ArcGIS® 9

Geoprocessing in ArcGIS

DATA CREDITS

Geographic data used in the quick-start tutorial provided courtesy of San Diego Association of Governments (SANDAG) and is used herein with permission.

Some of the illustrations in this work were made from data supplied by Collins Bartholomew Ltd.; IHS Energy; Riley County, Kansas, GIS; and SANDAG. They are used herein with permission.

AUTHOR

Jill McCoy

U.S. GOVERNMENT RESTRICTED/LIMITED RIGHTS

Contents

Introduction

1

IN THIS CHAPTER

- **Running tools via their dialog boxes**
- **Running tools at a command line**
- **Building models of your work flow**
- **Adding scripts to toolboxes**
- **Tips on learning how to perform geoprocessing in ArcGIS**

Geoprocessing is the processing of geographic information, one of the basic functions of a *geographic information system* (GIS). It provides a way to create new information by applying an operation to existing data. Any alteration or information extraction you perform on your data involves a geoprocessing task. It can be a task such as converting geographic data to a different format, or it can involve multiple tasks performed in sequence, such as those that clip, select, then intersect datasets.

Within ArcGIS® you can perform geoprocessing tasks in a number of ways.

Run a tool using its dialog box. Open a tool's dialog box, fill in parameter values, and click OK to run the tool.

Run tools at a command line. Type the tool name and its parameter values at a command line, then press Enter to run the tool.

Build and run a model that runs a sequence of geoprocessing tools. Create a model that runs the sequence of geoprocessing tools in your work flow. Alter parameter values, then rerun the model with a single click.

Create and run a script that runs geoprocessing tools. Use system batch processing scripts for repetitive tasks, such as those that run the same tool on multiple inputs, or create your own scripts that run geoprocessing tools.

The next few pages show you some of the geoprocessing tasks that can be performed within ArcGIS, and the quick-start tutorial in Chapter 2 will help you become familiar with the environment. As you start to use ArcGIS to perform your own geoprocessing tasks, you will discover even more.

Running tools via their dialog boxes

Your geoprocessing tasks can easily be performed by running the dialog box of a *system tool*—a tool installed with ArcGIS—or a *custom tool*—a model or a script you have created. Inside a tool's dialog box, simply enter parameter values and click OK to run the tool. Use the help available within the tool's dialog box to help you supply parameter values.

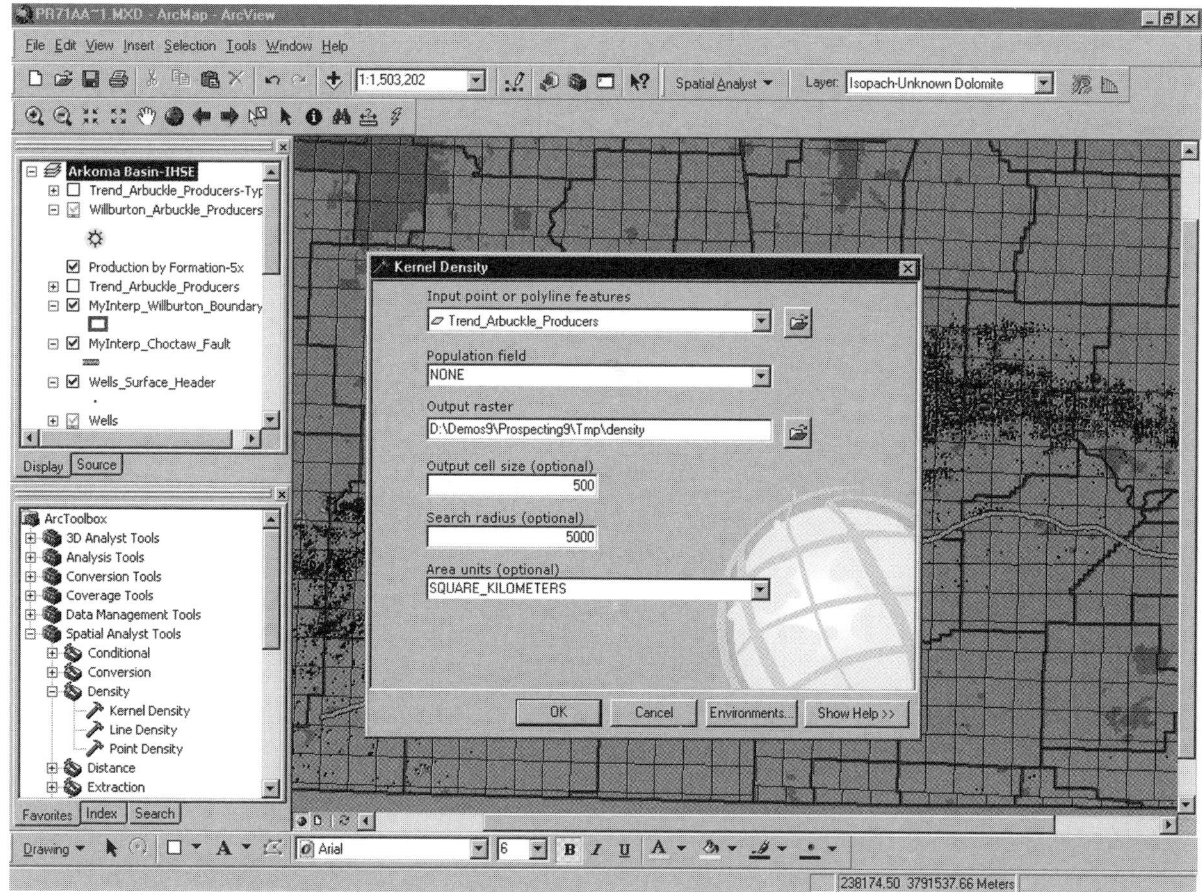

Run tools using a dialog box to supply parameter values. In the gas prospecting example, the Kernel Density system tool calculates the density of producing wells. The result from this tool shows areas that have a high density of producing wells, and this is the first step in finding areas similar to an existing gas field.

Running tools at a command line

Tools can be run at a command line. The tool name and its parameter values are typed as a string, and the tool executes when you press Enter. *Usage* is displayed as you type, helping you supply appropriate parameter values. You can edit the parameter values for a tool run at the command line in a dialog box, then reexecute the tool. Messages tell you the status of the processing.

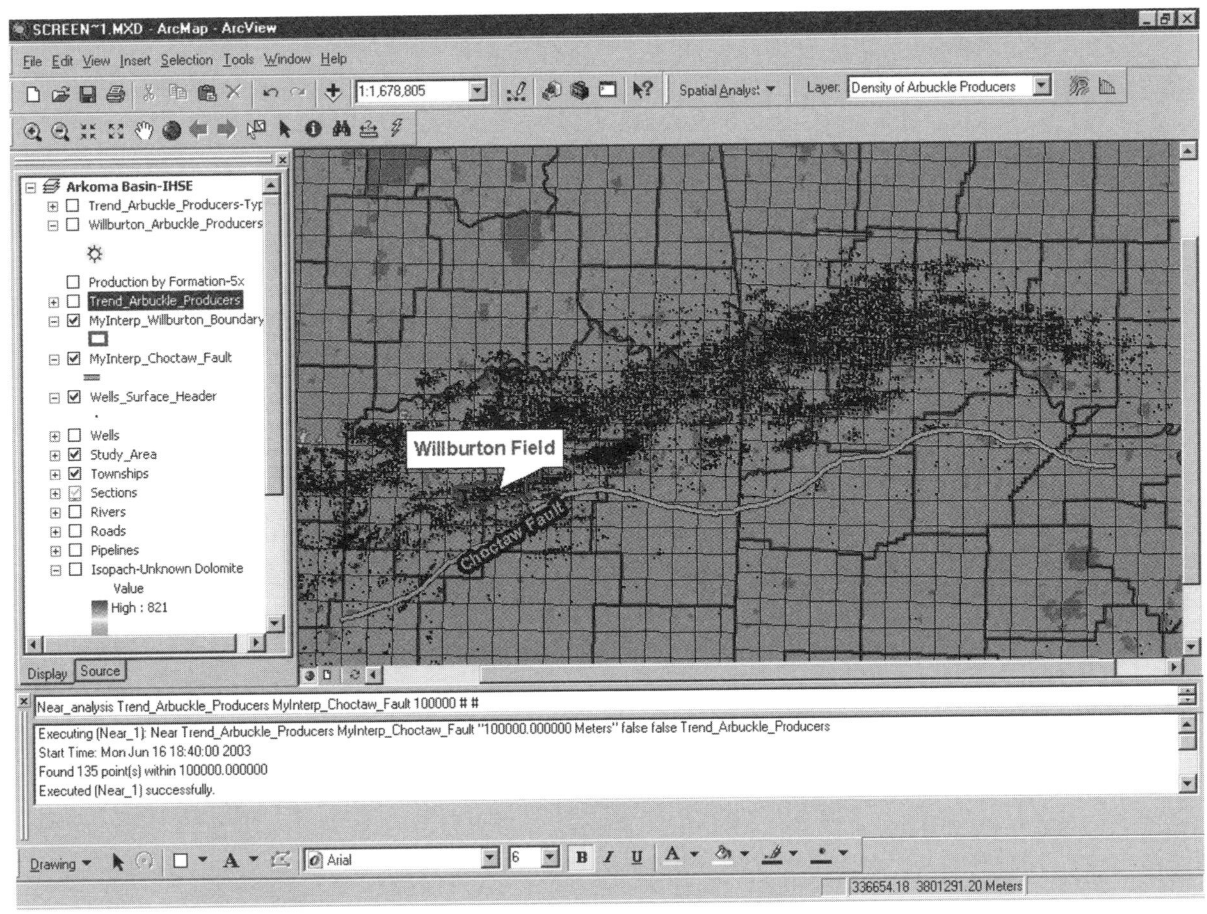

Run tools at a command line by typing the tool name and its parameter values. In the gas prospecting example, the Near tool is run to assess if the distance to a fault line is significant in the production of gas in the existing field.

Building models of your work flow

Create a *model* of your geoprocessing work flow by stringing processes together, then run the model with a single click. By setting *model parameters*, the user of your model can supply values for these parameters when your model is run via its dialog box or the command line.

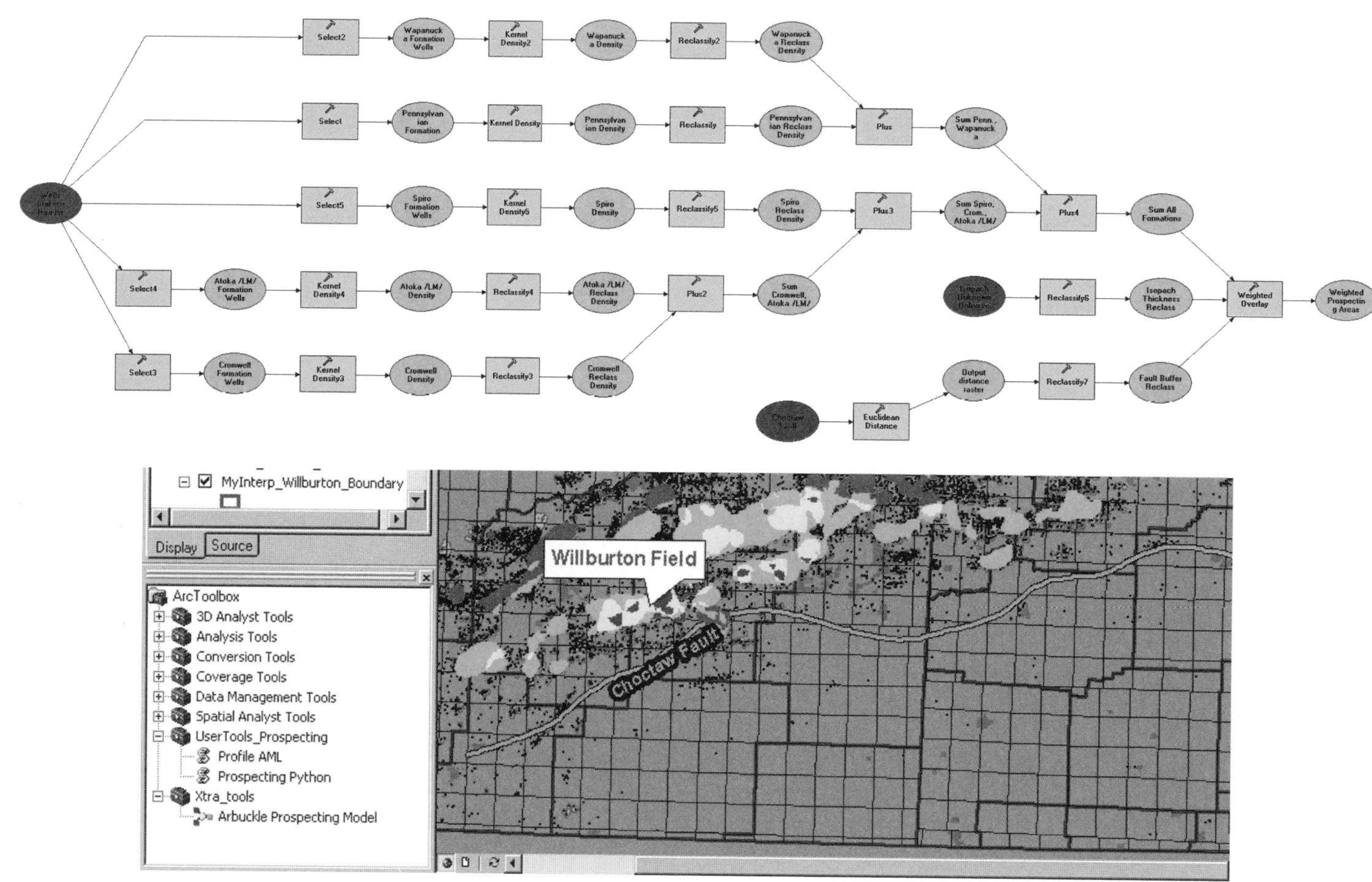

Build a model by adding tools to a ModelBuilder™ window, setting parameter values for each tool, and stringing processes together. The Arbuckle Prospecting model above locates areas with features similar to those within an existing gas field. The located areas (in red in the display) could have a high probability of being gas-producing areas.

Adding scripts to toolboxes

Scripts can be run from within their scripting application, or they can be added to a toolbox and run like any other tool from a dialog box, the command line, another model, or a script. Scripts can be written in any *Component Object Model (COM)*-compliant scripting language, such as Python, JScript, or VBScript, or they can be *ARC Macro Language (AML™)* scripts or executable files.

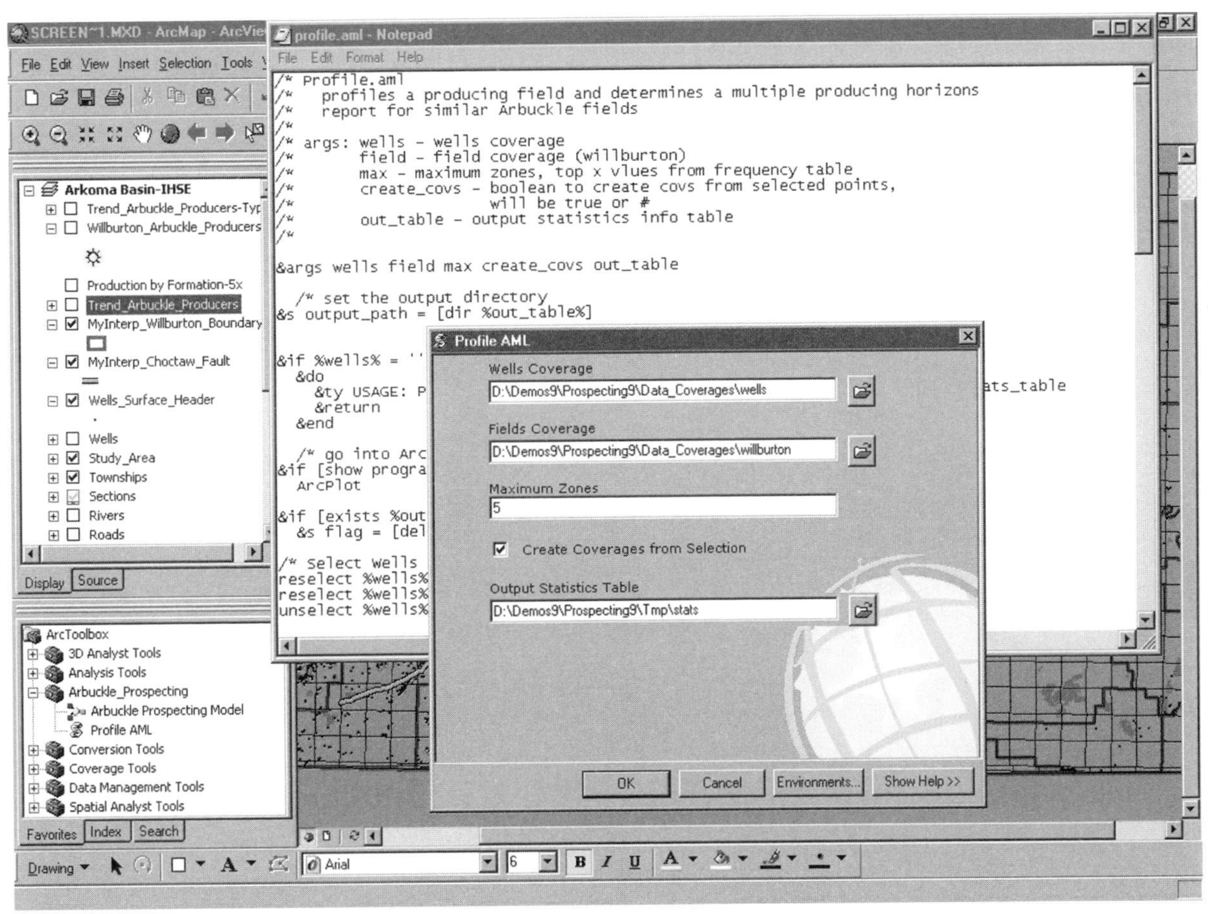

Add scripts to toolboxes and run them by supplying values for parameters. The profile AML tool determines the five shallowest gas-producing formations above the existing gas field.

Tips on learning how to perform geoprocessing in ArcGIS

If you're new to the concept of GIS, remember that you don't need to know everything about geoprocessing to get immediate results. Begin learning how to approach geoprocessing by reading Chapter 2, 'Quick-start tutorial'. This chapter introduces you to the methods available to perform geoprocessing tasks with ArcGIS and provides an excellent starting point as you begin to think about how to tackle your own particular geoprocessing tasks. ArcGIS comes with the data used in this tutorial, so you can follow along step by step at your computer.

If you prefer to jump right in and experiment on your own, flip to the appropriate chapter when you come across a task for which you need help. Use the index, the table of contents, and the 'About this book' section that follows to help you find the information you are looking for.

Finding answers to questions

Like most people, your goal is to complete your tasks while investing a minimum amount of time and effort in learning how to use the software. You want intuitive, easy-to-use software that gives you immediate results without having to read pages of documentation. However, when you do have a question, you want the answer quickly so you can complete your task. That's what this book is all about—helping you get the answers you need, when you need them.

This book describes the framework and methods for performing geoprocessing tasks in ArcGIS. Although you can read this book from start to finish, you'll likely use it more as a reference. You may also refer to the glossary in this book if you come across any unfamiliar GIS terms or need to refresh your memory.

About this book

This book is designed to help you use ArcGIS to perform your geoprocessing tasks. Topics covered in Chapter 2, 'Quick-start tutorial', assume you are familiar with the fundamentals of GIS and have a basic knowledge of ArcGIS. If you are new to GIS, you are encouraged to take some time to read the relevant ArcGIS documentation that you received in your ArcGIS package, such as *Getting Started With ArcGIS*, and other books that coincide with the ArcGIS applications you will be using, such as *Using ArcMap* and *Using ArcCatalog*. To learn more about writing scripts, *Writing Geoprocessing Scripts With ArcGIS* provides useful information. It is not necessary to read such documentation to continue with this book; simply use it as a reference.

Chapter 3, 'Geoprocessing basics', explains how geoprocessing fits within a GIS, explains the methods you can use to geoprocess your data, and describes the workspaces and data sources that can be geoprocessed. It also explains information you'll need to know when running tools, setting geoprocessing options, and sharing your tools. Chapter 4, 'Working with toolboxes', explains how to create and manage toolboxes, how to add and view documentation for toolboxes, and the rules for accessing toolboxes. Chapter 5, 'Working with toolsets and tools', explains how to manage toolsets and how to work with tools. It describes how to create your own model inside a toolbox and how to add a script to a toolbox. It also explains how to set parameters so the user of your model or script can supply values for parameters within the dialog box of your model or script, how to add documentation to your tools, and how to find the tools you are looking for. Chapter 6, 'Specifying environment settings', explains the hierarchy of environment settings that you can set and provides information for each environment setting that can be applied to your results from running tools. Chapter 7, 'Using the Command Line window', explains how to run commands at the command line and explains how to work with the message section of the Command Line window. Chapter 8, 'Introducing model building', gives conceptual information on building models, gives an overview of the ModelBuilder window, and explains the process of creating and building a new model. Chapter 9, 'Using the ModelBuilder window', gives more detailed information for using the ModelBuilder window to build models. The Appendix explains how the default stylesheets applied to tool dialog boxes can be altered to change the appearance of your dialog boxes.

Getting help on your computer

In addition to this book, you can use the online Help system to learn more about geoprocessing. See Chapter 4, 'Viewing documentation for toolboxes', and Chapter 5, 'Viewing documentation for tools', for information on getting help for toolboxes and tools.

For a list of the tools available with each product of ArcGIS (ArcView®, ArcEditor™ and ArcInfo™), type "Quick Reference Guide" in the Search tab of the online Help system and double-click the link to the Geoprocessing Commands Quick Reference Guide.

Contacting ESRI

If you need to contact ESRI for technical support, refer to 'Contacting Technical Support' in the 'Getting more help' section of the ArcGIS Desktop Help system. You can also visit ESRI on the Web at *www.esri.com* and *support.esri.com* for more information on geoprocessing and ArcGIS.

ESRI education solutions

ESRI provides educational opportunities related to geographic information science, GIS applications, and technology. You can choose among instructor-led courses, Web-based courses, and self-study workbooks to find education solutions that fit your learning style. For more information, go to *www.esri.com/education*.

Quick-start tutorial

2

IN THIS CHAPTER

- **Exercise 1: Finding coastal sage scrub near proposed roads**
- **Exercise 2: Joining fields to the vegetation data**
- **Exercise 3: Clipping data to the study area**
- **Exercise 4: Finding high-quality habitat patches**
- **Exercise 5: Finding habitat patches impacted by proposed roads**

This tutorial will take you on a tour of the geoprocessing functionality within ArcGIS as you find areas in San Diego County with the best potential to support the California gnatcatcher (Polioptila californica).

The California gnatcatcher is a small nonmigratory songbird that resides predominantly in southwestern California and northwestern Baja. The gnatcatcher population has been in decline over several decades because of extensive habitat loss due predominantly to urbanization. Most habitats are fragmented by roads; building tracts; and other barriers insurmountable to the small, weak-flying bird. The species is listed as threatened and is dependent upon certain vegetation types for its survival.

The goal of this tutorial is to find patches of potentially high-quality habitat—patches that contain features essential to the conservation of the species—then to identify which patches would be impacted by proposed roads. New roads will only serve to further fragment these fragile regions, so those high-quality habitat patches that will be impacted need to be identified.

In achieving the above goal, you will learn how to:

- Run geoprocessing tools through dialog boxes and the command line.
- Create models inside, and add scripts to, toolboxes.

By following the tutorial, you'll quickly learn how to perform your own geoprocessing tasks with ArcGIS. Included is an estimate of the time it will take to complete each exercise. You can work through the entire tutorial or complete each exercise one at a time. Although recommended, you don't have to complete the exercises in sequence.

Tutorial requirements

- It is assumed that you have installed ArcGIS Desktop before beginning this tutorial.
- You can perform Exercise 1 with an ArcView, ArcEditor, or ArcInfo license. The rest of the exercises require an ArcInfo license.
- The data required is included on the ArcGIS Desktop CD. After running the ArcGIS setup, on the Additional Installation Components dialog box, check to install the ArcGIS Tutorial Data. In the ArcGIS Tutorial Data Setup wizard, select to install the Geoprocessing data (the default installation path is arcgis\ArcTutor\Geoprocessing) on the drive where the tutorial data is installed.

  Climate.shp, futrds, majorrds.shp, study_quads.shp, vegtype, and vegtable.dbf were provided courtesy of the San Diego Association of Governments (SANDAG) for educational purposes only. The data has been manipulated for the purpose of the tutorial. Assumptions applied and actual outcomes may vary. ESRI is not inviting reliance on this data or the methodologies followed. You should always verify actual data and exercise your own professional judgment when interpreting any outcomes.

Data	Description
climate.shp	General climate zones.
elevlt250.shp	Areas where the elevation is less than 250 m. This shapefile was derived from a United States Geological Survey (USGS) digital elevation model (DEM).
futrds	A geodatabase feature class of all proposed freeways and roads.
habitat_analysis.mdb	A personal geodatabase containing the feature classes futrds and vegtype.
majorrds.shp	All major roads and freeways.
multi_clip.py	A Python script used to clip multiple feature classes and place them into a personal geodatabase.
query.txt	Includes an expression that can be copied and pasted into the expression box on the Select tool's dialog box used in Exercise 4. It can be used as an alternative to the Query Builder or typing the expression.
slopelt40.shp	Areas where the slope of the land is less than 40 percent. This shapefile was derived from a USGS DEM.
study_quads.shp	The study area. USGS 7.5 quad boundaries covering the towns of La Jolla and La Mesa in San Diego County.
vegtype	A geodatabase feature class of vegetation type information.

vegtable.dbf	A table containing fields that will be joined to a layer you'll create from vegtype. The fields include a vegetation description field and a field indicating whether the vegetation type is suitable for habitation by the gnatcatcher.

The *ArcCatalog™ tree* provides access to datasets stored on disk. In the ArcCatalog tree, click a dataset, then click the Metadata tab to explore the *metadata* (information) of each dataset in detail. Explore metadata for the tutorial folder to gain an overview of the data contained in the folder and how it was created.

Exercise 1: Finding coastal sage scrub near proposed roads

When working with ArcGIS you'll likely need to perform geoprocessing tasks at some stage in your work flow. In ArcGIS geoprocessing tasks are performed by running *tools*. Many times your work flow takes multiple steps to produce the desired result. You can perform each geoprocessing task individually by running a system tool via its dialog box or the command line, or you can string tasks together to build a model or a script.

In this exercise you'll copy the tutorial data locally using ArcCatalog, then in ArcMap™ you'll review and run an existing model that runs a sequence of system tools. The model identifies the vegetation types near proposed roads.

The interface you use to build and edit models is the *ModelBuilder window*. By adding a tool to a ModelBuilder window and supplying values for the tool's *parameters*, you construct a *process*. A model is built by connecting processes together. The model can be run with a single click. A model increases your efficiency because you can easily alter input data or other parameter values, then reexecute the model to produce different results.

The diagram that follows shows the model you'll run. In the model, the roads are first buffered using values from the Distance field in the roads attribute table. The value for the Distance field of each road segment depends on the width of the road. The wider the road, the farther the distance (buffer), due to the impact on the surrounding area. The output from the Buffer tool (Buffer Zones) is used to clip the vegetation data (Vegetation Types) to identify the vegetation types within the buffer zones (Vegetation Near Roads).

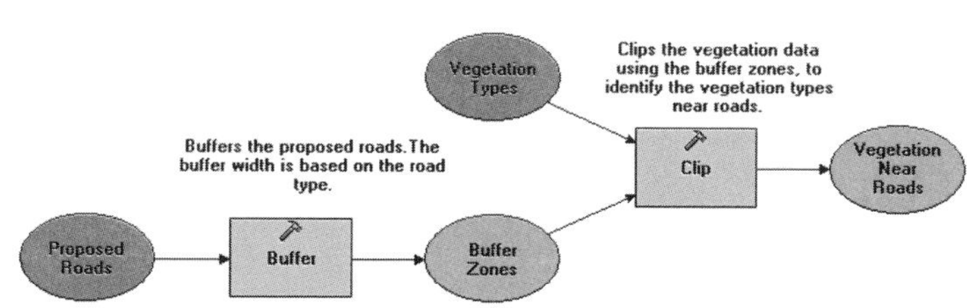

Identifying vegetation types near proposed roads
Vegetation Types
Clips the vegetation data using the buffer zones, to identify the vegetation types near roads.
Buffers the proposed roads. The buffer width is based on the road type.
Clip
Vegetation Near Roads
Proposed Roads
Buffer
Buffer Zones

The above diagram shows the model you'll run to identify vegetation types near proposed roads.

After running the model, you'll select all features in the final *output data* that represent coastal sage scrub vegetation to identify the patches of this vegetation type that would be impacted by the addition of the proposed roads.

This exercise will take approximately 15 minutes to complete.

Organizing your data

Before working with geoprocessing tools you'll first organize your tutorial data using ArcCatalog.

Starting ArcCatalog

1. Start ArcCatalog by either double-clicking a shortcut installed on your desktop or using the Programs list in your Start menu.

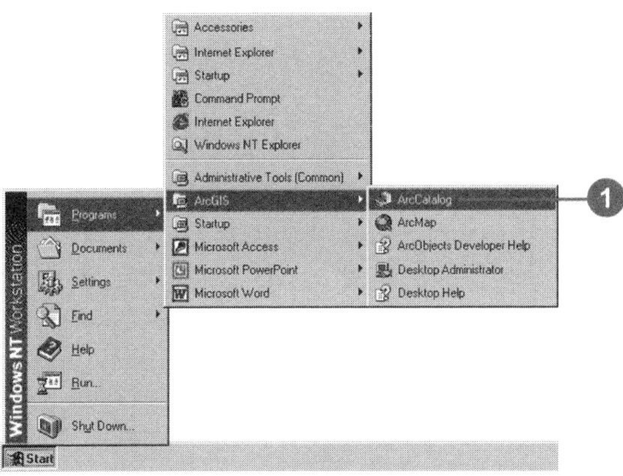

Locating the tutorial data

By connecting to a folder in ArcCatalog you can quickly see the folders and data sources it contains. You'll now begin organizing your tutorial data by establishing a folder connection to its location.

1. Click the Connect to Folder button.

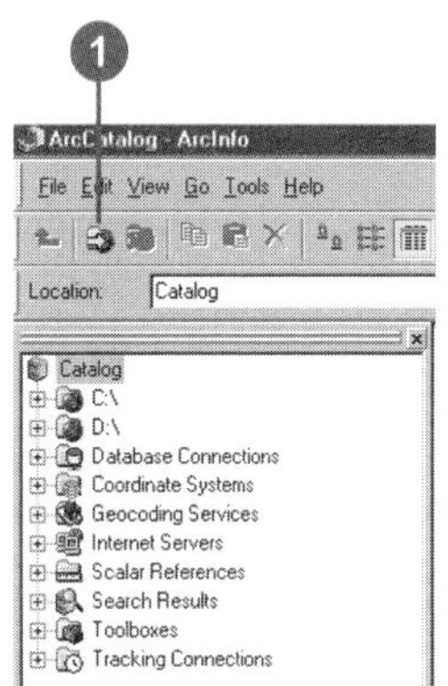

2. Type the path or navigate to the location where you installed the tutorial data. For example, if you installed ArcGIS on your C:\ drive, type "C:\arcgis\ArcTutor\Geoprocessing", then click OK to establish a folder connection.

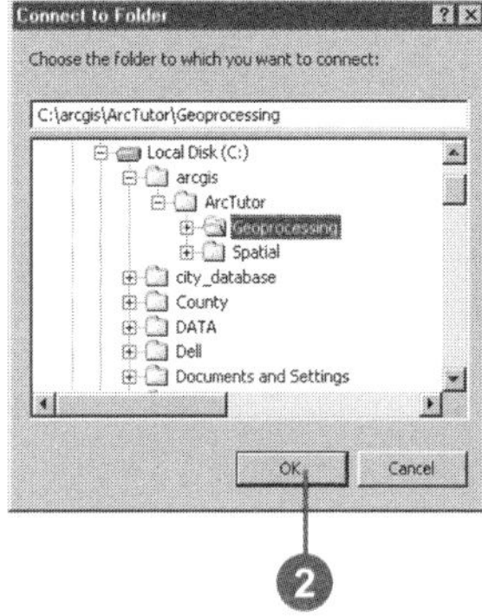

Your new folder connection to the tutorial data is listed in the ArcCatalog tree.

Creating a working copy of the tutorial data

You'll now copy the tutorial data into a folder on a local disk to maintain the integrity of the original data. Once it has been copied, you will then create a connection to the folder containing the data.

1. Click the connection to the tutorial data and click the San_Diego folder in the Contents tab.

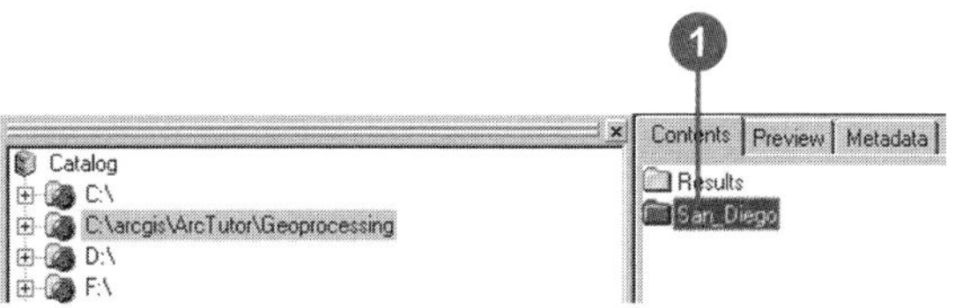

2. Click the Copy button on the Standard toolbar.

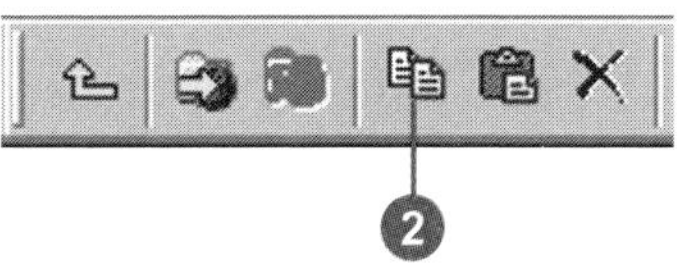

3. Click the C:\ (or an alternative drive) folder connection in the ArcCatalog tree, then click the Paste button.

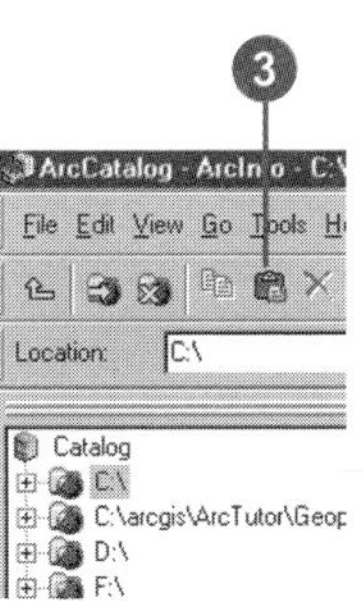

A new folder called San_Diego will appear in the Contents list.

4. Click the new San_Diego folder in the Contents tab to select it, then click again to rename the folder.
5. Type "GP_Tutorial" and press Enter.

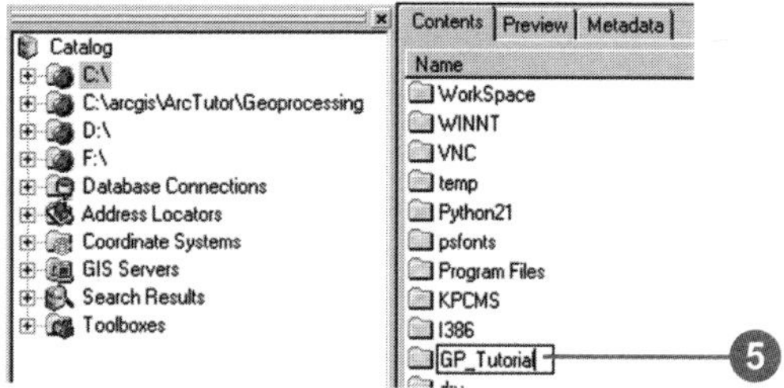

6. Click the Connect to Folder button again, create a connection to your GP_Tutorial folder, and click OK.

   Your new folder connection, for example, C:\GP_Tutorial, is now listed in the ArcCatalog tree. You will access this connection many times during this tutorial.

Reviewing an existing model

Models can be run within any ArcGIS Desktop application. In this exercise you'll work within ArcMap to identify vegetation types near proposed roads.

Opening the Vegetation Analysis.mxd

1. Click the connection to your GP_Tutorial folder in the ArcCatalog tree, click the Contents tab, then double-click the map document Vegetation Analysis to open it.

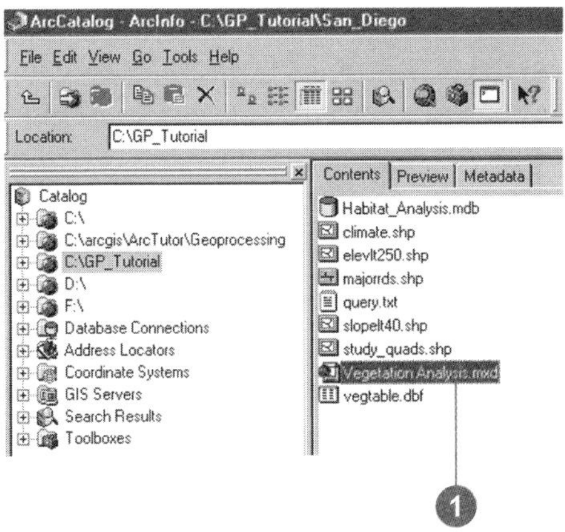

As you have finished using ArcCatalog in this exercise, close your ArcCatalog session.

2. Click ArcCatalog on the desktop taskbar to maximize it.
3. Click File, then Exit to exit the ArcCatalog session.

Opening the ArcToolbox window

The *ArcToolbox™ window* is dockable in any ArcGIS Desktop application. It provides access to tools you have stored on disk.

1. Click the Show/Hide ArcToolbox Window button on the Standard toolbar to open the ArcToolbox window.

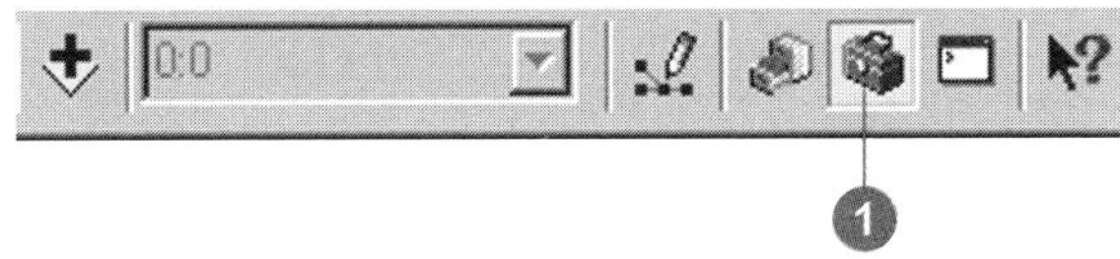

Toolboxes can contain system tools—tools installed by default—or they can contain custom tools that you have created, such as models or scripts, that might run a number of tools at one time. You'll now review a model that has already been built using two system tools.

2. Right-click the ArcToolbox folder in the ArcToolbox window and click Add Toolbox. Click the Look in dropdown arrow and click the connection to your local copy of the tutorial data—for example, C:\GP_Tutorial. Double-click Habitat_Analysis.mdb, then click the toolbox My_Analysis_Tools and click Open.
3. Expand the My_Analysis_Tools toolbox.
4. Right-click the model Find Vegetation Near Roads and click Edit.

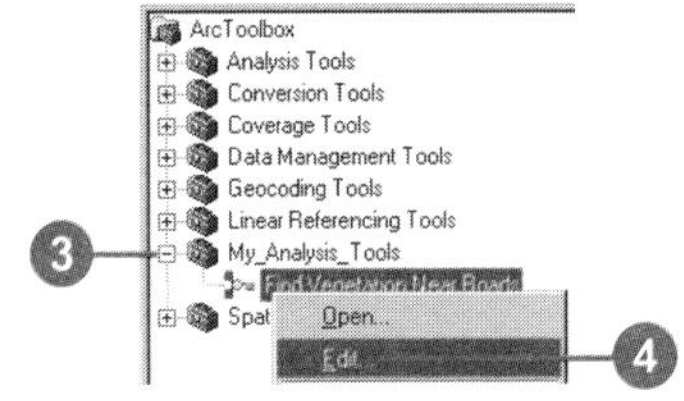

This opens a previously created model in its ModelBuilder window so you can review it.

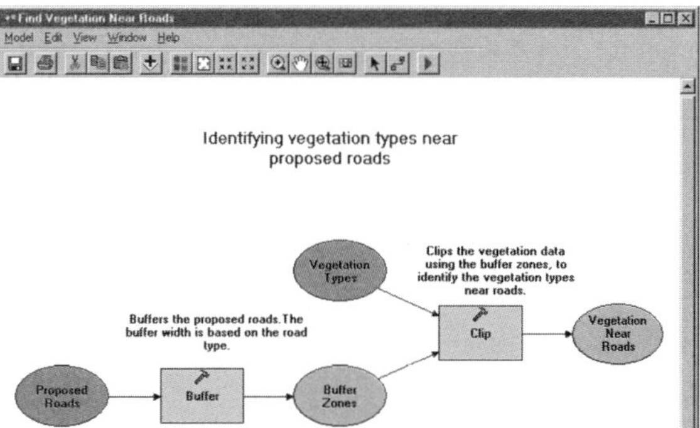

Input data elements that represent existing data appear as blue ovals in the model. Tool elements that represent tools stored on disk appear as yellow rectangles. *Derived data* elements, which represent data that will be created when the model is run, appear as green ovals.

5. Click the Search tab on the ArcToolbox window.
6. Type "Buffer", then click Search to locate the Buffer tool in the ArcToolbox window.

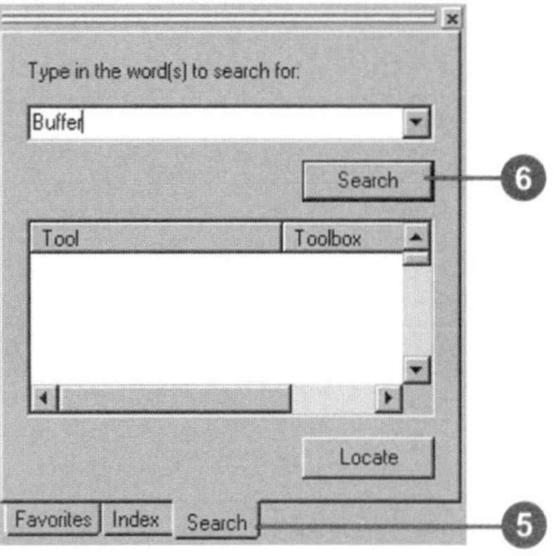

If you have the Analysis Tools toolbox and the Coverage Tools toolbox added to the ArcToolbox window, you'll see two Buffer tools in the list of found tools. The Buffer tool in the Coverage Tools toolbox only works with coverages as inputs. The tool in the Analysis Tools toolbox works with feature classes as inputs.

7. Click the Analysis Tools Buffer and click Locate.

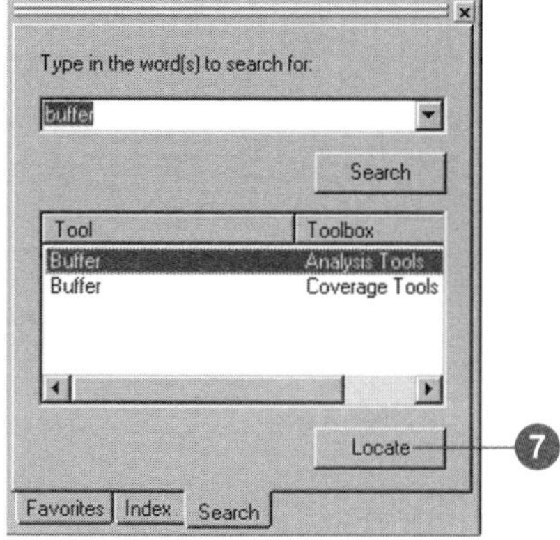

The tool is located in the ArcToolbox window.

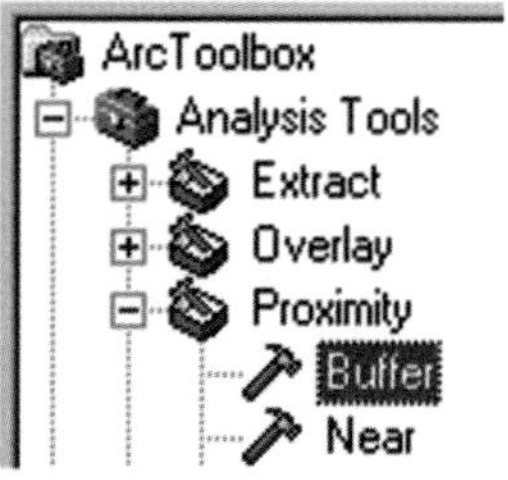

Rather than opening the Buffer tool's dialog box from the ArcToolbox window, you'll open it within the ModelBuilder window to view the parameter values set for the tool.

8. Right-click the Buffer tool element in the model and click Open to open the tool's dialog box.

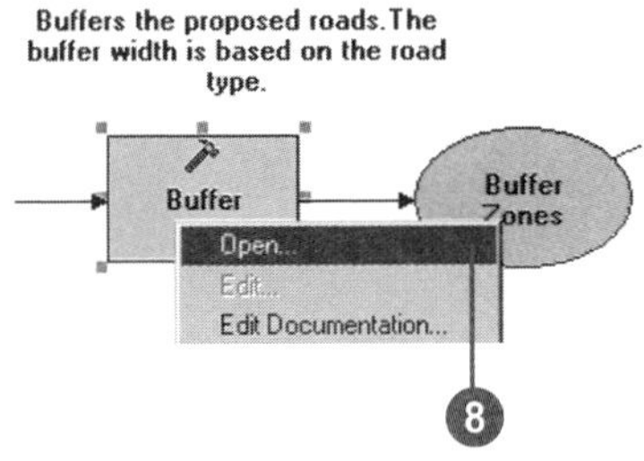

A description for the tool is displayed in the Help panel by default.

9. Click each parameter in the dialog box to view help in the Help panel for each.
10. Click Help to get more detailed help for the tool if desired.
11. Click Cancel on the Buffer dialog box.

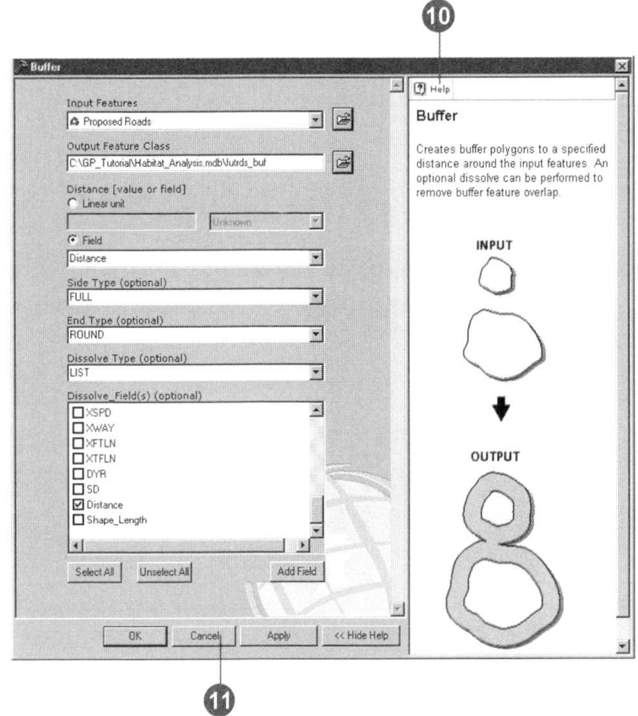

12. Right-click the Clip tool element and click Open to open the tool's dialog box.
13. Examine the parameter values set for the tool and view the help for each parameter as you did with the Buffer tool, then click Cancel.

In this model there are two processes, each consisting of a tool and its parameter values—one using the Buffer tool and the other using the Clip tool. The two processes are connected together to build the model.

Running the model

The model is colored in, meaning it is in a ready-to-run state, as all necessary parameter values have been supplied in each tool's dialog box.

1. Click Model, then Run.

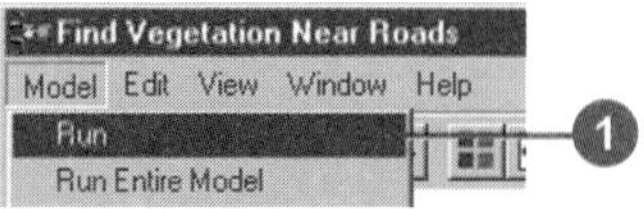

When the model runs, the tool of the process that is currently executing is highlighted in red. When each process has executed, a dropshadow appears behind the tool and its derived data element, indicating that the process has run and the derived data has been created on disk.

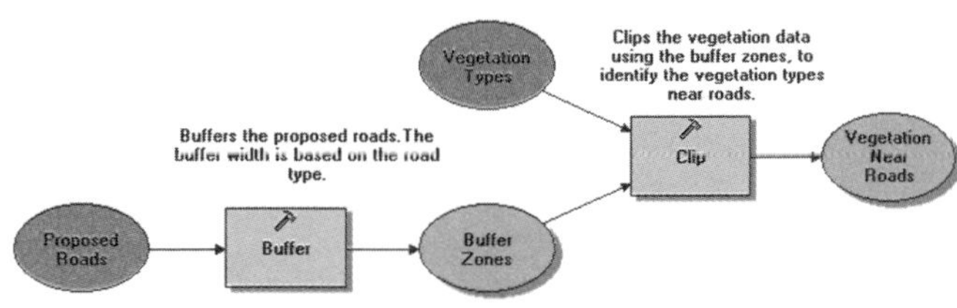

2. Click Close on the progress dialog box.

The final result is added to the ArcMap display by default, because the option to add derived data to the display is checked on the context menu of the derived data element Vegetation Near Roads.

3. Right-click the Vegetation Near Roads derived data element in the model to see that Add To Display is checked.

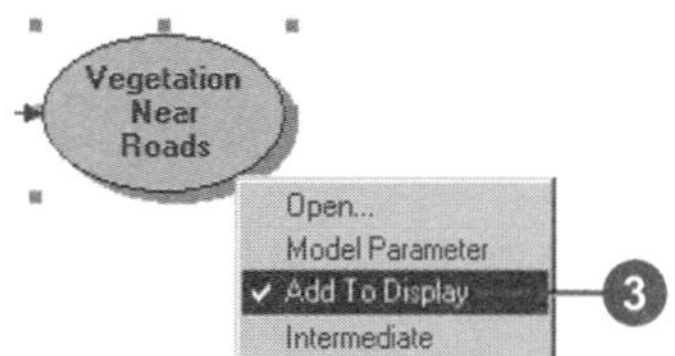

4. Click Model, then Close, then click Yes to save the changes you have made to the model.

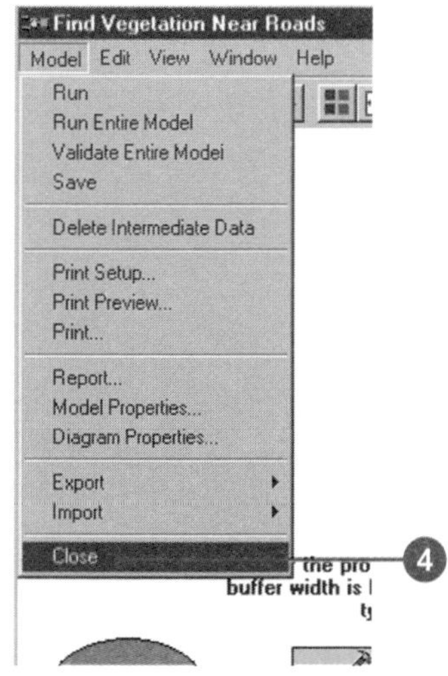

5. Examine the final result (vegtype_clip) in the ArcMap display. In the graphic below, the color used to symbolize vegtype_clip is purple. The color is randomly generated for the output feature class, so the color used in symbolizing your feature class might be different.

   All vegetation types near proposed roads have been identified.

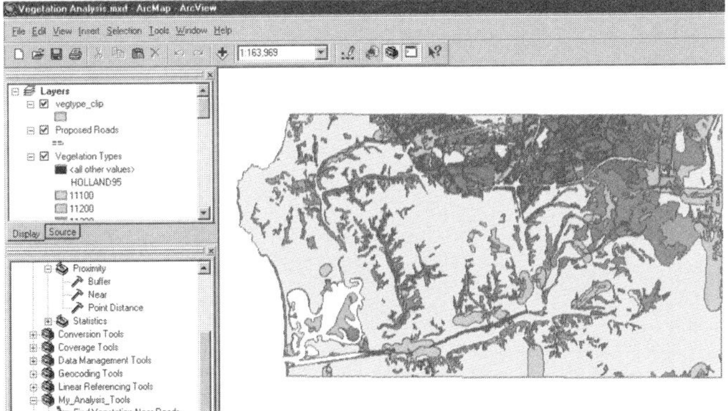

Selecting areas of coastal sage scrub

You'll now select areas where the vegetation type is San Diego coastal sage scrub to see how much of this particular vegetation type would be impacted by the proposed roads.

1. Click Selection on the Main menu and click Select By Attributes.

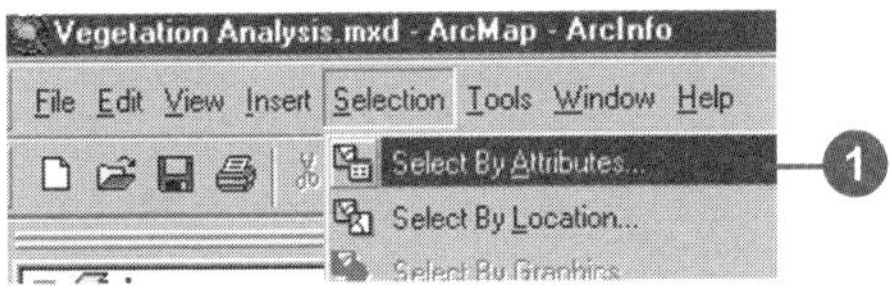

2. Click the Layer dropdown arrow and click vegtype_clip.
3. Double-click the field [HOLLAND95] to add it to the expression box.
4. Click = to add the equals operator to the expression box.
5. Click Get Unique Values then double-click the unique value '32500' to add it to the expression box.

   HOLLAND95 is a code for vegetation type. The value 32500 corresponds to San Diego coastal sage scrub.
6. Click Verify to check the syntax of the expression.
7. Click Apply to select the features.
8. Click Close.

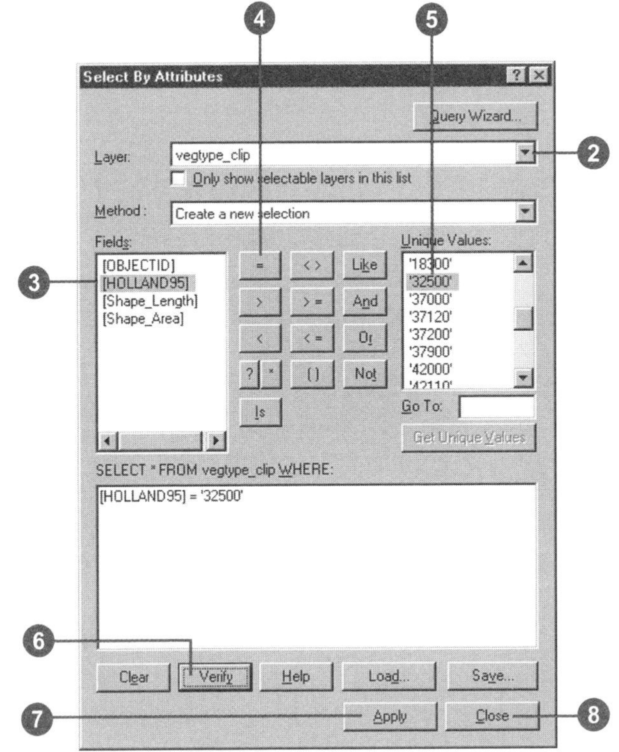

9. Examine the selection in the ArcMap display. All areas near proposed roads where the vegetation is San Diego coastal sage scrub are identified.

This brings you to the end of Exercise 1. This exercise has given a brief introduction to running models. The rest of this tutorial goes into more detail about geoprocessing in ArcGIS. You'll use more of the system tools, run a script to batch process multiple inputs, and build more complex models. You can continue on to Exercise 2 or stop and complete the tutorial at a later time. If you do not move on to Exercise 2 now, do not delete your working copy of the tutorial data, which is your GP_Tutorial folder, and do not remove the folder connection that accesses it from ArcCatalog.

10. Click File on the Main menu of ArcMap and click Exit. Click Yes to save your changes.

Exercise 2: Joining fields to the vegetation data

There are multiple system tools that you can access, such as dissolve or union, through *system toolboxes* that are installed with ArcGIS.

In ArcCatalog you can choose whether to access system tools via the ArcCatalog tree or the ArcToolbox window. If your toolboxes are located in different locations on disk, it is recommended that you use the ArcToolbox window to centralize the location of the toolboxes you use most frequently. In applications in which the ArcCatalog tree is not accessible, such as in ArcMap, you'll always work with tools using the ArcToolbox window.

Some manipulations you make to your data should be done as preparation steps, so you can avoid the repetition of unnecessary steps in your main work flow. In this exercise you'll first locate the system toolboxes on disk, then you'll open the ArcToolbox window. Once you have specified appropriate environment settings, you'll join two fields to a layer you'll create from the vegetation data (vegtype). This exercise will take approximately 20 minutes to complete.

Setting up

1. Start ArcCatalog by using the Programs list in your Start menu.

If you did the previous exercise, skip the next step and go to the next section ('Locating the system toolboxes').

2. Copy the GP_Tutorial folder from arcgis\ArcTutor\Geoprocessing\Results\Ex1 on the drive where you installed ArcGIS to a local drive, such as your C:\ drive.

Locating the system toolboxes

The Toolboxes folder in the ArcCatalog tree contains two folders: a My Toolboxes folder and a System Toolboxes folder. The *My Toolboxes folder* points to a location on disk that can be changed in the Geoprocessing tab of the Options dialog box. It is the location used for new toolboxes created inside the ArcToolbox window. The My Toolboxes folder also contains a History toolbox within which *history models* are created automatically for each session, recording the tools and parameter values used. For more information on the My Toolboxes folder, see 'Changing the default location of the My Toolboxes folder' in Chapter 4. For more information on history tools, see 'Keeping track of geoprocessing operations' in Chapter 3.

The System Toolboxes folder contains the toolboxes that come with ArcGIS. The Toolboxes folder is not added to the ArcCatalog tree by default.

1. Click the Tools menu and click Options.

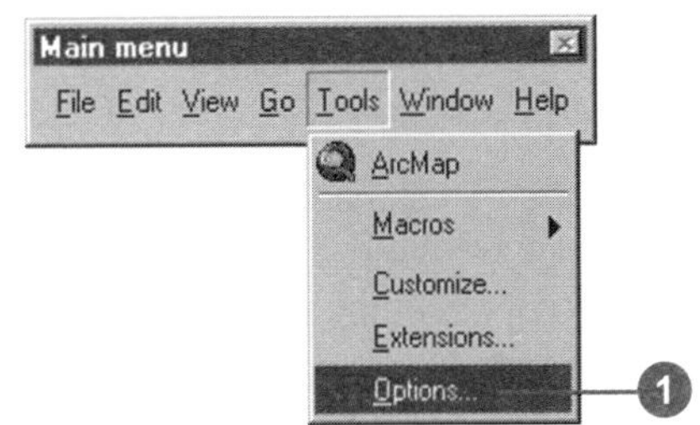

2. Click the General tab and check Toolboxes if it is not already checked.

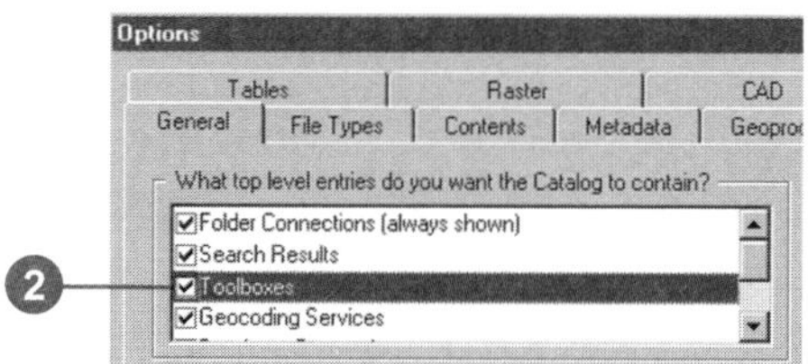

3. Click OK on the Options dialog box.
4. Expand the Toolboxes folder, then expand the System Toolboxes folder to view the toolboxes that are available.

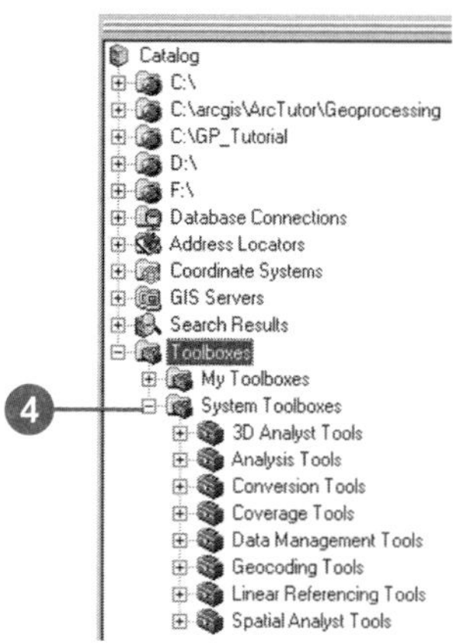

Each toolbox contains a set of system tools that can be used to geoprocess your data.

Working with the ArcToolbox window

The ArcToolbox window provides a way to centralize the location of the tools you use most frequently. You can remove toolboxes you don't use often from the ArcToolbox window and add them back at any time. The ArcToolbox window is dockable inside any ArcGIS Desktop application.

1. Click the Show/Hide ArcToolbox Window button on the Standard toolbar of ArcCatalog to open the ArcToolbox window.

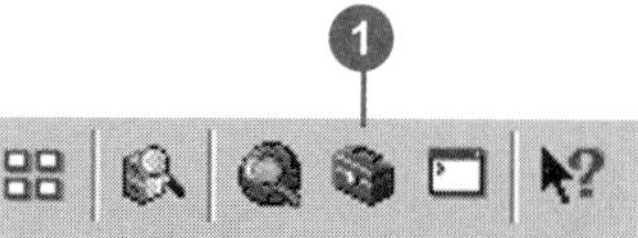

The window is placed within the application. Its position may vary depending on the windows you have open.

Docking the ArcToolbox window

The ArcToolbox window can be placed anywhere in the application or on the desktop.

1. Click and drag the bar at the top of the ArcToolbox window.
2. Place the ArcToolbox window over the ArcCatalog tree, then drop the panel. The outline of the ArcToolbox window helps you to place it in the desired position.

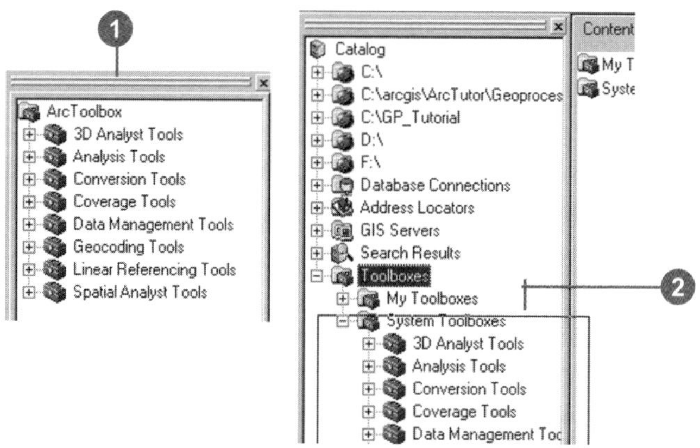

Your ArcToolbox window should now display under the ArcCatalog tree, as the following diagram shows.

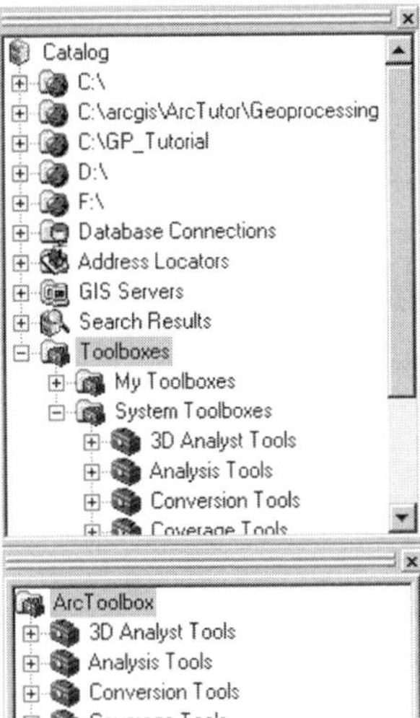

Note: If you hold down the Ctrl key and drag the ArcToolbox window, it will not dock itself to the application.

Setting the geoprocessing environment

Before you start to perform geoprocessing tasks on your data, you should set up any relevant *environment settings*. The application Environment Settings dialog box allows you to change the default application settings that are used when you run geoprocessing tools.

Note: You can specify environment settings for the application that will apply to all tools. Alternatively, you can set environment settings for a specific model or for a process within a model. Environment settings for the application are used by default unless they are specified for a model or for a process within a model. Settings for a process will override settings for the model and for the application. Settings for a model will override settings for the application. For more information on environment settings, see Chapter 6, 'Specifying environment settings'.

1. Right-click the ArcToolbox window and click Environments to open the Environment Settings dialog box for the application.

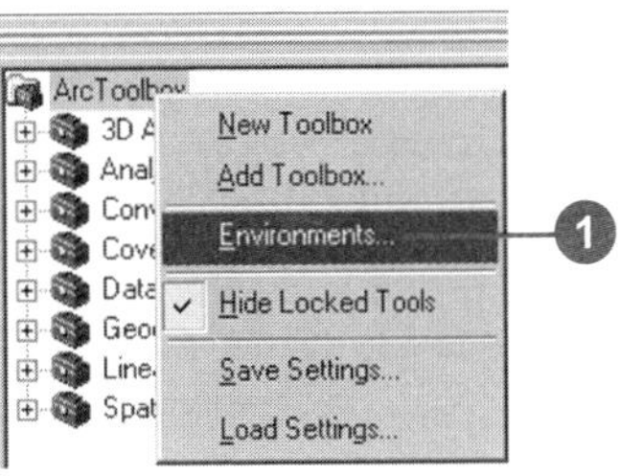

2. Click General Settings to expand its contents.
3. Click the Current Workspace parameter and type the path to your GP_Tutorial folder—for example, "C:\GP_Tutorial"—for its value. Alternatively, click the Browse button to the right of the parameter and navigate to this location.
4. Click OK.

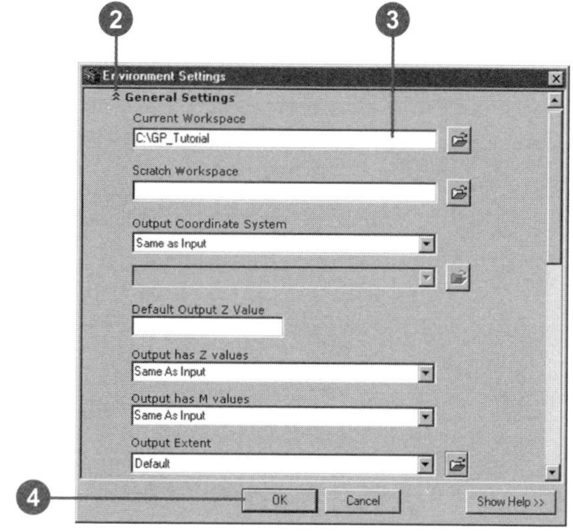

Now that the *current workspace* is set, you can simply type the name for the input and output parameter values. Data will be taken from, and placed in, this location, saving you time typing in or browsing to the location of data each time you run a tool.

Adding fields to the vegetation data

If you performed Exercise 1, you selected areas where the vegetation type was San Diego coastal sage scrub—polygons with a value of 32500 for the HOLLAND95 field in the attribute table of vegtype.shp. This vegetation type is preferred by the California gnatcatcher. However, there are other vegetation types that are also suitable. You'll now join two fields from a dBASE® Format (DBF) table (vegtable.dbf) to a layer you'll create from your vegetation feature class, vegtype, located inside Habitat_Analysis.mdb. The fields you'll join to the layer are HABITAT—a Boolean field indicating whether or not the vegetation type is suitable for habitation by the gnatcatcher—and VEG_TYPE—a description for the HOLLAND95 code.

As when joining fields in ArcMap, the Add Join tool only accepts in-memory *layers* that reference geographic data stored in a data source. You'll create a layer—vegtypelayer—and join vegtable.dbf to it. Then you'll create a new layer from the result of the Add Join tool, so you can rename the fields to be more meaningful and make visible those fields that you want to see in the attribute table of the output layer. You'll then copy the features in the layer to a new feature class that you'll save inside Habitat_Analysis.mdb.

Creating a layer

1. Expand your GP_Tutorial connection, then click the plus sign next to the Habitat_Analysis.mdb to view its contents.
2. In the ArcToolbox window, expand the Data Management Tools toolbox, then expand the Layers and Table Views toolset.
3. Right-click Make Feature Layer and click Open or double-click the tool.

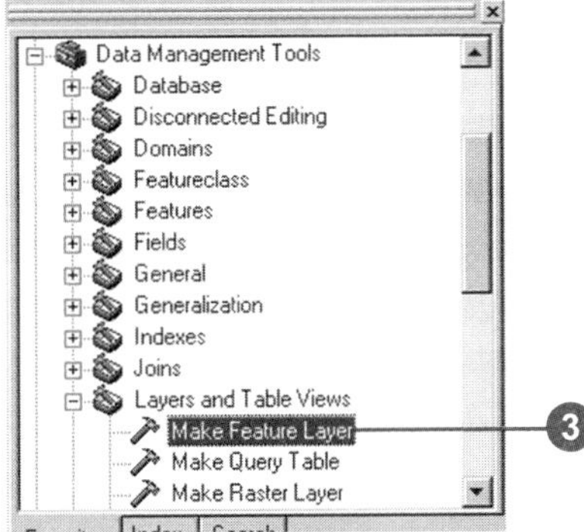

4. Drag the feature class vegtype from within the Habitat_Analysis geodatabase into the Input Features text box.
5. Type vegtypelayer for the name of the in-memory layer to create.
6. Click OK.

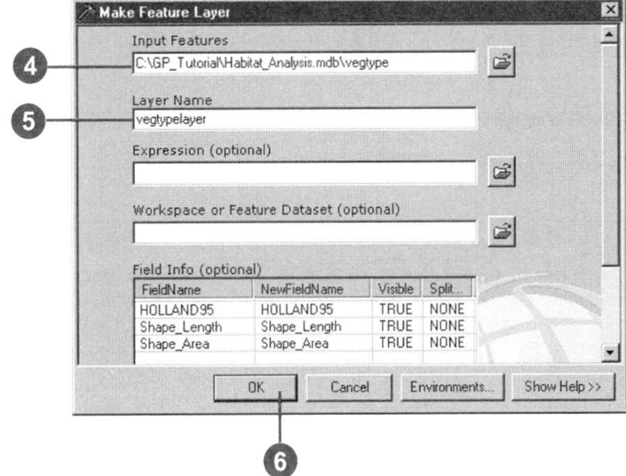

An in-memory layer is created. You cannot see this layer in the ArcCatalog tree, but it can be used as input to geoprocessing tools that support it.

7. Click Details on the progress dialog box to view the execution messages produced.
8. Check Close this dialog when completed successfully on the progress dialog box, then click Close.

Joining the DBF table to vegtypelayer

1. In the ArcToolbox window, expand the Data Management Tools toolbox, then expand the Joins toolset.
2. Right-click Add Join and click Open.

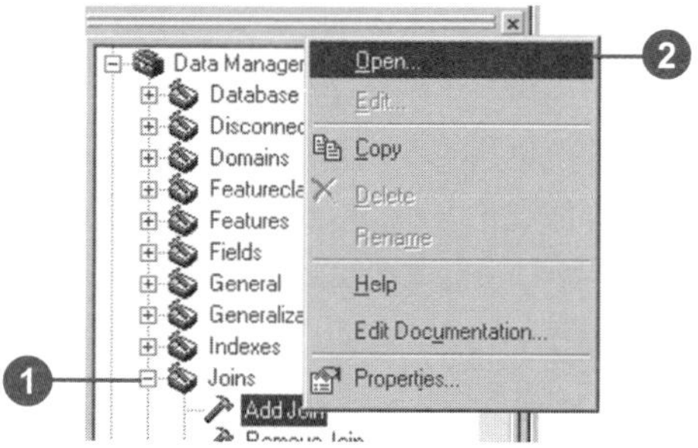

Notice that all required parameters in the Add Join dialog box—except Keep All, as this is an optional parameter—need values set for them. There is a green circle next to these parameters, indicating that a value needs to be specified.

3. Click the dropdown arrow and click vegtypelayer to use the created in-memory layer as the value for the Layer Name or Table View parameter.

   The Add Join tool appends attributes from one table to another using a common field. Here, the common field is named the same in the layer and in the table (HOLLAND95).

4. Click the dropdown arrow for the Input Join Field parameter and click HOLLAND95.
5. Type "vegtable.dbf" for the value of the Join Table parameter.

   The path to the current workspace is supplied for you when you click another parameter.

6. Click the dropdown arrow for the Output Join Field parameter and click HOLLAND95.
7. Leave the default Keep All to retain all records in the result.
8. Click OK.

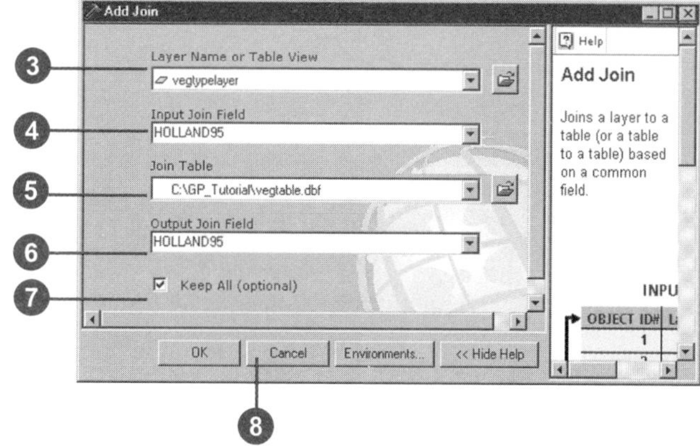

Vegtypelayer will now contain two fields: HABITAT and VEG_TYPE.

Renaming and choosing visible fields

You'll now create an in-memory layer from vegtypelayer with a subset of fields. You'll also rename the fields to remove the concatenated text introduced by joining fields.

1. In the ArcToolbox window, expand the Layers and Table Views toolset located inside the Data Management Tools toolbox.
2. Right-click Make Feature Layer and click Open.
3. Click the Input Features parameter dropdown arrow and click the in-memory layer vegtypelayer.
4. Type "vegtype_new" for the value of the parameter Layer Name. This is the name of the output layer to create. The output layer is an in-memory layer that references the data source of the layer set for the Input Features parameter.
5. In the Field Info table, click and widen the NewFieldName column.
6. In the Field Info table, click vegtype.HOLLAND95 in the NewFieldName column and remove the text vegtype.. Remove the text vegtype. from vegtype.Shape_Length and vegtype.Shape_Area. Remove the text vegtable. from vegtable.HABITAT and vegtable.VEG_TYPE.
7. Click TRUE in the Visible column for both vegtable.OID and vegtable.HOLLAND95 and change it to FALSE. You don't need the OID field, and you have two HOLLAND95 fields, so one can be removed from view.
8. Click OK.

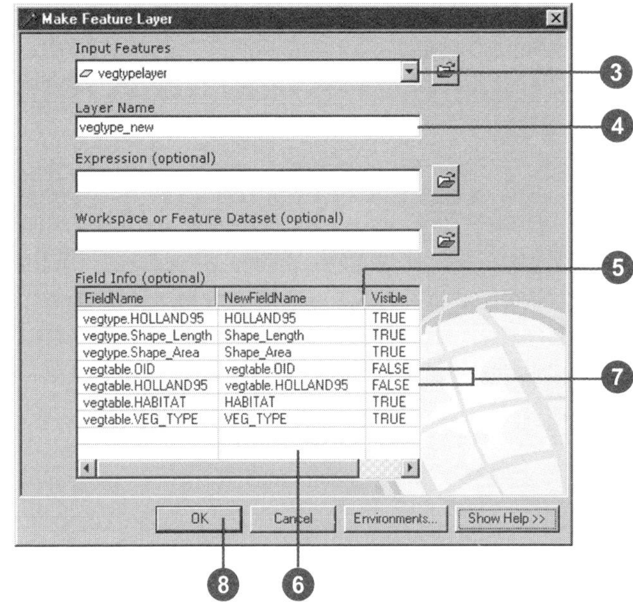

Creating a feature class from the layer

You'll now copy the features in the layer to a feature class using the Copy Features tool.

1. In the ArcToolbox window, expand the Features toolset located inside the Data Management Tools toolbox.
2. Right-click Copy Features and click Open.

3. Click the dropdown arrow for the Input Features parameter and click the vegtype_new layer you created.
4. Click the Browse button next to the parameter Output Feature Class.
5. Click the Look in dropdown arrow and click the connection to the tutorial data (C:\GP_Tutorial). Double-click Habitat_Analysis.mdb, then type "vegtype2" for the name of the output feature class to create inside the geodatabase.
6. Click Save on the Output Feature Class dialog box, then click OK.

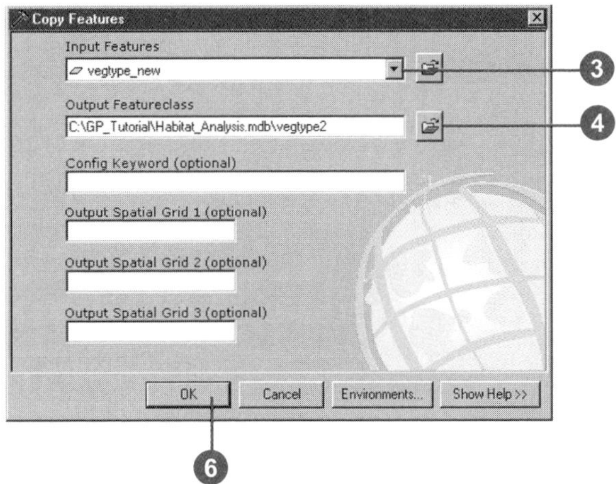

You'll now delete the feature class vegtype from your Habitat_Analysis geodatabase, as it is no longer needed.

7. In the ArcCatalog tree, right-click vegtype inside the Habitat_Analysis geodatabase and click Delete.
8. Right-click vegtype2 and click Rename. Type "vegtype" and press Enter.

You can continue on to Exercise 3 or stop and complete the tutorial at a later time. If you do not move on to Exercise 3 now, do not delete your working copy of the tutorial data or the folder connection that accesses it in ArcCatalog.

Exercise 3: Clipping data to the study area

You can complete many tasks using the system tools as you have learned in Exercise 2; however, the process of running single tools can be time-consuming. When you want to perform *batch processing* (running the same tool on multiple inputs), such as when converting multiple datasets to a different format, running the same tool multiple times is not ideal. To automate your work flow so you can, for instance, perform a task on multiple inputs in one step, you can add your own scripts to a toolbox. Scripts you add to a toolbox work as any system tool. They can be run via a dialog box, at the command line, inside a model, or inside another script.

This exercise will take approximately 15 minutes to complete. You'll add a previously created script to a toolbox that will clip the tutorial datasets located in your GP_Tutorial folder and place them inside your Habitat_Analysis personal geodatabase. If you would like to create the script from scratch before starting this exercise, follow the example 'Creating a New Script Module' in Chapter 2 of *Writing Geoprocessing Scripts With ArcGIS*.

Setting up

1. If you don't have ArcCatalog open, start it by using the Programs list in your Start menu.

If you did the previous exercise, skip the next step and go to the next section, 'Adding a new toolbox to Habitat_Analysis.mdb'.

2. Copy the GP_Tutorial folder from arcgis\ArcTutor\Geoprocessing\Results\Ex2 on the drive where you installed ArcGIS to a local drive, for example, your C:\ drive.

Adding a new toolbox to Habitat_Analysis.mdb

You'll first create a new toolbox to hold the tools you'll create in this exercise.

1. Right-click Habitat_Analysis.mdb in your GP_Tutorial folder, point to New, and click Toolbox.

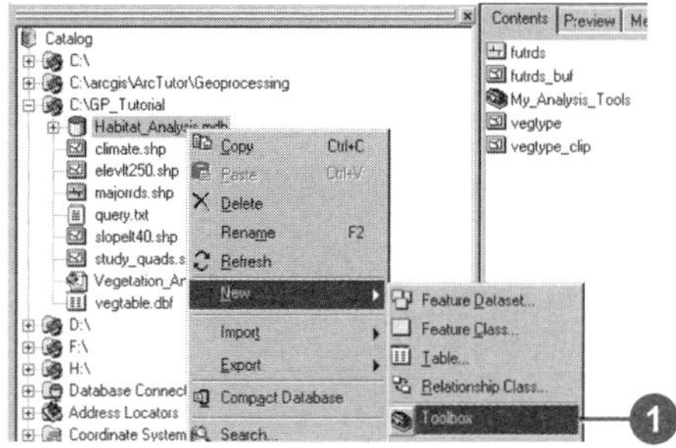

A new toolbox is created inside the geodatabase.

2. Expand your Habitat_Analysis geodatabase to see the created toolbox in the ArcCatalog tree.
3. Right-click the toolbox and click Rename. Type "My_Management_Tools" and press Enter.

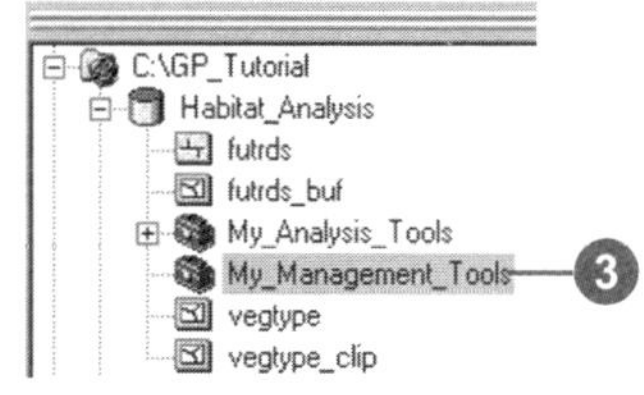

Creating a shortcut to the toolbox

You'll create a shortcut to your My_Management_Tools toolbox in the ArcToolbox window. The toolbox shortcut will point to your My_Management_Tools toolbox stored inside your Habitat_Analysis geodatabase on disk.

1. Right-click your My_Management_Tools toolbox and click Add to ArcToolbox.

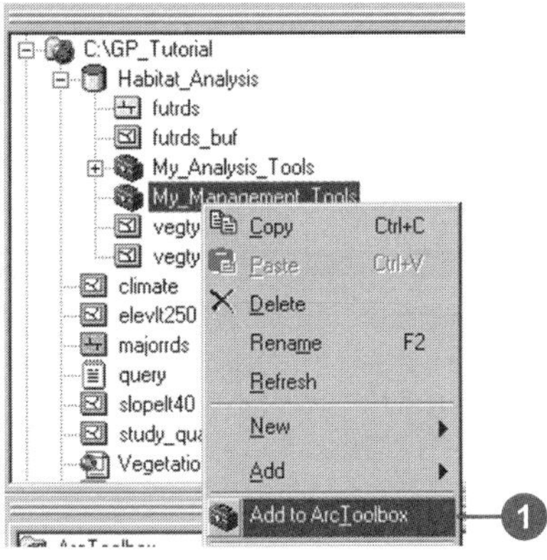

2. If the ArcToolbox window is not displayed, click the Show/Hide ArcToolbox Window button on the Standard toolbar, then collapse all the toolboxes inside the ArcToolbox window so you can see your My_Management_Tools toolbox.

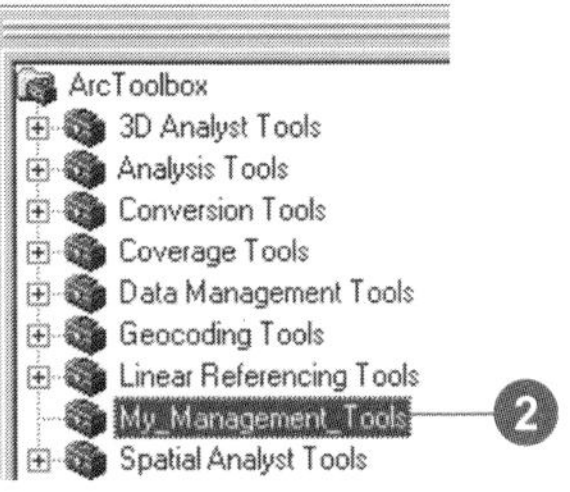

Adding a script to your toolbox

1. Right-click your My_Management_Tools toolbox in the ArcToolbox window, point to Add, and click Script.

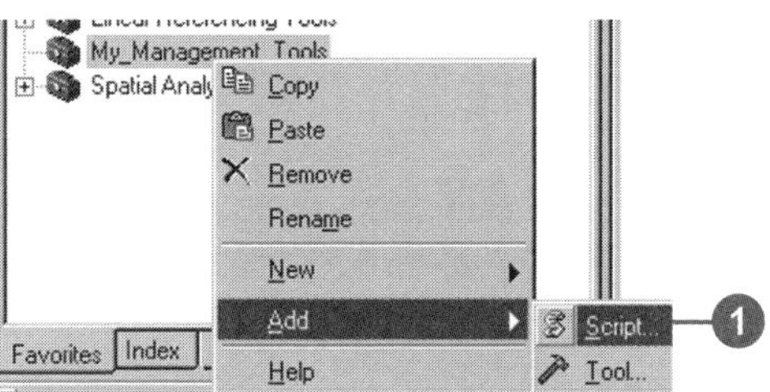

2. Type "Multi_Clip" for the name of the script and type "Clip Feature Classes" for the label.
3. Optionally, add a description to explain what the script will do.

   The script's name can be used when running the script inside another script or at the command line. The label is used as the script's display name in the user interface.

   The default stylesheet will be applied to the dialog box of the script unless specified otherwise. Here you'll use the default, so leave the Stylesheet text box blank.
4. Check Store *relative path* names.

   All pathnames referenced by the script will be stored relative to the location of the toolbox containing the script. Storing relative paths avoids the hassle of repairing pathnames to data sources if the toolbox and associated data sources are moved to a new location. Provided the same directory structure is used at the new location, the tools contained within the toolbox will still be able to find their data sources by traversing the relative paths.

5. Click Next.

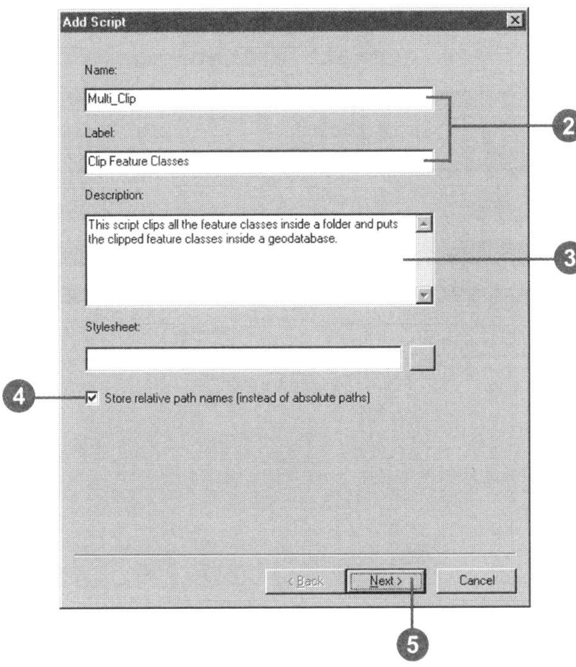

6. Click the Browse button to the right of the Script File text box and navigate to your GP_Tutorial folder.

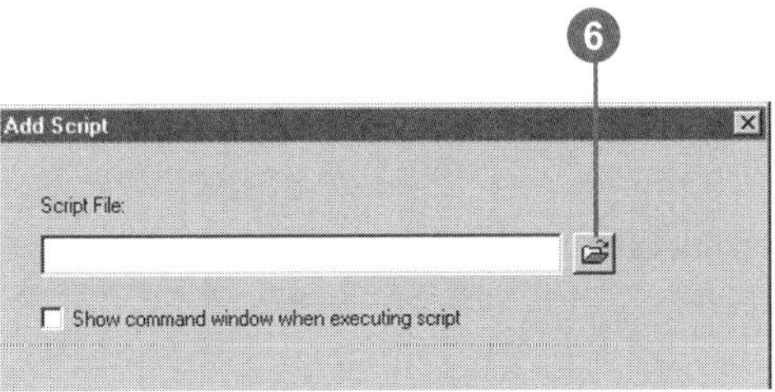

7. Click multi_clip.py and click Open to add the script.

   A *script* is a simple text file that can be created in any text editor. It holds a loosely typed (meaning variable declaration is not required) scripting language, such as JScript, VBScript, or Python. Refer to *Writing Geoprocessing Scripts With ArcGIS* to learn more about writing geoprocessing scripts.

8. Click Next.

   Leave the parameters and their properties as they are at this point.

9. Click Finish.

   The script Clip Feature Classes has been added to your My_Management_Tools toolbox.

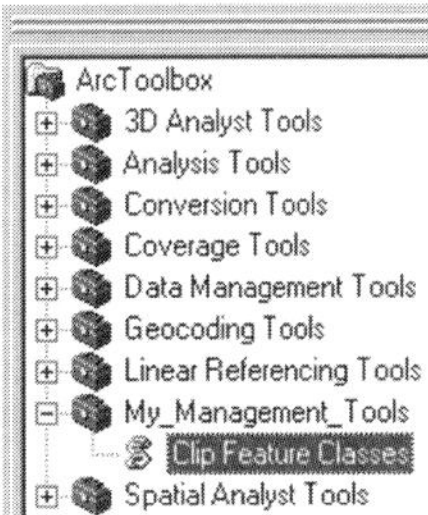

Editing a script

1. Right-click the script in the ArcToolbox window and click Edit.

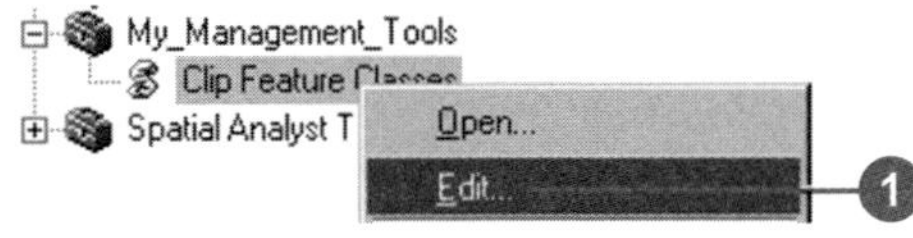

This opens the Python script in the PythonWin application so you can view or modify the contents of the script.

2. Examine the code. Notice that the values set for the variables gp.workspace, clipFeatures, out_workspace, and clusterTolerance are set as system arguments (sys.argv [ ]). Four parameters need to be defined for these system arguments in the Properties dialog box of the script so the parameters will display on the script's dialog box.

   For more information on the contents of this particular script, see 'Creating a New Script Module' in Chapter 2 of *Writing Geoprocessing Scripts With ArcGIS.*

   Note: Instead of using system arguments for the values of variables, you could hard code the value. However, hard coding the value means that only the value specified in the script can be used—for example, the path to a workspace containing input data.

3. Close the PythonWin application.

Setting script parameters

1. Right-click your Clip Feature Classes script in the ArcToolbox window and click Open to open its dialog box.

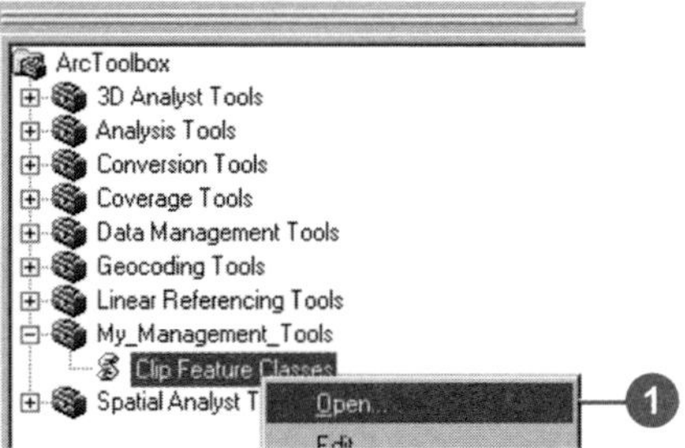

   Notice that there are no parameters on the dialog box. This is because they have not been defined for the four system arguments (sys.argv [ ]) you saw in the code.

2. Click Cancel.

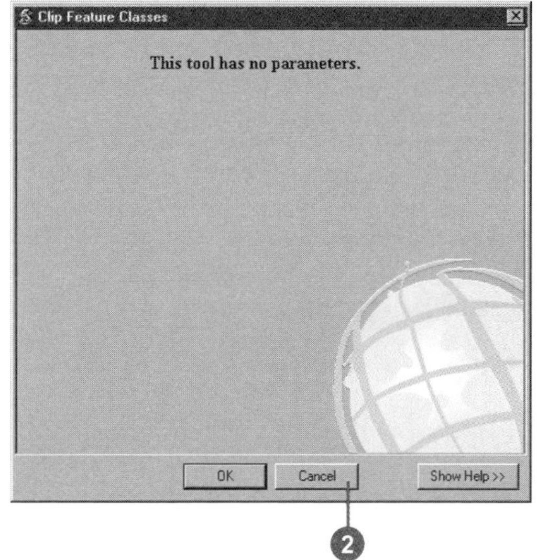

3. Right-click the Clip Feature Classes script and click Properties.

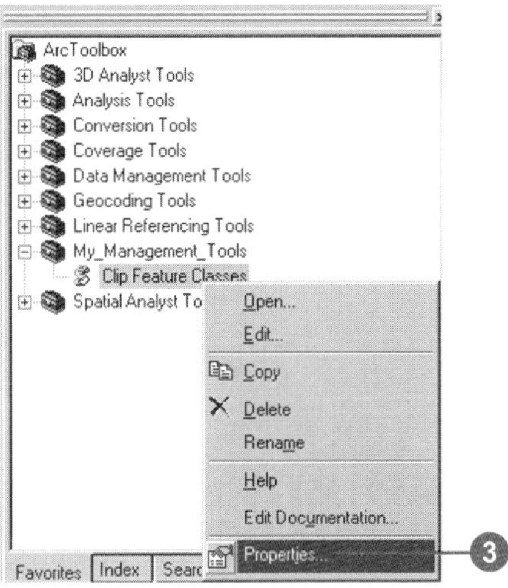

4. Click the Parameters tab.

Setting the Input Folder parameter

1. In the Parameters section, click in the first row under the column heading Display Name.
2. Type "Input Folder" for the name of the parameter.

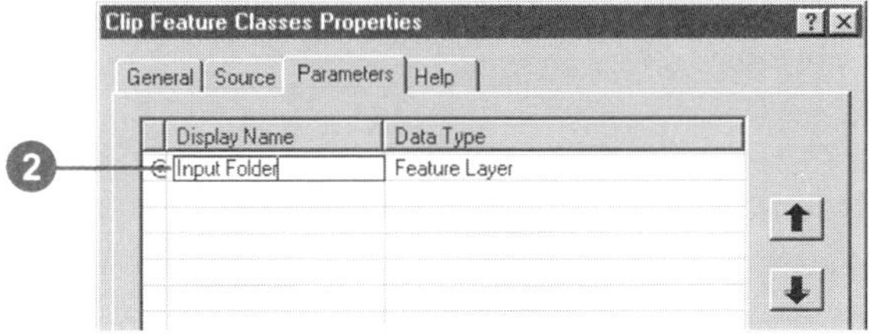

You must specify the type of data a parameter can store.

3. Click the same row under the column heading Data Type to specify the *data type* for the parameter.
4. Scroll through the list of data types and click Folder.

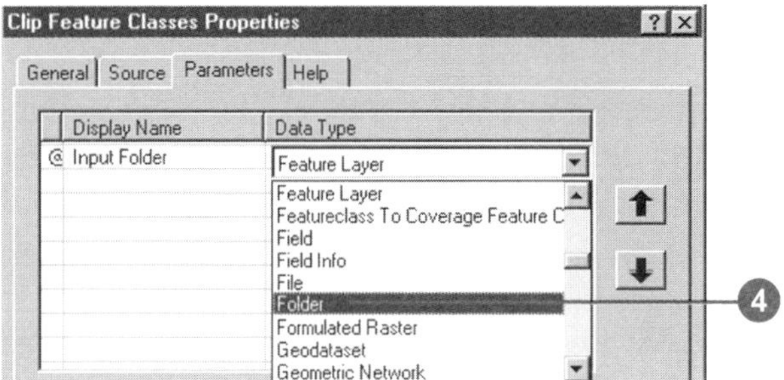

5. Examine the property values set for the Input Folder parameter in the Parameter Properties section of the dialog box.

   The values that could be set for the property Type are either Required or Optional. As it is required to set an input folder, leave the default value, Required.

   The values that can be set for the property Direction are either Input or Output. As this parameter is an input parameter—that is, the folder contains the input feature classes that will be clipped—leave the default value, Input.

   If the input parameter accepted multiple values, such as multiple folders, you would set the MultiValue property value to Yes. However, here you can only process one folder at a time, so leave the default value, No.

   You can specify a default value that will display for the parameter on the tool's dialog box. For the input folder, you want the default value to be the path to your GP_Tutorial folder.

6. Type "C:\GP_Tutorial" for the value of the property Default if your GP_Tutorial folder is on your C:\ drive. Alternatively, type the path to your GP_Tutorial folder.

   This value will be displayed and used by this parameter by default.

   Don't set a value for the Environment property; instead, accept the default. This property will be explained when the cluster tolerance parameter property values are defined.

   Attribute domains are used to constrain the values allowed for the parameter. There are two different types of attribute domains: range domains or coded value domains. A range domain specifies a valid range of values, and a coded value domain specifies a valid set of values for a parameter. For this parameter, an attribute domain is not necessary.

   Some parameters are dependent on the value of other parameters for information, such as a parameter that lists the fields in an input table. The fields can only be displayed once the input data parameter has a value. The input folder is not dependent on any other parameter, so you'll use the default.

   The values for the properties of the parameter Input Folder should now be set up as shown in the graphic that follows.

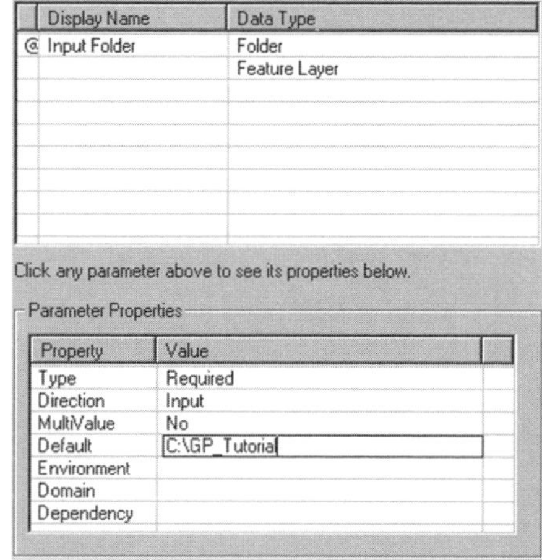

Setting the Input Clip Feature Class parameter

1. Click in the second row under the column heading Display Name.
2. Type "Input Clip Feature Class".
3. Click the same row under the column heading Data Type to specify the data type for the parameter.
4. Scroll up through the list of data types and click Feature Class.
5. Leave the defaults for the values of the properties Type, Direction, and MultiValue, then type the location for GP_Tutorial\study_quads.shp, such as "C:\GP_Tutorial\study_quads.shp", for the value of the property Default. This shapefile will be displayed and used as the value for the parameter by default.

   Your Properties dialog box should now resemble the graphic that follows.

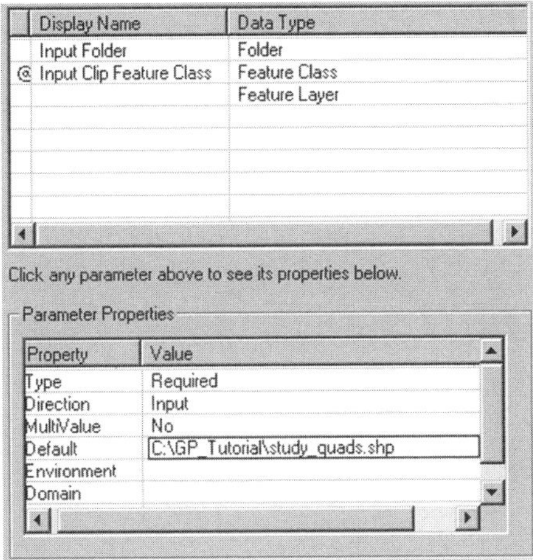

Setting the Output Workspace parameter

1. Click in the third row under the column heading Display Name.
2. Type "Output Workspace".
3. Click the same row under the column heading Data Type to specify the data type for the parameter.
4. Scroll through the list of data types and click Workspace.
5. As this parameter is an input parameter (an existing workspace into which the outputs will be placed) accept the default value of input for the property Direction.
6. For the value of the property Default, type the location for GP_Tutorial\Habitat_Analysis.mdb, such as "C:\GP_Tutorial\Habitat_Analysis.mdb". This personal geodatabase will be displayed and used as the value for the parameter by default. All clipped outputs will, by default, be placed in this location.

The values set for the properties of the parameter Output Workspace should resemble those in the graphic that follows.

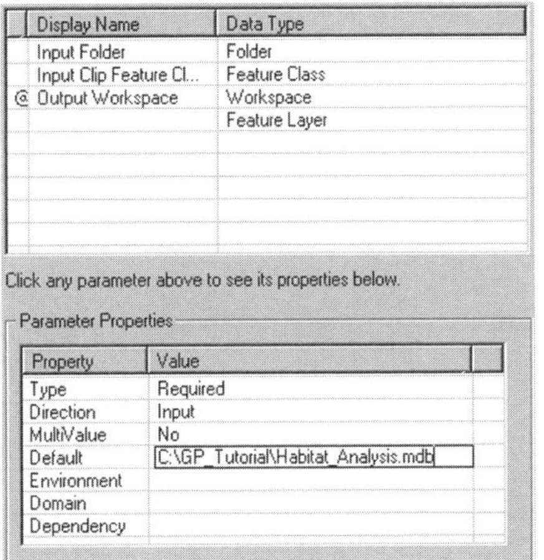

Setting the Cluster Tolerance parameter

1. Click in the fourth row under the column heading Display Name.
2. Type "Cluster Tolerance".
3. Click the same row under the column heading Data Type to specify the data type for the parameter.
4. Scroll through the list of data types and click Linear unit.
5. Click the row for the property Type in the Value column to open the dropdown list of values. Click Optional. It is not required to specify a value for the cluster tolerance parameter. A default value will be used if there is no value set.

   Leave the defaults for the properties Direction, MultiValue, and Default. You could type a value to use as the default. Instead, you'll take the value set for cluster tolerance from the Environment Settings dialog box to mimic the behavior of the system tools.
6. Click the row for the property Environment in the Value column to open the dropdown list of environment setting values. Click Cluster Tolerance. The value set for the cluster tolerance in the Environment Settings dialog box will be displayed as the value for the parameter by default. If the value set in the Environment Settings dialog box is changed, the value specified there will be used by the Cluster Tolerance parameter.

   Leave the default for the Domain property. If you wanted to limit the range of values that could be set for the parameter, you could set up a domain. The cluster tolerance depends on the input data, so you won't set a domain for this parameter.

The values set for the properties of the Cluster Tolerance parameter should resemble those in the graphic that follows.

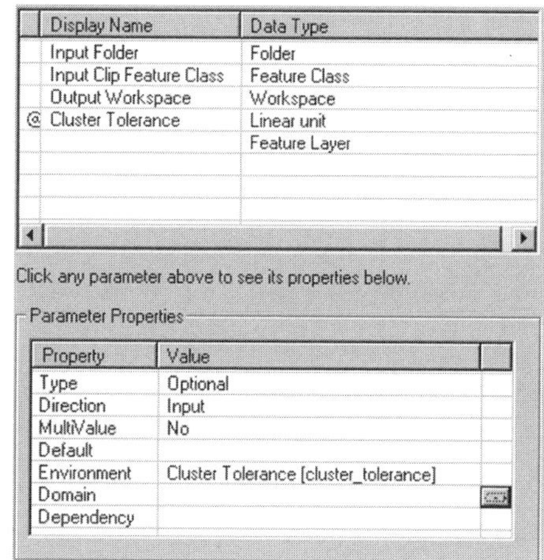

7. Click OK.

Running the script

1. Double-click the Clip Feature Classes script in the ArcToolbox window to open it. The parameters you defined are displayed on its dialog box.

   Notice how each parameter contains a default value. For the Input Folder, Input Clip Feature Class, and Output Workspace parameters, these are the values you set up for the Default property of each parameter.

   The *cluster tolerance* is the distance that determines the range in which features are made coincident. You'll leave the default (blank) and the units Unknown for the Cluster Tolerance parameter. You set the value for the Cluster Tolerance parameter to be taken from the Environment Settings dialog box when you set up the properties for this parameter. As you haven't specified a value for this parameter in the Environment Settings dialog box, a default value will be calculated. The default value calculated will ensure the precision of the data, as it always integrates the data at the smallest possible cluster tolerance value.

   The units for the Cluster Tolerance parameter are set to Unknown. The units given to the spatial reference of the output feature classes will be used by the Cluster Tolerance parameter. If you were to set the units to be anything other than Unknown, they would be converted to the same units as the output. The output feature classes get their units from the spatial reference set for the first input, that is, the first shapefile in the GP_Tutorial folder to be clipped, provided you haven't specified otherwise in the Spatial Reference section of the Environment Settings dialog box.

2. Click OK.

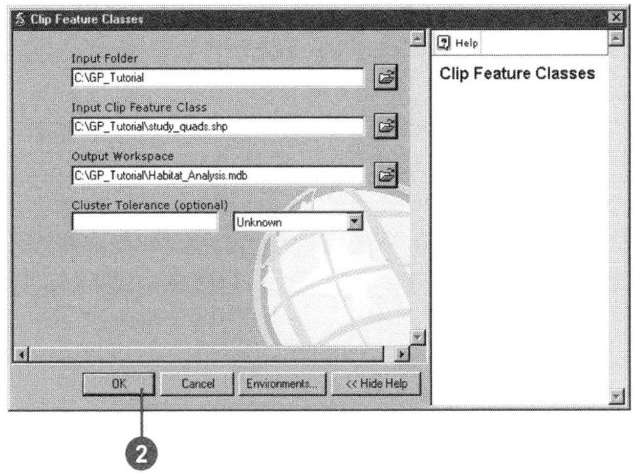

The clipped feature classes are added to your Habitat_Analysis geodatabase. You will use these feature classes in the next exercise to locate potential high-quality habitat areas for the California gnatcatcher.

Deleting unnecessary datasets

You can now delete the shapefiles and the DBF table you copied locally to your GP_Tutorial folder as they are not needed for the rest of this tutorial.

1. In the ArcCatalog tree, click the connection to the tutorial data, which is your GP_Tutorial folder. In the Contents tab, click each shapefile and the DBF file; press Ctrl to select more than one item.
2. Right-click the selection and click Delete.

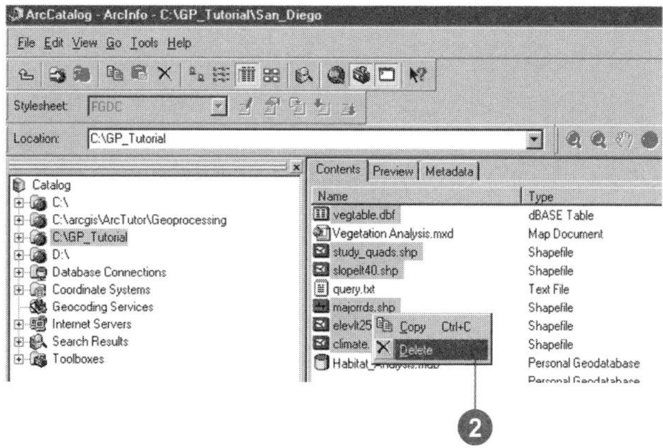

Closing ArcCatalog

1. Click the File menu and click Exit to close ArcCatalog.

   You can continue on to Exercise 4 or stop and complete the tutorial at a later time. If you do not move on to Exercise 4 now, do not delete your GP_Tutorial folder, and do not remove the folder connection that accesses it.

Exercise 4: Finding high-quality habitat patches

Creating a model provides a way to run a sequence of geoprocessing tools by enabling you to string your tools together in a visual diagram. A model can be run again and again, and you can alter parameter values to experiment with different outcomes.

Your task in this exercise is to locate high-quality habitat patches—areas that have the best potential to support the California gnatcatcher. You will add data from your Habitat_Analysis geodatabase to the table of contents of your ArcMap session. You'll then create a new model to find potential habitat areas, incorporating the following five criteria:

1. Roads form a barrier for the gnatcatcher. The impact of roads on gnatcatcher habitat increases proportionately with the road size.
2. The gnatcatcher prefers to live in areas where the vegetation is of a particular type, most preferably, San Diego coastal sage scrub, though other vegetation types are also suitable.
3. Parent gnatcatchers drive fledglings off their territory once they are ready. Fragmentation of existing habitat has meant young gnatcatchers cannot move very far away. For the species to survive long term, they need sizable patches of suitable vegetation, though home ranges tend to be smaller nearer the coast and larger in the drier, sparser inland areas. Nearer the coast—within the Maritime (zone 1) and Coastal (zone 2) climate zones—vegetation patches greater than and equal to 25 acres (1,089,000 ft^2) will be considered suitable. In the inland areas—within the Transitional climate zone (zone 3)—vegetation patches greater than and equal to 50 acres (2,178,000 ft^2) will be considered suitable.
4. Gnatcatcher sightings have been highly correlated with elevation. Areas below 250 m in elevation have contained the majority of all documented sightings in the past. Only areas where the elevation is less than 250 m above sea level will be considered as suitable habitat locations.
5. The gnatcatcher prefers to nest in areas where the slope of the terrain is minimal. Habitat on slopes that are steeper than 40 percent have low potential for nesting. Only areas where the slope of the terrain is less than 40 percent will be considered as suitable habitat locations.

Setting up

If you did the previous exercise, skip the next step and go to the next section ('Starting ArcMap and adding data').

1. Copy the GP_Tutorial folder from C:\arcgis\ArcTutor\Geoprocessing\Results\Ex3 to your C:\ drive or an alternative drive.

This exercise will take approximately 40 minutes to complete.

The following diagram shows the work flow you'll go through in this exercise to identify high-quality habitat patches.

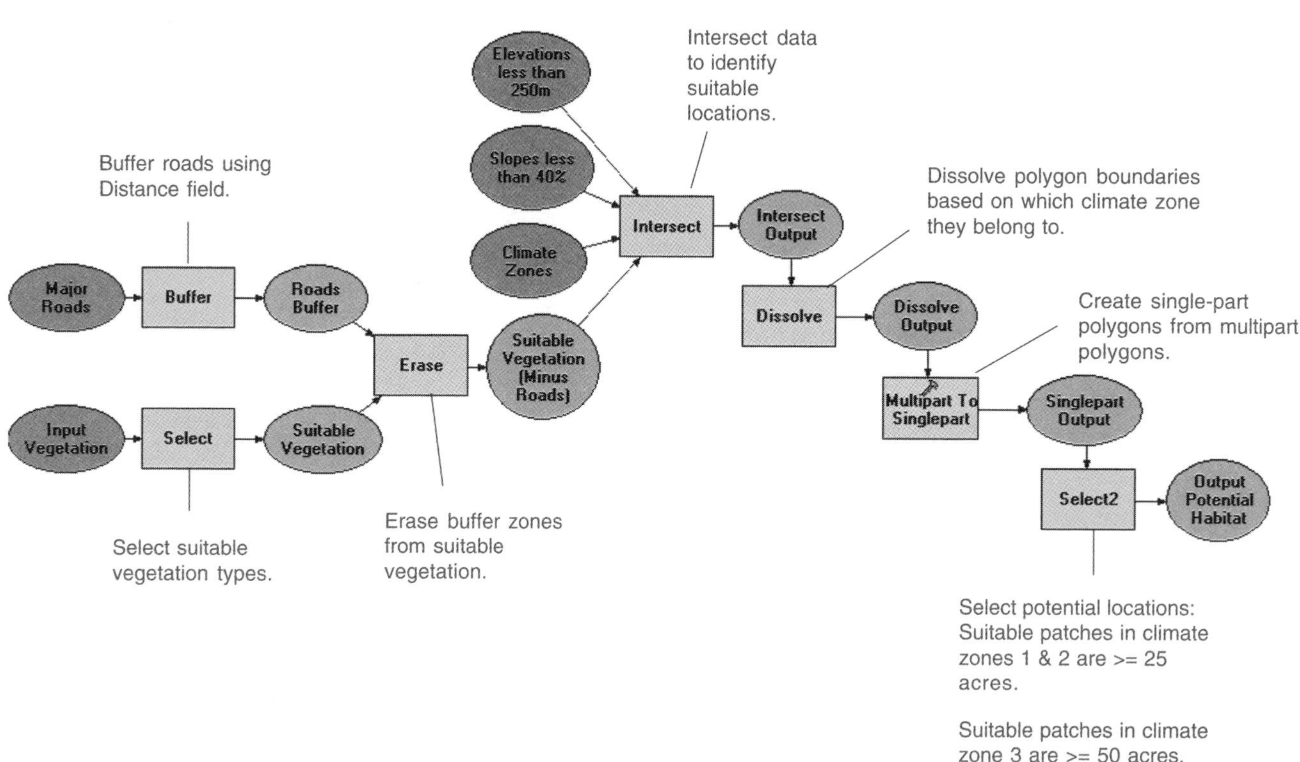

Starting ArcMap and adding data

You'll start by opening ArcMap and adding the feature classes you created in the Habitat_Analysis geodatabase.

1. Start ArcMap by using the Programs list in your Start menu.
2. Click OK to open a new, empty map.
3. Click the Add Data button on the Standard toolbar, then double-click the folder connection to your local copy of the tutorial data, for example, C:\GP_Tutorial.
4. Double-click the Habitat_Analysis geodatabase to open it and select all feature classes except vegtype_clip and futrds_buf; use the Ctrl key to select multiple feature classes.
5. Click Add.

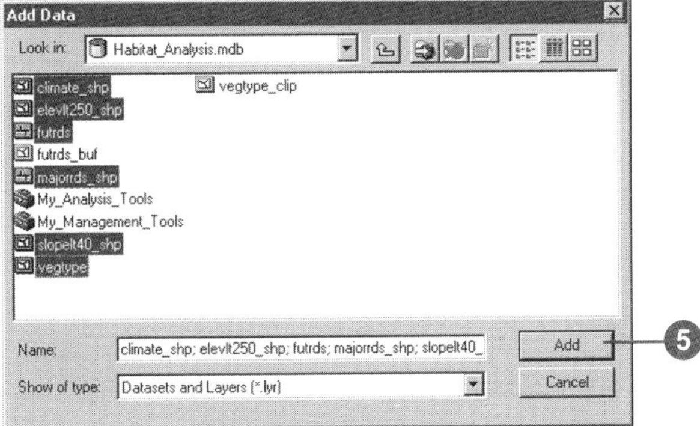

The feature classes are added to your ArcMap session as layers.

Opening the ArcToolbox window

When working in ArcMap, you access toolboxes through the ArcToolbox window. By saving the map document you save any changes made to the ArcToolbox window. This allows you to have different versions of the ArcToolbox window with each map document you create, depending on which tools you want to use for each project.

1. Click the Show/Hide ArcToolbox Window button on the Standard toolbar to open the ArcToolbox window if it is not already open.

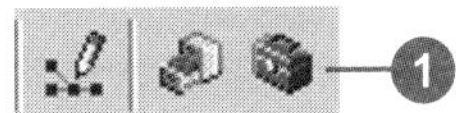

2. If the ArcToolbox window does not display under the ArcMap table of contents, click and drag the bar at the top of the ArcToolbox window and place it over the table of contents, then drop the panel.

Creating a personal geodatabase to hold intermediate results

The *Command Line window* consists of a command line and a message section. All tools added to the ArcToolbox window can be executed from the command line as well as from their dialog boxes. The *command line* provides a quick way to execute tools if you are familiar with their parameters.

You'll now create a personal geodatabase (Temp_Results.mdb) by running a tool at the command line. This personal geodatabase will hold intermediate data that you'll create to produce the final result. Results you want to keep will be placed inside the existing geodatabase, Habitat_Analysis.mdb.

1. Click Show/Hide Command Line Window on the Standard toolbar to open the window.

As with the ArcToolbox window, the Command Line window can be docked anywhere inside any ArcGIS Desktop application.

2. Press and hold the Ctrl key while you click and drag the bar at the top of the Command Line window and position it at the lower-left corner of your ArcCatalog application. Holding the Ctrl key prevents the window from docking to the application.
3. Release the Ctrl key and drop the panel.

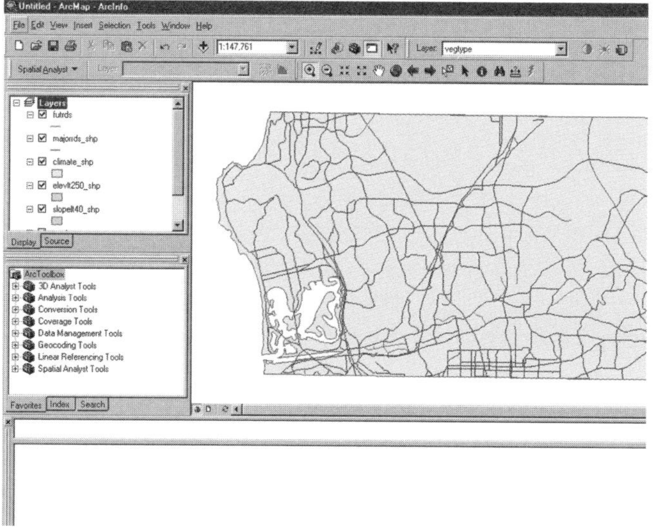

The window will dock at the bottom of your application. If you find that it is not docked in the position shown, click the Command Line title bar to snap the window into position. If the Draw toolbar is present, click View, then Toolbars, and uncheck Draw.

The Command Line window contains a prompt (a command line) where you can run tools and a message window, showing all the messages associated with running tools—at the command line, through a tool's dialog box, by running a model from within a ModelBuilder window or any messages returned after running a standalone script.

AutoComplete enables you to type part of the tool name until you see the name of the tool displayed in the dropdown list. You can then press the Space key to automatically complete the rest of the tool name. Once the tool name is entered, the tool's usage appears as a ToolTip. It prompts you for the next parameter value that must be specified by highlighting the parameter in bold.

4. Type "CreatePersonalGDB", then press the Space key.

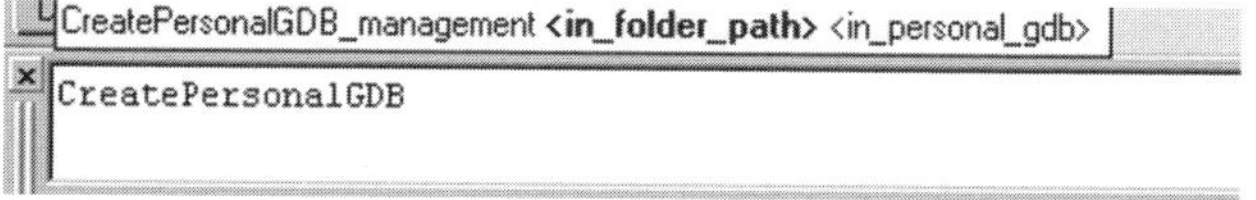

Note: The Data Management Tools toolbox must be added to the ArcToolbox window to run tools from this toolbox at the command line. If the toolbox is not added, right-click the ArcToolbox window and click Add Toolbox. Navigate to the Toolboxes folder at the root level of the ArcCatalog tree and locate the Data Management Tools toolbox, then add it to the ArcToolbox window.

5. Type “C:\GP_Tutorial” (or your alternative drive) for the value of the <in_folder_path> parameter, then type a space.
6. Type “Temp_Results.mdb” for the value of the <in_personal_gdb> parameter. This is the personal geodatabase to create inside your GP_Tutorial folder.

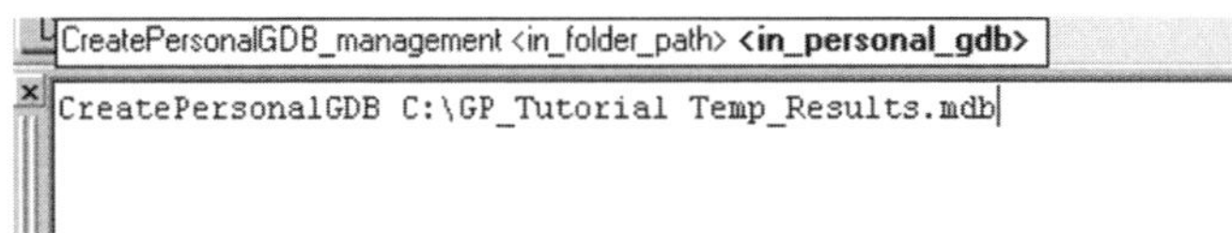

7. Press Enter to run the CreatePersonalGDB tool.

   A new personal geodatabase called Temp_Results.mdb is placed in your GP_Tutorial folder.

Setting workspaces

You'll set the current workspace to point to your Habitat_Analysis personal geodatabase. You'll also set up a *scratch workspace* to point to your Temp_Results personal geodatabase. When building a model it is useful to set up your workspaces in this manner. In each tool's dialog box, you can simply leave the default path provided (to your scratch workspace) for most outputs. For those outputs you want to keep separate, such as the final result from running the model, you can blank out the default path and type a name for the output from the tool. Such outputs will be placed in the location set for your current workspace. You'll see this as you build the model in this exercise.

1. Right-click the ArcToolbox folder and click Environments.

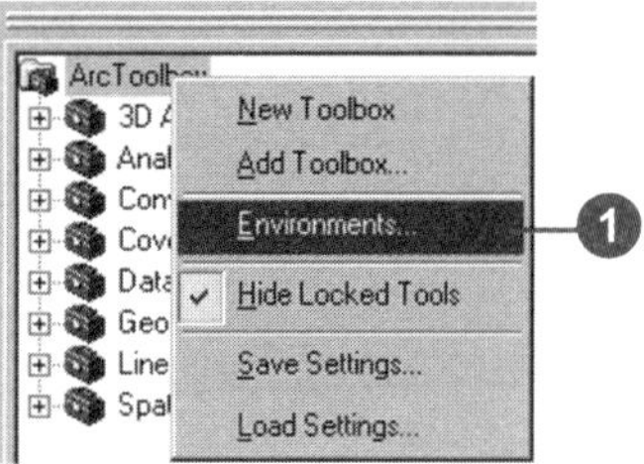

2. Click the General Settings dropdown arrow to expand its contents. For the value of the Current Workspace environment setting, type “C:\GP_Tutorial\Habitat_Analysis.mdb”, or type your alternative location. Alternatively, click the Browse button to navigate to this location.
3. For the value of the Scratch Workspace environment setting, type “C:\GP_Tutorial\Temp_Results.mdb”, or type your alternative location. Alternatively, click the Browse button to navigate to this location.
4. Click OK.

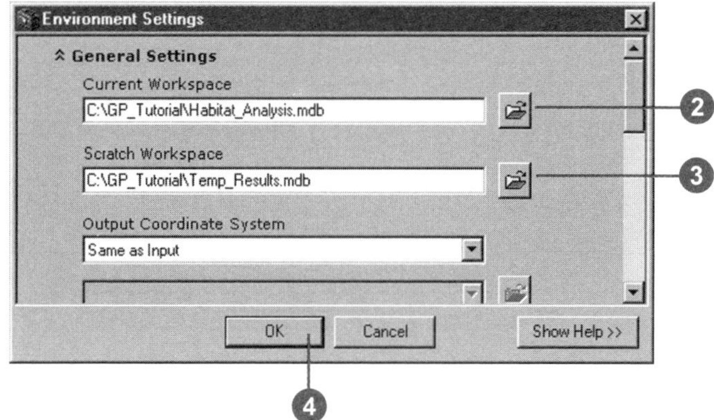

Adding the My_Analysis_Tools toolbox

You'll first add your My_Analysis_Tools toolbox to the ArcToolbox window within your ArcMap session.

1. Right-click the ArcToolbox window and click Add Toolbox.

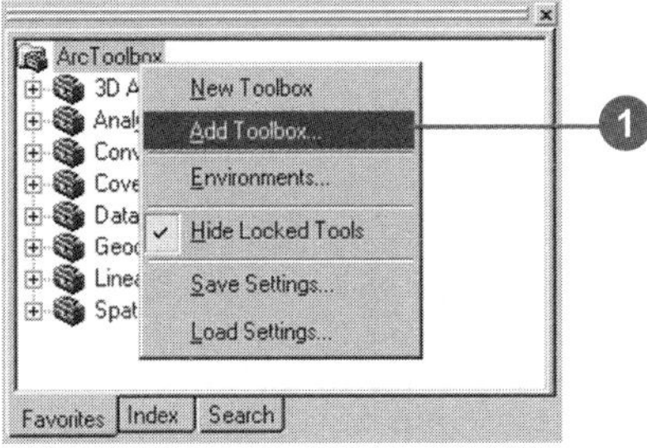

2. Click the Look in dropdown arrow and navigate to Habitat_Analysis.mdb in your GP_Tutorial folder.
3. Click My_Analysis_Tools and click Open.

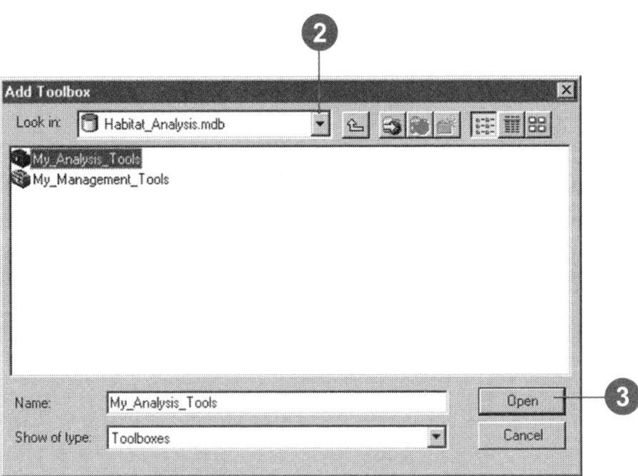

The toolbox is added to the ArcToolbox window.

Saving the map document

Geoprocessing settings include environment settings, the state of the ArcToolbox window, and variables you have created at the command line. These settings are saved in ArcMap when you save a map document. There are times when you want to switch between applications, have different settings for different projects, or use the default settings. In these cases you can save the geoprocessing settings by right-clicking the ArcToolbox window and clicking Save Settings. The default settings or saved settings in a file can be loaded at any time by right-clicking the ArcToolbox window and clicking Load Settings.

1. Click File on the Main menu and click Save As.
2. Click the Save in dropdown arrow and navigate to your GP_Tutorial folder.
3. Type "Habitat_Analysis.mxd" for the filename and click Save.

Creating a new model

You will now create a model by building processes and connecting them together to find areas with the best potential to support the gnatcatcher.

1. Right-click your My_Analysis_Tools toolbox in the ArcToolbox window, point to New, and click Model.

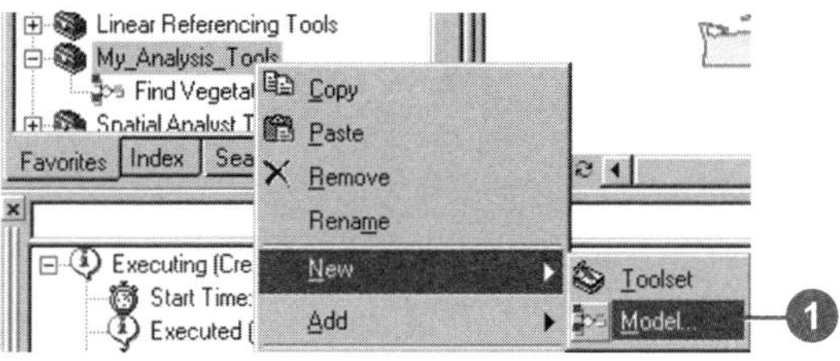

An empty ModelBuilder window will open.

Selecting suitable vegetation types

You'll select from the vegetation feature class (vegtype) those vegetation types preferred by the gnatcatcher. You'll use the Select tool to select all suitable vegetation types (polygons with a Habitat value of 1).

1. Expand the Extract toolset located inside the Analysis Tools toolbox.
2. Click and drag the Select tool into the display window.

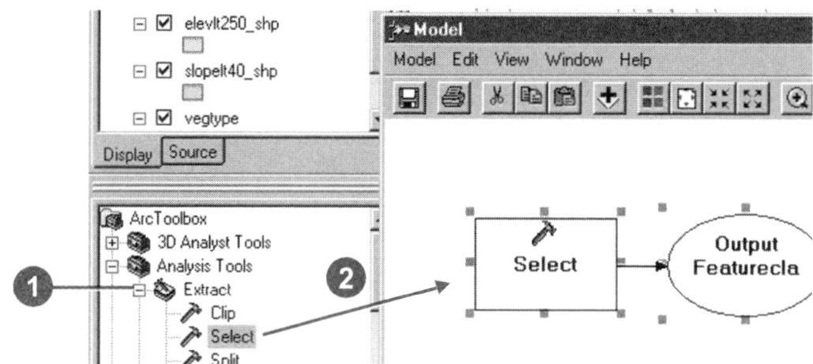

An element that references the Select tool is created in the display window.

3. Right-click the Select tool element and click Open.

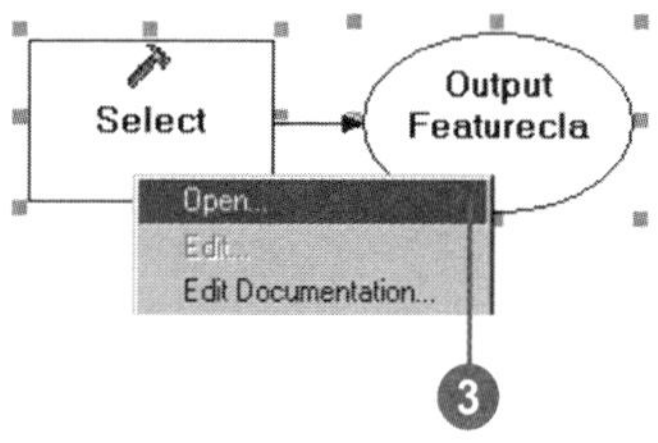

4. Click the Input Features dropdown arrow and click the layer vegtype.

   Notice that layers from the ArcMap table of contents have a yellow layer icon to the left of them to identify them as layers.

5. Since the Scratch Workspace was set earlier, the default path given for the Output Feature Class parameter is set to this location. Leave the default location and type "select_output" for the name.

6. Type "[HABITAT] = 1" for the value of the Expression parameter, or click the Browse button next to the Expression parameter and use the Query Builder to build the expression.

   When all parameters have valid values, the green circles to the left of each parameter will be removed, indicating that a valid value has been set for each parameter. If you have entered an invalid value for a parameter, a red cross icon will display to the left of the parameter. By placing the cursor over such icons, you can obtain information to help solve the problem with the current parameter value that is set.

7. Click OK.

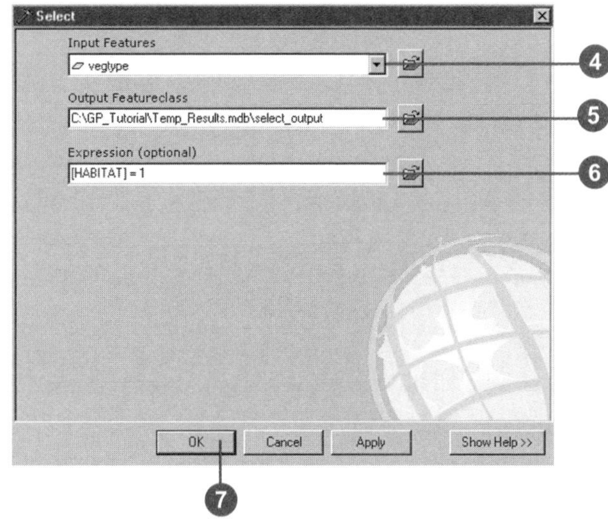

8. Click Auto Layout then click Full Extent to apply the current diagram properties to the elements and to place the elements within the display window.

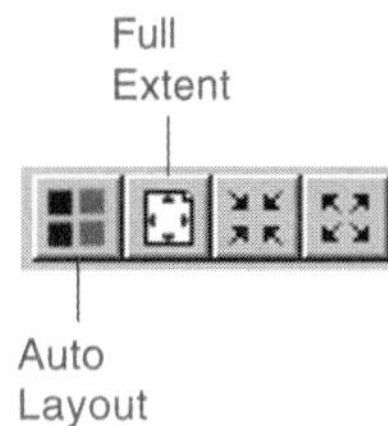

Notice that the process is now colored in, meaning it is ready to run.

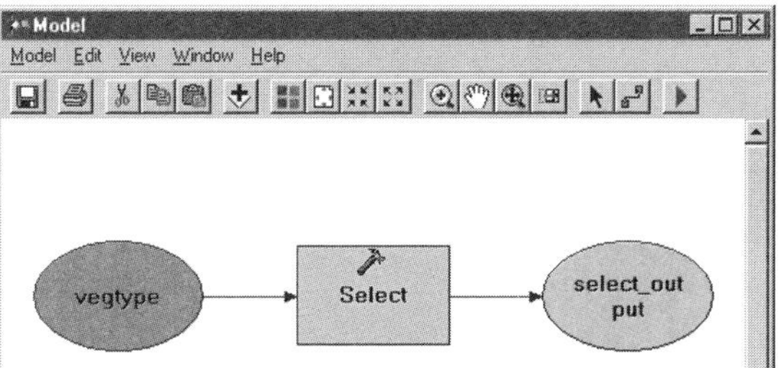

9. Right-click the select_output derived data element and click Rename.

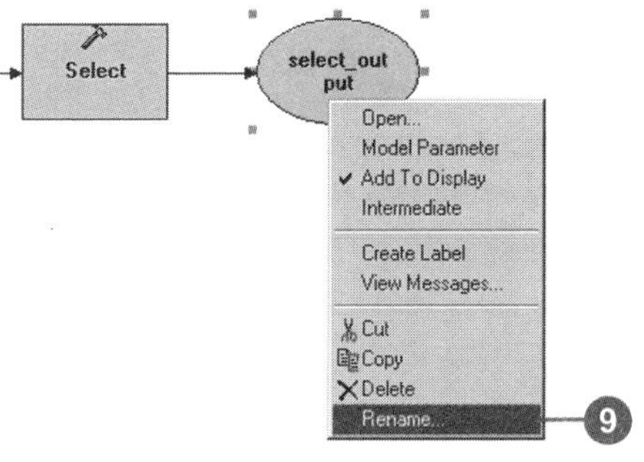

10. Type "Suitable Vegetation" and click OK.

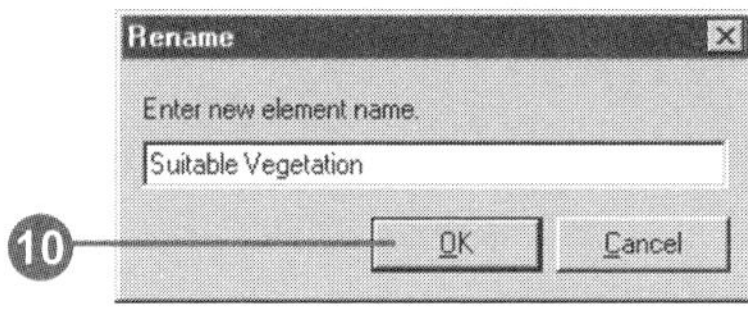

11. Right-click the Suitable Vegetation data element and click Add To Display.

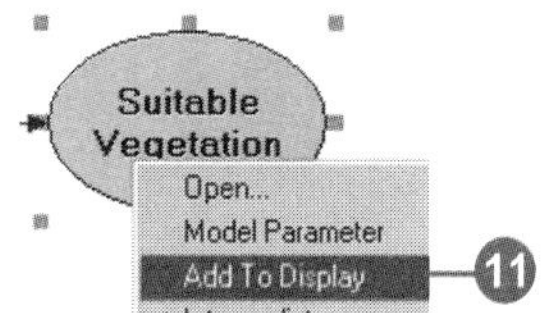

With the Add To Display property checked, the derived data referenced by this element will be added to the display each time the model is run.

12. Right-click the Select tool element and click Run to run the process.

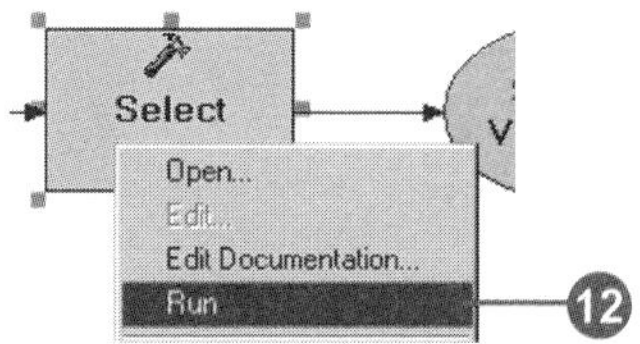

Notice that as the process runs, its progress is documented in the message section of the Command Line window, and the tool element that references the tool is highlighted in red. When the process has finished, the tool and its derived data element become shaded, indicating that the process has run and the derived data has been created on disk.

13. Check 'Close this dialog when completed processing' if the Select progress dialog box is present, then click Close.

14. Examine the layer select_output in your ArcMap display. All vegetation types that are preferable for the gnatcatcher, including areas of San Diego coastal sage scrub, have been selected from the vegetation data.

Excluding the roads

You'll now exclude the areas near roads and freeways as potential habitat locations because roads form a barrier for the gnatcatcher, and there are impacts on the habitat patches near roads. You'll buffer the roads using the Distance field in the majorrds_shp attribute table, then erase the resultant buffer zones from the suitable vegetation patches.

Buffering the roads

1. Expand the Proximity toolset located inside the Analysis Tools toolbox.
2. Click and drag the Buffer tool into the display window.

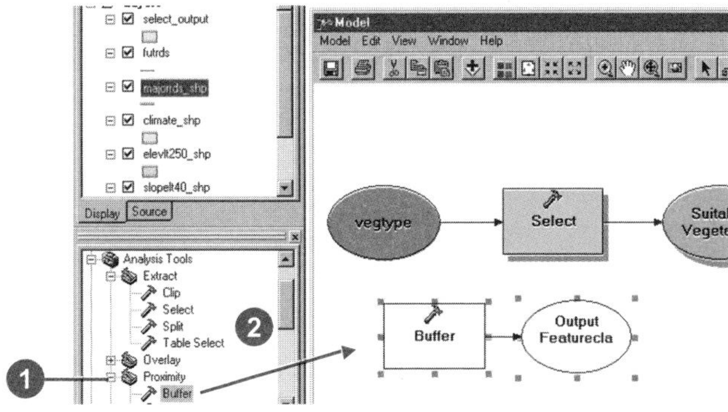

3. Right-click the Buffer tool element that represents the Buffer tool in the model and click Open.

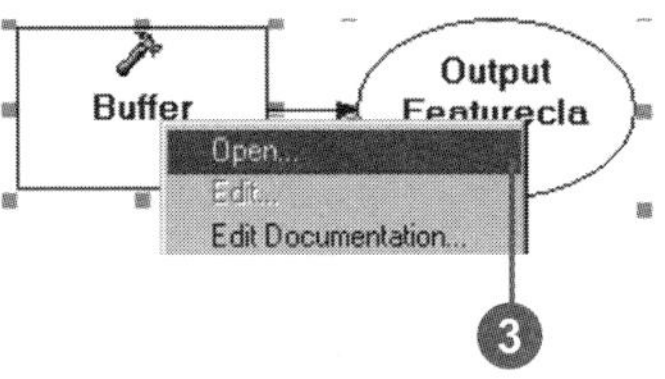

4. Click the Input Features dropdown arrow and click the layer majorrds_shp.
5. Leave the default location for the Output Feature Class parameter and type "buffer_output" for the name.
6. Click Field, click the Field dropdown arrow, and click Distance. The values in this field will be used for the width of the buffer.
7. Click the Dissolve Type dropdown arrow and click LIST to list the dissolve fields.
8. Check Distance in the Dissolve_Field(s) list. All polygons that share the same distance value will be dissolved.
9. Leave the defaults for the rest of the parameter values and click OK.

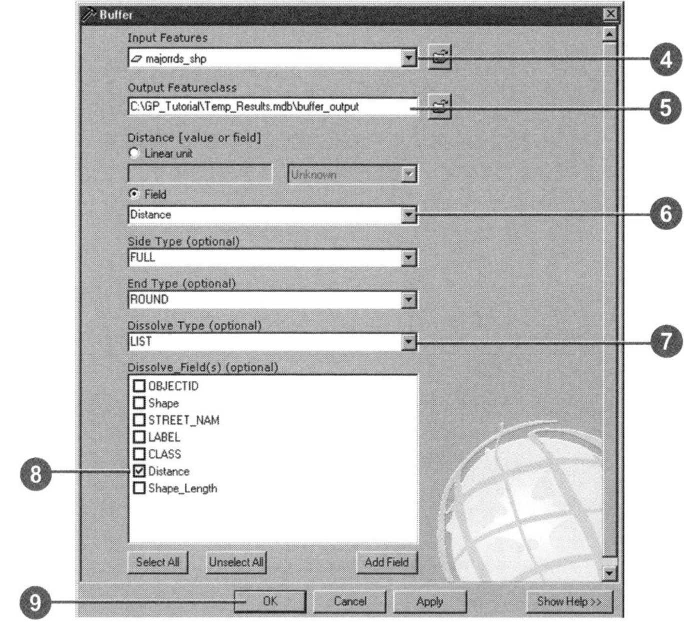

10. Click Auto Layout then click Full Extent. Right-click the buffer_output derived data element and click Rename. Type "Roads Buffer" and click OK.
11. Right-click Roads Buffer and click Add To Display.
12. Right-click the Buffer tool element and click Run to run the process.

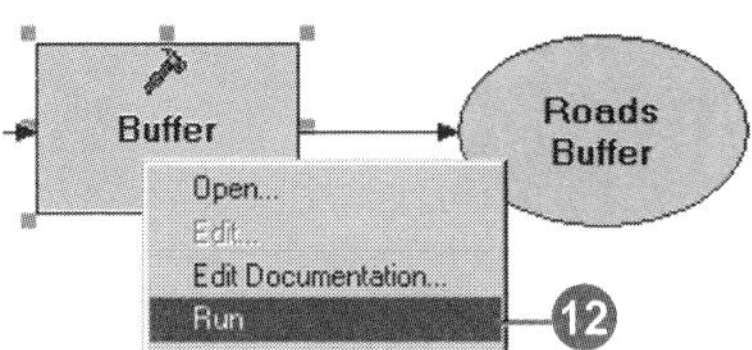

13. Examine the layer buffer_output in your ArcMap display.

Erasing the roads

You'll now erase the buffer zones from the suitable vegetation patches to exclude these areas from the analysis.

1. Expand the Overlay toolset located inside the Analysis Tools toolbox.
2. Click and drag the Erase tool into the display window.

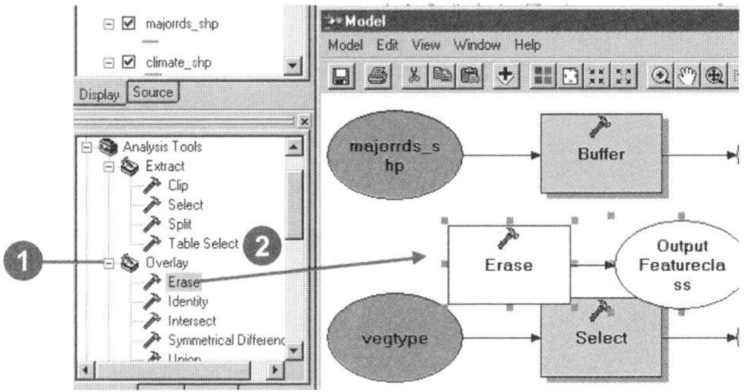

3. Right-click the Erase tool in the model and click Open.
4. Click the Input Features dropdown arrow.

   Notice that you now have four variables at the top of the list of inputs as well as all the layers in the table of contents to choose from.

   When you drag data into a ModelBuilder window or when you set the data referenced by input or derived data elements within a tool's dialog box, the elements created in the ModelBuilder window are variables that can be shared between processes.
5. Click the variable Suitable Vegetation from the dropdown list. Note the blue icon indicating that it is a variable.
6. Click the Erase Features dropdown arrow and click the variable Roads Buffer.
7. Leave the default location for the value of the Output Feature Class parameter and type "erase_output" for the name.

   Leave the default value (blank) for the Cluster Tolerance parameter and leave the units as Feet. An appropriate default value will be calculated using the units of the spatial reference set for the output (in this case, in feet, as the input spatial reference units are in feet).

   All lines within the default cluster tolerance calculated will be considered coincident.
8. Click OK.

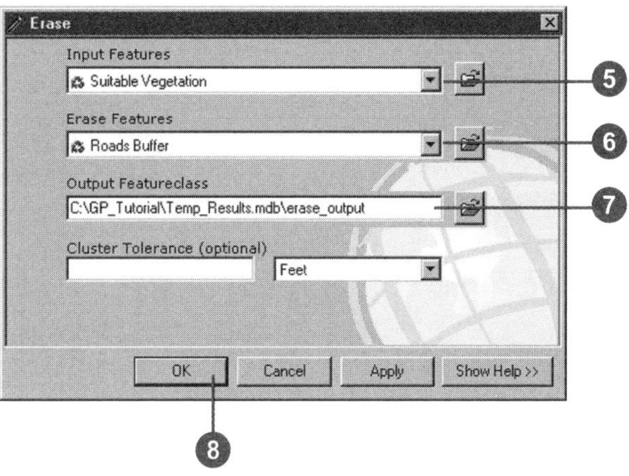

9. Click Auto Layout then Full Extent to display all model elements in their default position within the display window.
10. Right-click the erase_output derived data element and click Rename. Type "Suitable Vegetation (Minus Roads)" and click OK.
11. Click the element Suitable Vegetation (Minus Roads) to select it, then click the blue handles and drag them to resize the element so all the text is displayed.
12. Right-click the Suitable Vegetation (Minus Roads) derived data element and click Add To Display.
13. Right-click the Erase tool and click Run.

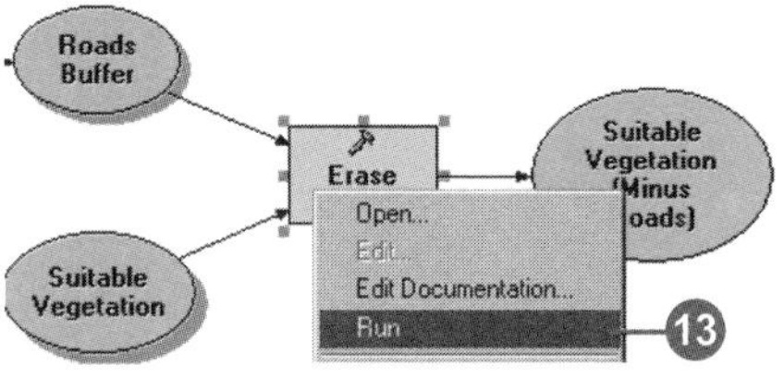

It is good practice to save your changes to the model as you are building it.

14. Click the Model menu and click Save.

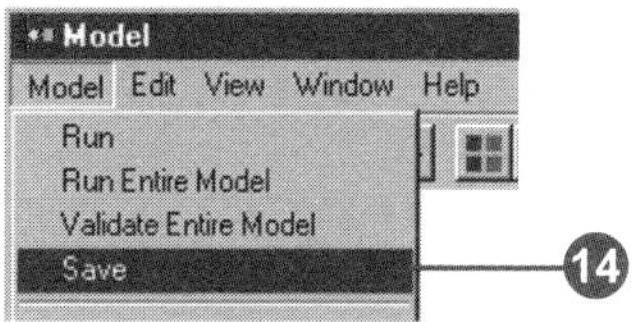

The model is saved with its default name. You'll rename the model later.

15. Examine the layer erase_output in your ArcMap display. It shows all vegetation patches that are away from existing roads and are suitable for habitation by the gnatcatcher.

Selecting potential habitat locations

You'll now use four tools to find potential habitat locations, incorporating the data referenced by the Suitable Vegetation (Minus Roads) element you have derived. You want to find locations where the terrain is less than 40 percent in slope and less than 250 m in elevation as well as where the area of the suitable locations is greater than or equal to 25 acres in climate zones 1 and 2 and greater than or equal to 50 acres in climate zone 3.

Intersecting feature classes

You'll first intersect slopelt40_shp, elevlt250_shp, climate_shp, and Suitable Vegetation (Minus Roads) to derive a dataset of suitable patches.

1. Expand the Overlay toolset located inside the Analysis Tools toolbox.
2. Click and drag the Intersect tool into the display window.

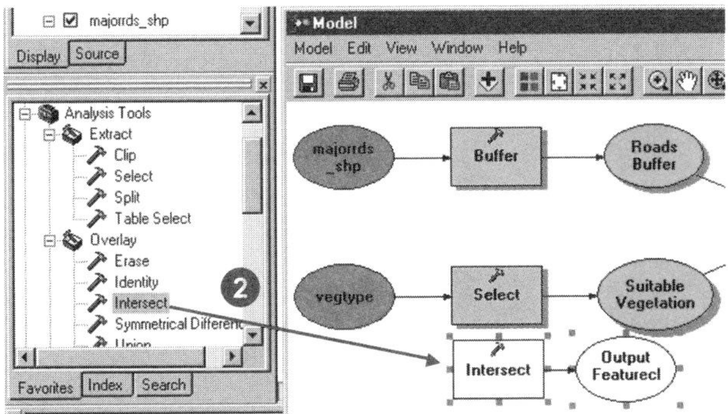

3. Right-click the Intersect tool and click Open.
4. Click the Input Features dropdown arrow and click the layer climate_shp. The layer is added to the features list.
5. Click the Input Features dropdown arrow again and click the layer slopelt40_shp.
6. Click the Input Features dropdown arrow again and click the layer elevlt250_shp.
7. Click the Input Features dropdown arrow again and click the variable Suitable Vegetation (Minus Roads).

   You'll use the default and will not assign any ranks to the input feature classes. The default takes an average if the distance between feature classes is less than the cluster tolerance. You would assign ranks if you knew that some of your feature classes were more integrally sound than others. Features in the feature classes with lower ranks will snap to features in the feature classes with higher ranks; the highest rank is 1, and lower ranks go up in value.
8. Leave the default location for the parameter Output Feature Class and type "intersect_output" for the name.

   Leave the default to join all attributes.

   Leave the default for the value of the Cluster Tolerance parameter and the units as Feet. The default (blank) means that an appropriate default value will be calculated, using the units of the spatial reference set for the output, in this case, in feet, as the input spatial reference units are in feet.

   All lines within the default cluster tolerance will be considered coincident.

   Leave the default for the Output Type so the geometry of the output feature class will be the same as the input with the lowest geometry type (polygon in this case).
9. Click OK.

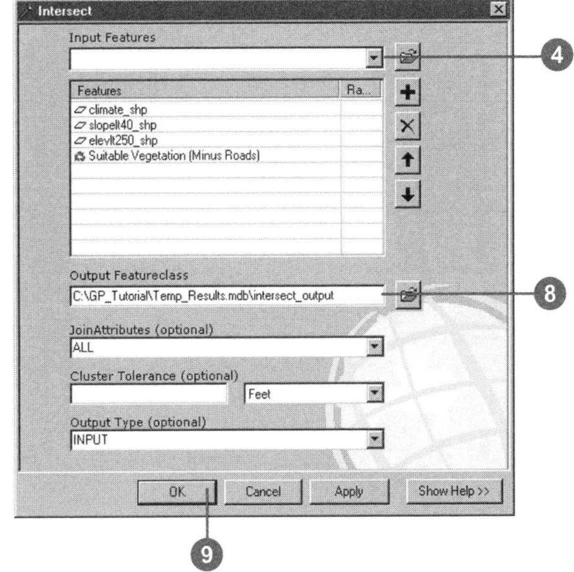

10. Click Auto Layout, then Full Extent to display all the model elements in their default position in the display window.

11. Right-click the intersect_output derived data element and click Rename. Type "Intersect Output" and click OK.

    Your model should now resemble the following graphic.

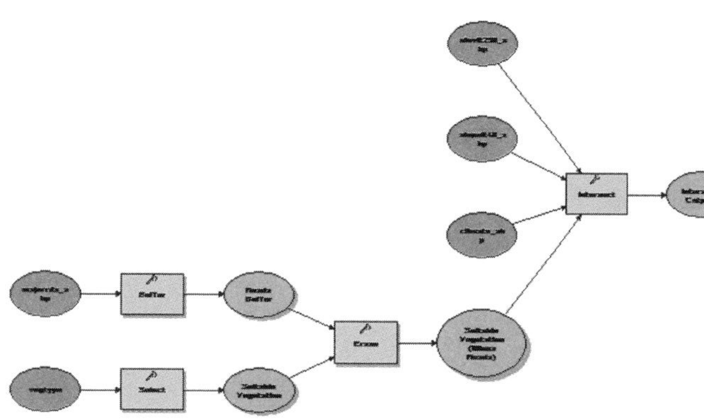

12. Right-click the Intersect Output derived data element and click Add To Display.

13. Right-click the Intersect tool and click Run.

14. Examine the layer intersect_output in your ArcMap display.

15. Click the Model menu and click Save.

Dissolving polygons

You'll now dissolve the polygons of intersect_output to remove unnecessary boundaries between polygons. In the final tool you'll select polygons based on the climate zone they belong to. Zones 1 and 2 both have a CLIMATE_ID of 2 as they are both close to the coast, and zone 3 has a CLIMATE_ID of 3, representing areas farther inland. You'll dissolve all polygons with the same CLIMATE_ID.

1. Expand the Generalization toolset located inside the Data Management Tools toolbox.

2. Click and drag the Dissolve tool into the ModelBuilder window.

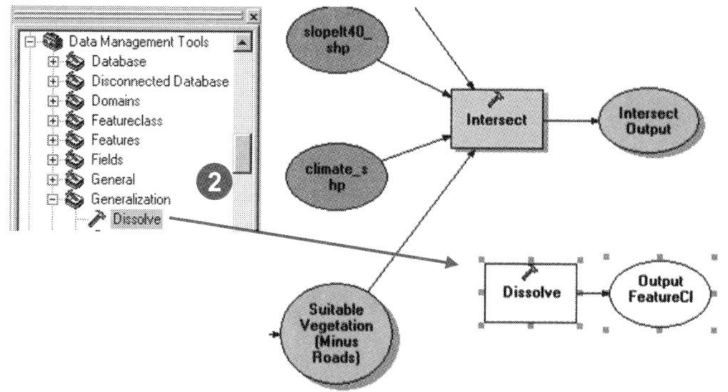

3. Right-click the Dissolve tool and click Open.

4. Click the Input Features dropdown arrow and click the variable Intersect Output.
5. Leave the default location for the Output Feature Class parameter and type "dissolve_output" for the name.
6. Check CLIMATE_ID in the Dissolve_Field(s) list to use CLIMATE_ID as the dissolve field.

   The Dissolve tool gives you the option to create a statistics field, but you don't need to create one at this time, so leave the Statistics Field(s) parameter blank.
7. Click OK.

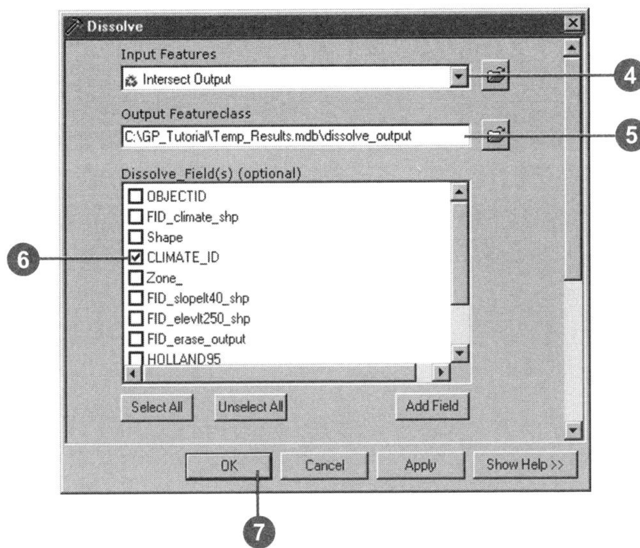

8. Click Auto Layout, then Full Extent to display all the model elements in their default position in the display window.
9. Right-click the dissolve_output derived data element and click Rename. Type "Dissolve Output" and click OK.
10. Right-click the Dissolve Output derived data element and click Add To Display.
11. Right-click the Dissolve tool and click Run.

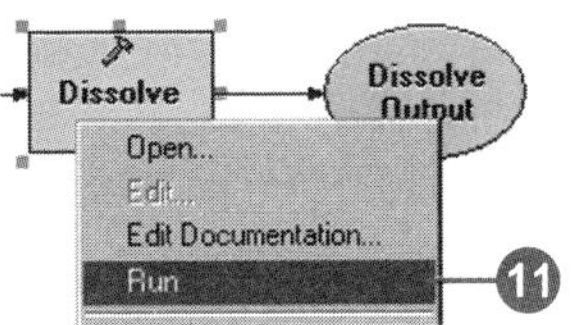

12. Examine the layer dissolve_output in your ArcMap display.
13. Right-click this layer in the table of contents of ArcMap and click Open Attribute Table. There are now only two rows in the table. The polygons have been dissolved into two groups, those with a CLIMATE_ID of 2 (climate zones 1 and 2) and those with a CLIMATE_ID of 3 (which includes climate zone 3).
14. Close the table.
15. Click the Model menu and click Save.

Creating single-part polygons

The Dissolve tool creates multipart polygons—separate polygons sharing the same attributes. Before selecting the required information using the Select tool, you'll first convert the multipart polygons to single-part polygons in order to extract information about each polygon.

1. Expand the Features toolset located inside the Data Management Tools toolbox.
2. Click and drag the Multipart To Singlepart tool into the ModelBuilder window.

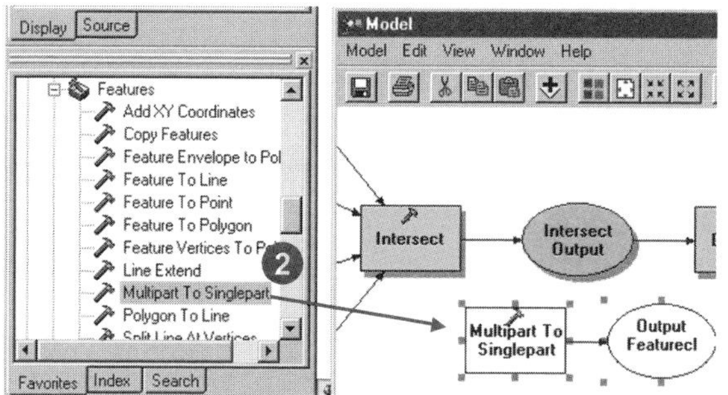

3. Right-click the Multipart To Singlepart tool and click Open.
4. Click the Input Features dropdown arrow and click the variable Dissolve Output.
5. Leave the default for the location of the Output Feature Class parameter and type "singlepart_output" for the name.
6. Click OK.

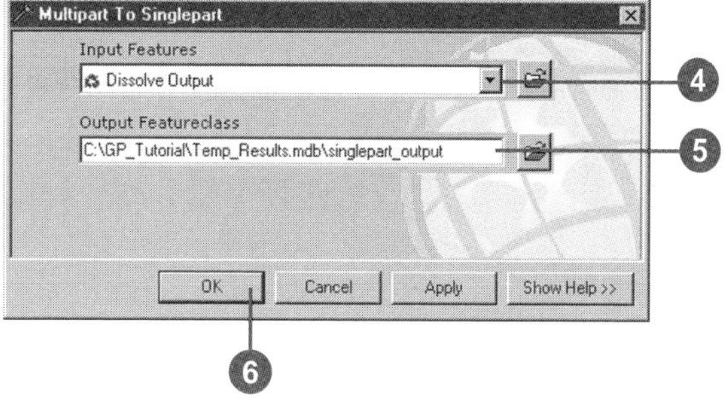

7. Click Auto Layout, then Full Extent to display all the model elements in their default positions in the display window.
8. Right-click the singlepart_output derived data element and click Rename. Type "Singlepart Output" and click OK.
9. Right-click the Singlepart Output derived data element and click Add To Display.
10. Right-click the Multipart To Singlepart tool and click Run.
11. Examine the output layer in your ArcMap display and the layer's attribute table.
12. Click the Identify tool and click a polygon.

    Notice that each polygon now contains its own information for climate zone, length, and area.

Selecting suitable locations

You'll now use the Select tool to extract the suitable locations.

1. Expand the Extract toolset located in the Analysis Tools toolbox.
2. Click and drag the Select tool into the ModelBuilder window.

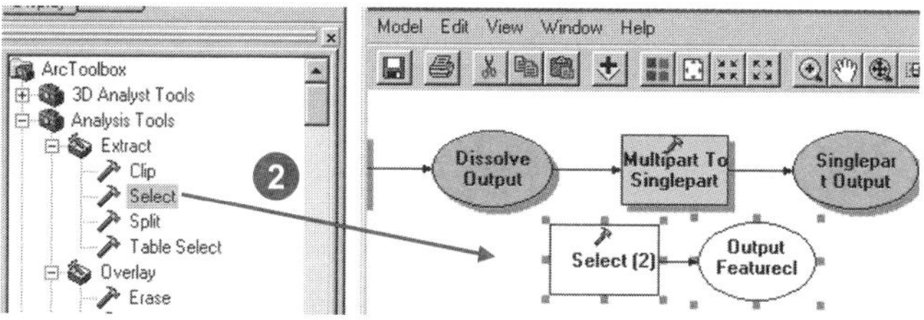

Note that the tool added will be called Select (2) as you have already added a Select tool.

3. Right-click the Select (2) tool element and click Open.
4. Click the Input Features dropdown arrow and click the variable Singlepart Output.
5. You'll put the final output inside your Habitat_Analysis geodatabase. Clear the default path and type "locations" for the value of the Output Feature Class parameter. The path to your current workspace will automatically be added when you click the next parameter in the dialog box.
6. Type the following expression for the value of the Expression parameter or click the browse button to the right of the Expression parameter and build the expression using the Query Builder dialog box.

   [CLIMATE_ID] = 2 AND [Shape_Area] >= 1089000 OR [CLIMATE_ID] = 3 AND [Shape_Area] >= 2178000

   Alternatively, browse in your operating system to query.txt located in your GP_Tutorial folder, copy the query, and paste it into the Expression parameter's input box.

   The expression extracts all areas where the climate zone is 1 or 2 (polygons with a CLIMATE_ID of 2) and the area of the vegetation patch is greater than or equal to 25 acres (1,089,000 ft^2), or where the climate zone is 3 (polygons with a CLIMATE_ID of 3) and the area of the vegetation patch is greater than or equal to 50 acres (2,178,000 ft^2).
7. Click OK.

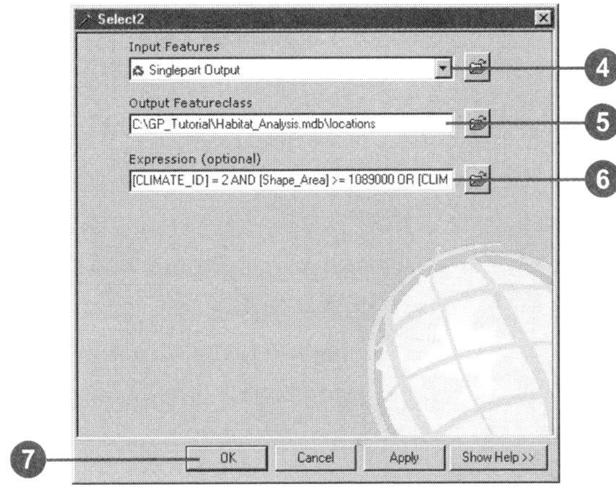

8. Click Auto Layout, then Full Extent to display all the model elements in their default position in the display window.
9. Right-click the element locations and click Rename. Type "Output Habitat Locations", then click OK.
10. Right-click the element Output Habitat Locations and click Add To Display.
11. Right-click the Select (2) tool element and click Run.

12. Click all results except locations in the table of contents (press the Ctrl key to select more than one layer), right-click, and click Remove.

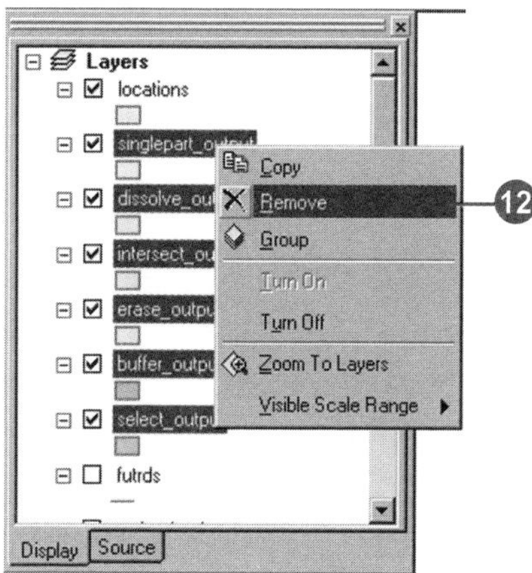

13. Examine the layer locations in your ArcMap display.

    It shows areas capable of supporting the California gnatcatcher based on the criteria set. Potential patches contain suitable vegetation types that are far enough away from roads, where the elevation is lower than 250 m, and where the slope of the terrain is less than 40 percent. The size of the vegetation patch is greater than or equal to 25 acres in climate zones 1 and 2 and greater than or equal to 50 acres in climate zone 3.

Saving and renaming the model

Any new model you create has a default name (Model). You can change the name of the model in the Model Properties dialog box.

1. Click Model and click Model Properties.

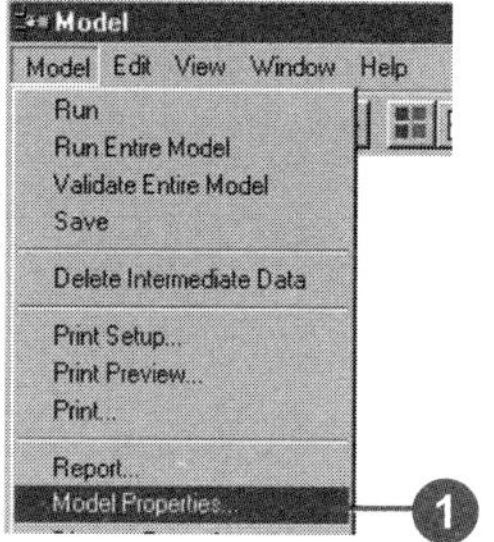

2. Click the General tab.
3. Specify a new name for your model, "Habitat_Analysis", and a label, "Find Potential Habitat". You would use the name if you were running the model at the command line or inside a script. The label is the display name for the model.
4. Add a description for the model, describing the model's contents.

   This description will appear in the Help panel of the model's dialog box. You'll view it later in this tutorial.
5. Check Store relative path names so that all paths for sources of information referenced by the tool are saved relative to the location of the toolbox. If the toolbox and its data are moved, paths to data sources will be altered accordingly.
6. Click OK.

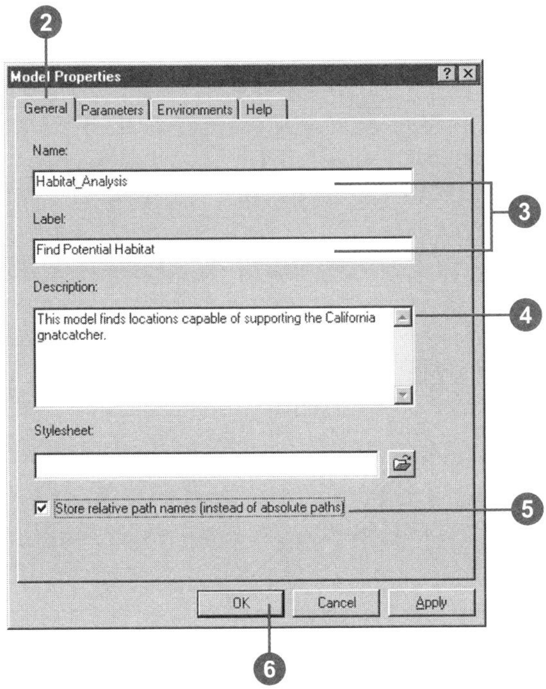

7. Click Model, Close, and Yes to save your changes.
8. Right-click your Find Potential Habitat model in the ArcToolbox window and click Open.

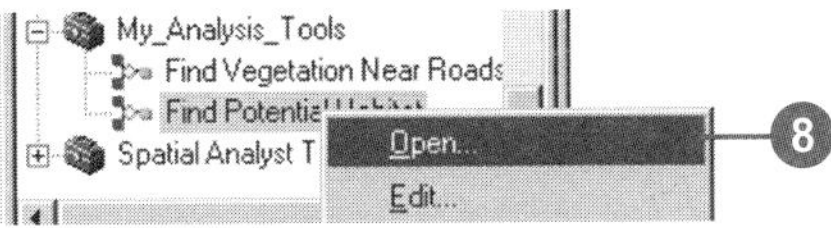

Notice that there are no parameters available in the model's dialog box. This is because you have not set any variables as model parameters.

9. Click Cancel.

Setting model parameters

In a similar way to setting variables as parameters for the dialog box of a script, as you did in Exercise 3, you must set the variables in the model as model parameters that you want to display in your model's dialog box.

1. Right-click the Find Potential Habitat tool and click Edit to open the model.
2. Right-click the element vegtype and click Model Parameter.

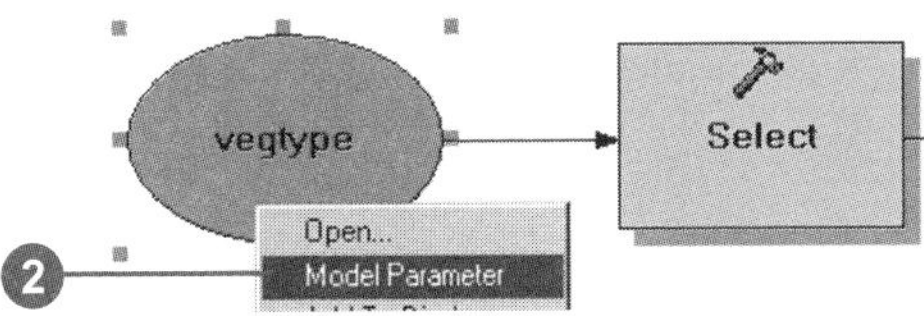

A 'P' icon will appear next to the element indicating it is set as a model parameter.

3. Right-click vegtype and click Rename. Type "Input Vegetation" and click OK. This is the name that will be displayed for the input data parameter in the model's dialog box.
4. Right-click Output Habitat Locations and click Model Parameter.
5. Click the Model menu and click Save, then click the Model menu again and click Close.

The Properties dialog box provides a way to change the name of the model, set parameters (an alternative way to setting them within the model), change their order, and set the tool's environment settings. You'll now examine the parameters that are set to display in the dialog box of your new Find Potential Habitat model.

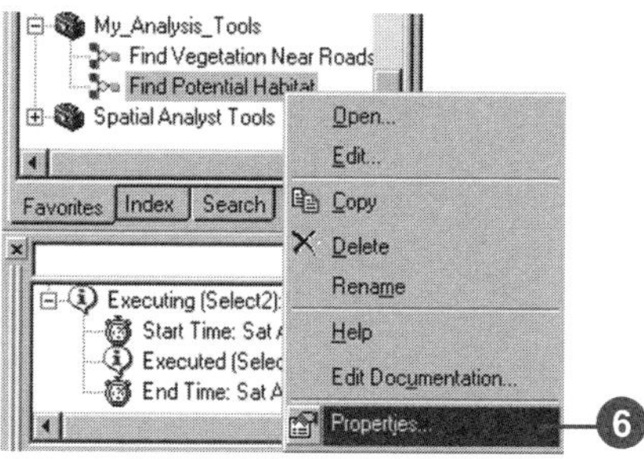

6. Right-click your Find Potential Habitat model and click Properties.
7. Click the Parameters tab.

   Notice the two variables set as model parameters for the user to specify values in the model's dialog box. The order of the parameters can be changed by clicking a parameter, then clicking Move Up or Move Down. The order for the parameters in the model's dialog box mimics the order seen in this list.

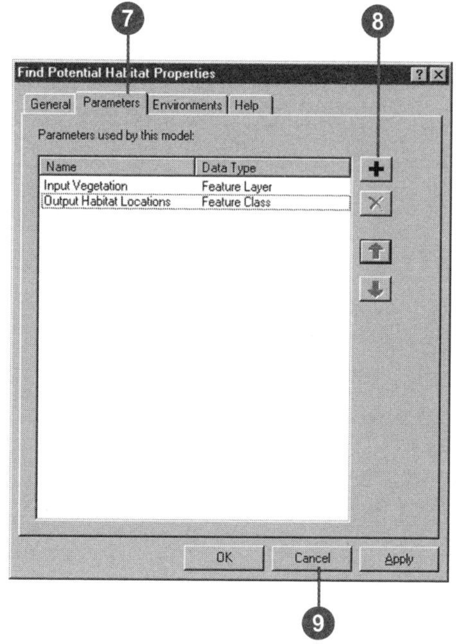

8. Click Add to see the variables that could also be set as parameters, then click Cancel on the Add Model Parameter dialog box.
9. Click Cancel.

Overwriting outputs

By default, results from running tools are not overwritten. You'll specify to overwrite results from running tools (via their dialog boxes) so the existing locations output will be overwritten.

1. Click Tools on the Main menu.
2. Click the Geoprocessing tab and check to overwrite the outputs of geoprocessing operations.
3. Click OK.

Running the model from its dialog box

1. Right-click the Find Potential Habitat model in the ArcToolbox window and click Open.

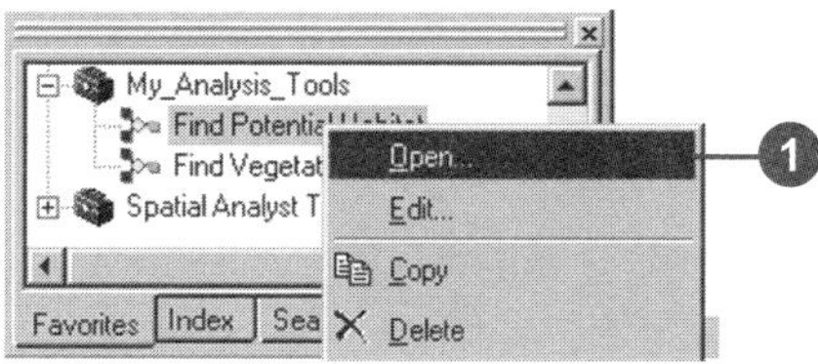

You now have two parameters displayed on the model's dialog box.

2. Click the Show Help button to see the description for the model that you wrote earlier.
3. Leave the default value for the Input Vegetation parameter to use vegtype.
4. Leave the default value for the Output Habitat Locations parameter to create an output called locations. This output will replace the output inside your current workspace, Habitat_Analysis.mdb.

   Although you already executed each individual tool while building the model, these tools will be executed again, and the results will be overwritten.

   Note: When you run a model from its dialog box, unlike when running the model within the ModelBuilder window, the intermediate data (data referenced by derived data elements that are flagged as intermediate) will be deleted after the model has executed.

5. Click OK to run the model.

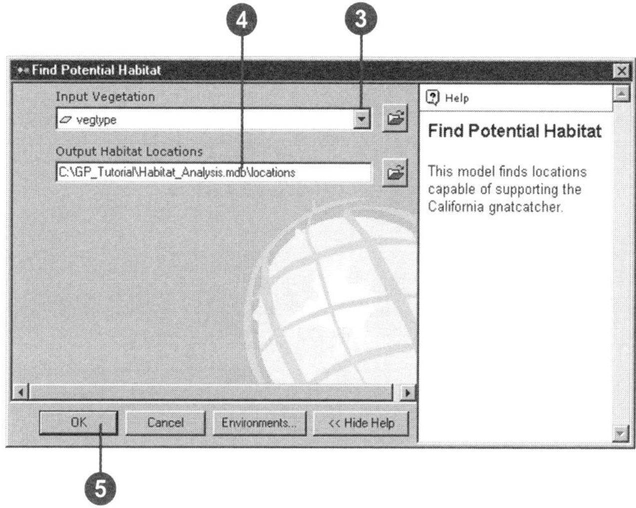

6. Examine the status messages in the message section of the Command Line window and examine the result that is added to your display.

   Note: The final result is added to the display by default. This is because the option to add results of geoprocessing operations to the display is checked on by default in the Geoprocessing tab of the Options dialog box, accessed via the Tools menu on the Main menu.

Note also that the 'locations' result created earlier by running the model from within the ModelBuilder window has been overwritten because the option to overwrite the outputs from geoprocessing operations is turned on in the Options dialog box. Also, the result produced is permanent, because the option to create temporary results from tools is turned off by default in the Options dialog box.

7. Click File on the Main menu of ArcMap and click Save.

In this exercise you created a new model that finds potential habitat locations for the California gnatcatcher. In the next exercise you will find out which of these locations will be impacted by proposed roads.

You can continue on to Exercise 5 or stop and complete the tutorial at a later time. If you do not move on to Exercise 5 now, click File, then Exit to close your ArcMap session, but do not delete your GP_Tutorial folder.

Exercise 5: Finding habitat patches impacted by proposed roads

Once you have created models within, or added scripts to, a toolbox, they can be used just like any of the system tools. They can be added to the ModelBuilder window of another model, run from the command line, or run from another script.

In this exercise you'll create a new model and add to it the Find Potential Habitat model you created in Exercise 4. You'll clip the result from the Find Potential Habitat model (locations) by using buffer zones you'll create around the proposed roads to find potential habitats that may be impacted by the proposed roads.

This exercise will take approximately 16 minutes to complete.

Setting up

If you didn't do the previous exercises, follow the next eight steps. If you did Exercise 4, go to the next section, 'Creating a new model'.

1. Open ArcCatalog. In the ArcCatalog tree copy the folder GP_Tutorial from C:\arcgis\ArcTutor\Geoprocessing\Results\Ex4 to your C:\ drive or an alternative drive, then close ArcCatalog.
2. Open ArcMap and open the map document Habitat_Analysis.mxd from your GP_Tutorial folder. Right-click the ArcToolbox window and click Add Toolbox. Navigate to your My_Analysis_Tools toolbox and click Open.
3. Right-click the ArcToolbox window and click Environments.
4. Click the General Settings dropdown arrow to expand its contents and type "C:\GP_Tutorial\Habitat_Analysis.mdb", or type your alternative location, for the value of the Current Workspace environment setting. Alternatively, click the Browse button to navigate to this location.
5. Type "C:\GP_Tutorial\Temp_Results.mdb", or type your alternative location, for the value of the Scratch Workspace environment setting. Alternatively, click the Browse button to navigate to this location, then click OK.
6. Click Tools on the Main menu and click Options.
7. Click the Geoprocessing tab.
8. Check Overwrite the outputs of geoprocessing operations. Then click OK on the Options dialog box.

Creating a new model

1. If you have closed ArcMap, open it again and open the map document Habitat_Analysis.mxd from your GP_Tutorial folder.
2. Right-click your My_Analysis_Tools toolbox in the ArcToolbox window, point to New, and click Model.

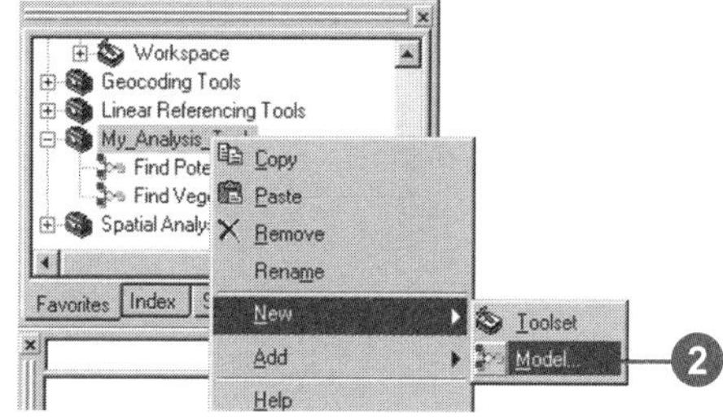

Adding the Find Potential Habitat model

1. Click the Find Potential Habitat model inside your My_Analysis_Tools toolbox and drag it into the ModelBuilder window of the new model.

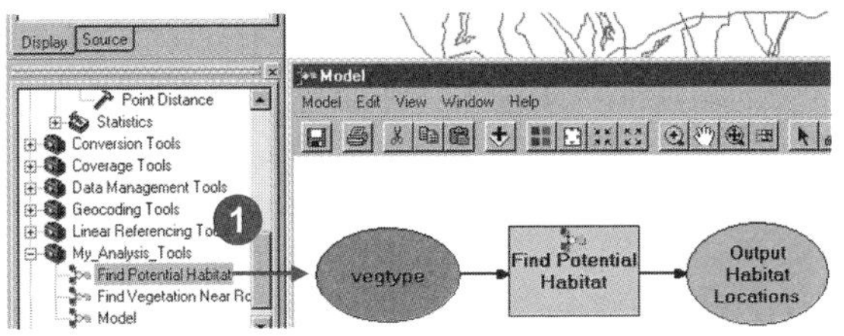

Notice the icon on the Find Potential Habitat tool element indicates that the tool is another model. You can right-click this element and click Edit to open the model inside its ModelBuilder window. You hide the complexity of a larger model by breaking it down into smaller submodels.

Buffering the proposed roads

1. Expand the Proximity toolset located inside the Analysis Tools toolbox.
2. Click and drag the Buffer tool into the display window.
3. Right-click the Buffer tool element that represents the Buffer tool in the model and click Open.
4. Click the Input Features dropdown arrow and click the layer futrds.

   Leave the default location and name for the Output Feature Class parameter.
5. Click Field, click the dropdown arrow, and click Distance. The values in this field will be used for the width of the buffer.
6. Click the Dissolve Type dropdown arrow and click LIST to list the dissolve fields. Check Distance in the Dissolve_Field(s) list. All polygons that share the same distance value will be dissolved.
7. Click OK.

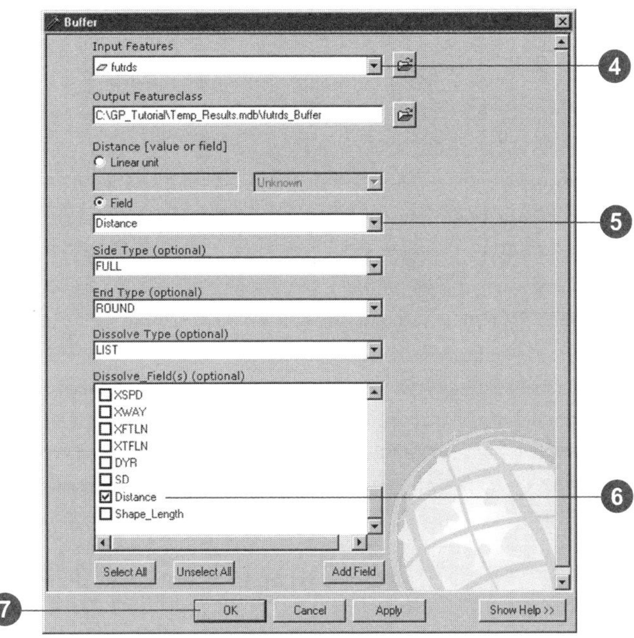

8. Click the Full Extent tool to display all elements in the display window.
9. Right-click the derived data element futrds_Buffer and click Rename.
10. Type "Buffer Zones" and click OK.

Clipping habitat patches using buffer zones

1. Expand the Extract toolset located inside the Analysis Tools toolbox.
2. Click the Clip tool and drag it into the ModelBuilder window.

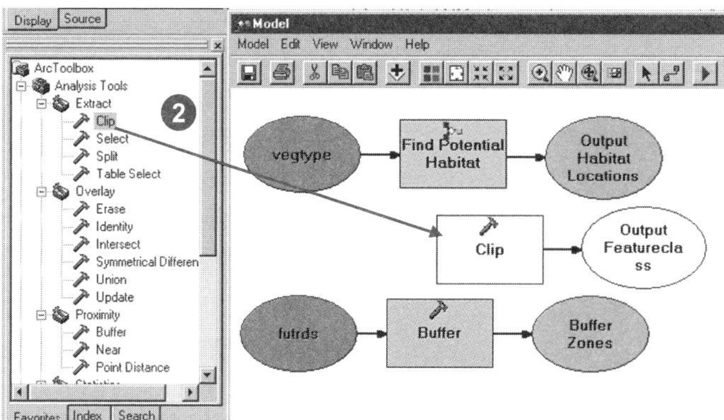

3. Click the Add Connection tool.

4. Click the element Output Habitat Locations, then click the Clip tool element.

   A *connector* is drawn between the two elements, connecting the output from the Find Potential Habitat tool as input for the Clip tool.

5. Click the element Buffer Zones, then click the Clip tool.

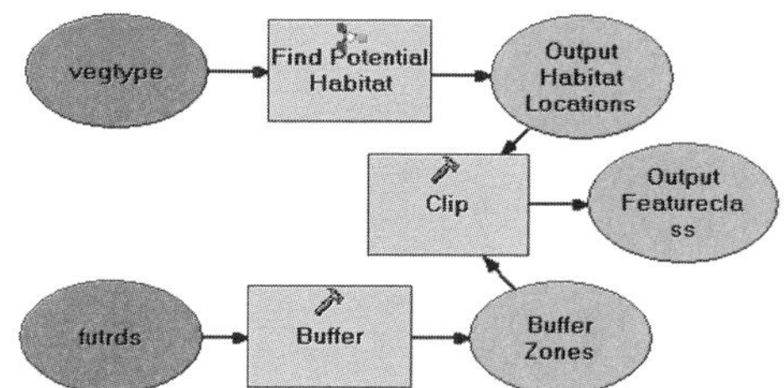

6. Double-click the Clip tool element to view the parameter values set in the tool's dialog box.

   Notice that by connecting the input data elements to the tool element, the values for the Input Features parameter and the Clip Features parameter are set inside the tool's dialog box.

7. As the result from the Clip tool is the final output from the model, clear the default path and type "impacted_habitat" for the name of the Output Feature Class. The path to the current workspace will be supplied by default.
8. Leave the default value (blank) for the Cluster Tolerance parameter and the units as Feet. An appropriate default value will be calculated using the units of the spatial reference set for the output, in this case, in feet, as the input data spatial reference units are in feet.
9. Click OK.

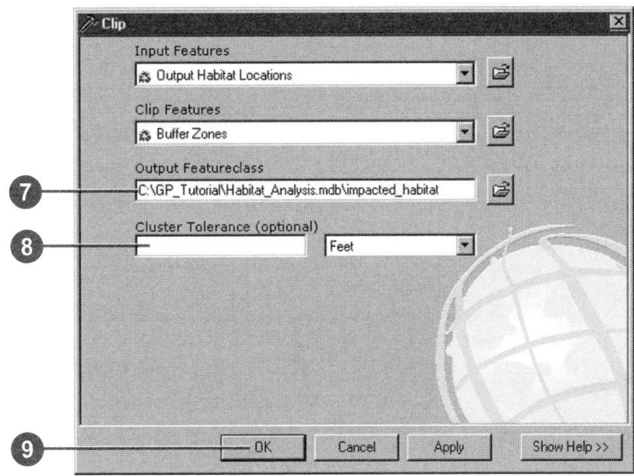

10. Click Auto Layout, then click Full Extent to display all the model elements in the display window in their default position.

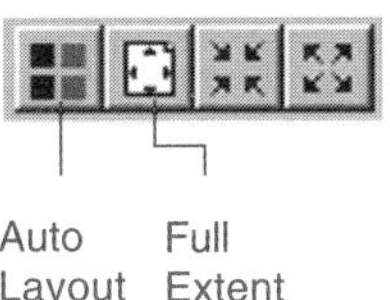

Renaming elements

1. Right-click the element impacted_habitat and click Rename.
2. Type "Output Impacted Habitat" and click OK.
3. Right-click the element vegtype and click Rename.
4. Type "Input Vegetation" and click OK.
5. Right-click the element futrds and click Rename.
6. Type "Input Proposed Roads" and click OK.

Setting model parameters

You want the user of this model to be able to specify the following inputs: Input Vegetation and Input Proposed Roads, as both of these could change over time. You also want the user to be able to set the path and name for the Output Habitat Locations output and the final output, Output Impacted Habitat.

So the user can specify values for these parameters, you'll make them model parameters.

1. Right-click the element Output Impacted Habitat and click Model Parameter.
2. Right-click the element Output Habitat Locations and click Model Parameter.
3. Right-click the element Input Vegetation and click Model Parameter.
4. Right-click the element Input Proposed Roads and click Model Parameter.

   Your model should resemble the graphic that follows.

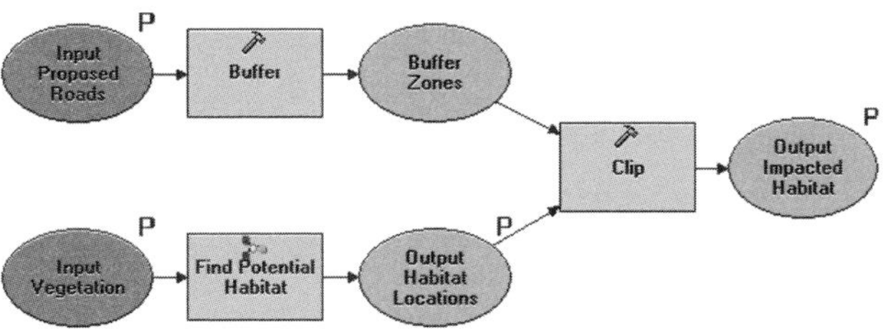

5. Click Model and Save, then click Model, then Close.

Renaming the model

Models can be renamed in the Model Properties dialog box within the ModelBuilder window, as you saw earlier, or from the Properties option on the model's context menu.

1. In the ArcToolbox window, right-click Model inside your My_Analysis_Tools toolbox and click Properties.
2. Click the General tab and type "impacted_habitat" for the name and "Find Impacted Habitat" for the label.
3. For the description of the model, type, "This model finds potential gnatcatcher habitat patches that might be impacted by proposed roads."
4. Check Store relative path names to save all paths for sources of information referenced by the tool relative to the location of the toolbox. If the toolbox and its data are moved, the model will still work without having to repair the paths to derived data.

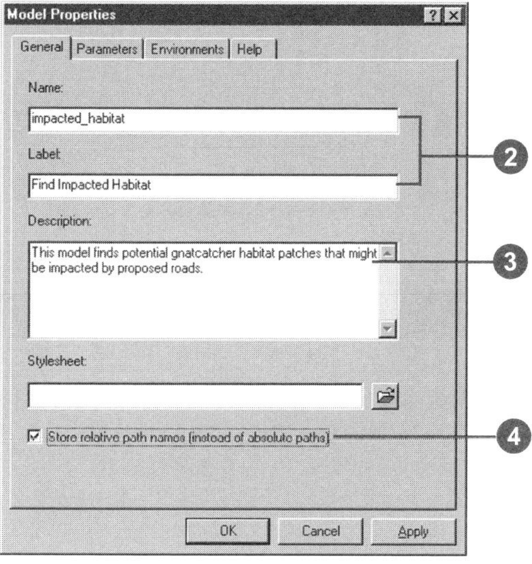

Changing the order of parameters

1. Click the Parameters tab.
2. Click Input Proposed Roads and click the Move Up arrow to place this parameter at the top of the list. Use the Move Up or Move Down arrows to place Input Vegetation after Input Proposed Roads, then Output Habitat Locations, and Output Impacted Habitat last.
3. Click OK.

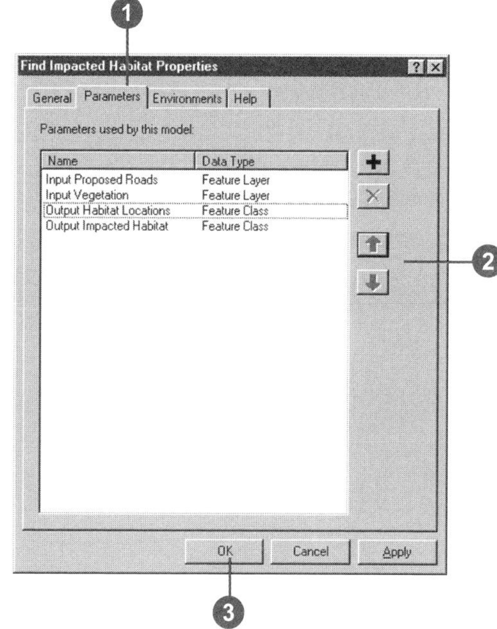

Running the model

1. In the ArcToolbox window, double-click the Find Impacted Habitat model to open its dialog box.
2. Click OK to run the model.

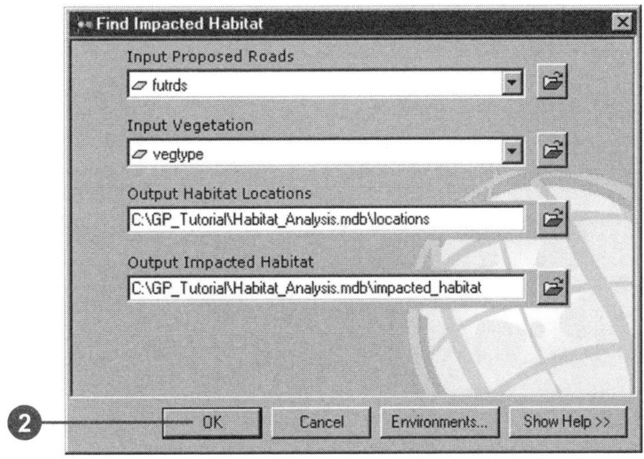

Displaying the results

1. Examine the status messages in the Command Line window.
2. Turn off all layers in the table of contents except impacted_habitat.
3. Examine the result in the ArcMap display.

These polygons represent habitat locations that would be impacted by the introduction of new roads.

4. Turn on the layers majorrds_shp and futrds in the table of contents and zoom in on the impacted areas.
5. Change the symbology of the layers if desired.

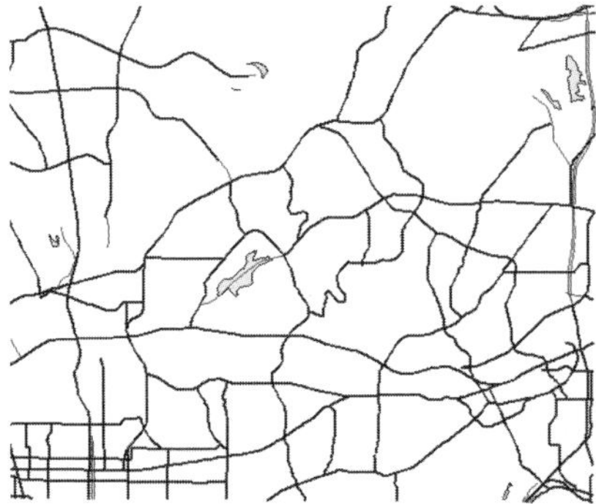

There is actually only one area of impacted habitat that a proposed road will go directly through. The rest of the impacted areas are close to new roads, but the roads will not travel directly through them. These areas fall within a certain distance from the proposed roads (the buffer distance assigned). They will be impacted by proposed roads, but only by noise and fume pollution.

6. Turn on the locations layer.
7. Change the order of the layers in the table of contents so they are in the following order from the top of the table of contents: futrds, majorrds_shp, impacted_habitat, then locations. The order of the rest of the layers does not matter.
8. Zoom to the full extent of the map to see all high-quality habitat locations, including those that are impacted by proposed roads.

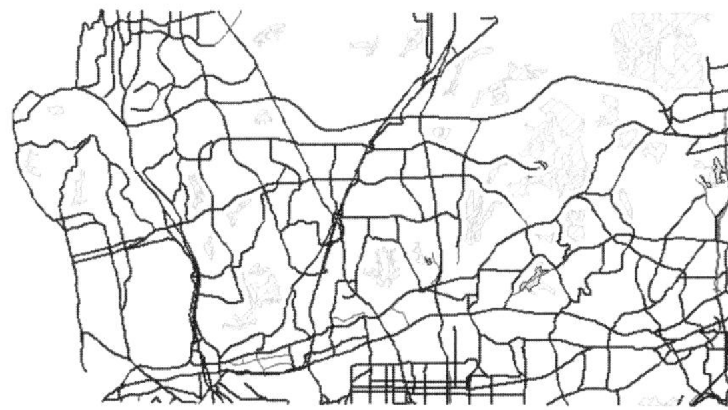

After an investigation in the field to see if there are indeed gnatcatchers present in the impacted habitat locations, the proposed roads that are planned to be routed through or near these locations will need to be assessed to see if they can be rerouted away from areas of high-quality habitat.

Saving the map document

1. Click File on the Main menu of ArcMap and click Save.
2. Click File again and click Exit to close the ArcMap session.

This brings you to the end of this tutorial. You should now have the experience to start using your own data and to get your geoprocessing tasks done with greater efficiency and ease. The rest of this book explains how to use ArcGIS to perform geoprocessing in more depth, so use it as a guide as you begin performing your own specific geoprocessing tasks.

Geoprocessing basics

3

IN THIS CHAPTER

- **Geoprocessing: The big picture**
- **Introducing geoprocessing methods**
- **Working with data**
- **Results from running tools**
- **Geoprocessing settings**
- **Saving and loading geoprocessing settings**
- **Keeping track of geoprocessing operations**
- **Sharing your geoprocessing work**

A GIS provides a spatial framework to support decisions to help manage the natural and man-made environment and resources of the earth. Geoprocessing allows you to define, manage, and analyze the information used to form these decisions.

The concept of geoprocessing is based on a framework of data transformation. A typical geoprocessing operation takes an input dataset, performs an operation on that input dataset, and returns the result of the operation as an output dataset. Geoprocessing your data within ArcGIS enables you to perform geoprocessing operations using many different data types and a number of different methods, and it gives you access to hundreds of geoprocessing tools.

This chapter discusses:

- How geoprocessing is a component of a GIS
- The different methods to perform geoprocessing tasks
- The different workspaces and data sources that are supported
- How to change the default settings that are applied to your results
- About geoprocessing settings and how to load and save them between sessions
- How to keep track of the geoprocessing tasks you perform
- How to share your geoprocessing work with others

Geoprocessing: The big picture

Geoprocessing is one of the most powerful components of a GIS. A GIS is composed of a GIS database and software tools used to manipulate the data from that GIS database. The results produced by using geoprocessing tools can be powerful aids in real-world decision making.

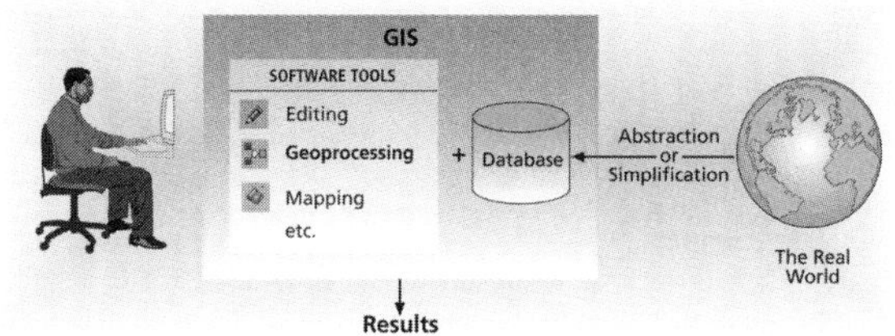

Geoprocessing tools manipulate data to produce results that model the real world.

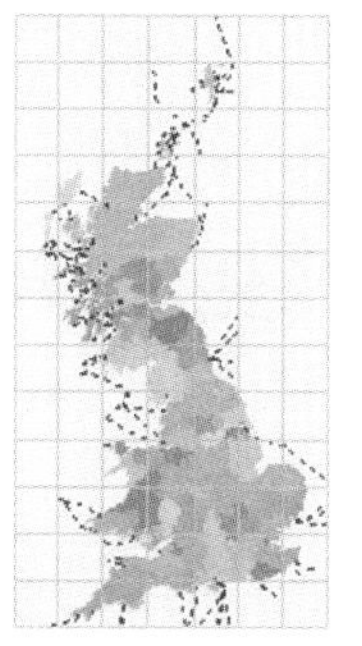

Map of Great Britain with the county of Greater Manchester highlighted

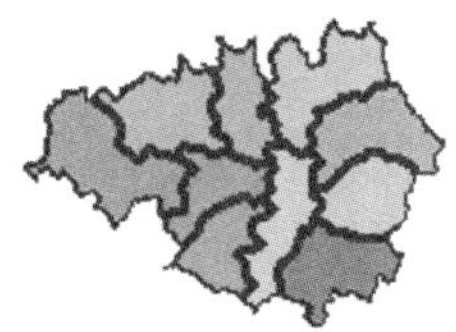

A new dataset of the county of Greater Manchester, created by selecting districts that fall into this county

What problems can geoprocessing help solve?

Geoprocessing can help you with an extensive range of tasks from preparing datasets, such as extracting a subset from a larger dataset or converting datasets to a different format, to performing analyses—finding possible answers to multiple spatial questions, such as "Where is the best location?" or "Whom will this affect?" The following examples show typical geoprocessing tasks.

Example 1

From a collection of datasets of Great Britain, you want to do some analysis in the county of Greater Manchester. You only have district data, but it includes which county each district falls into.

Using the Select tool you can select those districts that fall within the county. You can then clip the rest of your datasets to the extent of this new dataset. The result is a collection of datasets for the county of Greater Manchester.

Example 2

You are performing a study to find areas susceptible to flooding. Part of the study requires that you identify the soil types within each land parcel in each district. The Union tool combines information from separate datasets into one dataset, enabling you to extract the information you need.

Inputs to Union — *Output from Union*

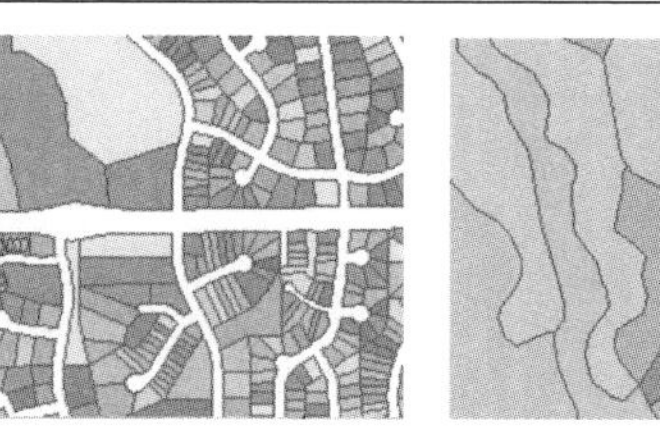

Land parcels

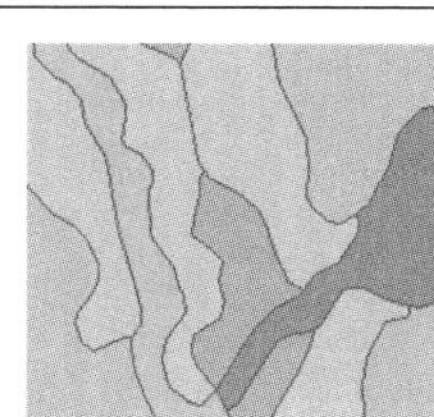

Soil types

Land parcels and soil types

Introducing geoprocessing methods

Geoprocessing forms a vital part of the work many companies do with a GIS. Countless geoprocessing tasks may be performed on a daily basis. Such tasks include converting data from one format to another or performing analysis by creating buffers or overlaying datasets.

Within ArcGIS, there are different methods available for performing geoprocessing tasks. The method you choose depends on the method best suited to the particular task and your personal preference.

You can perform geoprocessing tasks by running a tool via its dialog box, at the command line, or within a script or a model.

A *dialog box* is a form on which you supply the parameter values for the tool, then click OK to run the tool. At the command line, you type the tool name and its parameter values, then press Enter to run the tool. Alternatively, you can create your own models inside, or add scripts to, toolboxes. Models you create may run a chain of tools in sequence, and scripts are useful for batch processing multiple inputs, such as when converting multiple datasets to a different format. Scripts can be written in any COM-compliant scripting language, such as Python, JScript, or VBScript, and they do not have to be added to toolboxes. They can be run directly from within the appropriate scripting application. Each of these methods used to perform geoprocessing tasks is introduced in the sections following 'About toolboxes'.

For more information on toolboxes, see Chapter 4. For more information on creating a model and adding a script to a toolbox, see Chapter 5. See Chapter 7 for more information on using the command line. For more information on building models, see Chapters 8 and 9. For information on creating scripts, see *Writing Geoprocessing Scripts With ArcGIS.*

About toolboxes

Toolboxes can contain toolsets and tools. *Toolsets* can also contain toolsets and tools. You can access toolboxes from the ArcToolbox window or the ArcCatalog tree. There is a collection of system toolboxes that you can use, or you can create your own toolboxes to store a collection of system tools (tools installed with ArcGIS) or custom tools (models or scripts) that you have created. System tools are organized into *system toolsets*, making it easier to find the tools you are looking for. By creating *custom toolsets*, you can easily organize your tools.

To view system toolboxes in the ArcCatalog tree, click Tools on the Main menu and click Options. Click the General tab and check Toolboxes in the list of top-level entries.

The collection of system toolboxes are added by default as shortcuts to your ArcToolbox window from the Toolboxes folder in your ArcCatalog tree. The ArcToolbox window provides a shortcut to the toolboxes you have stored in folders or geodatabases on disk—those that can be seen in the ArcCatalog tree. It provides a way to centralize the location of toolboxes that might be spread out in different folders and geodatabases on your computer's disk. The ArcToolbox window can be docked to any ArcGIS Desktop application, such as ArcMap, ArcGlobe, or ArcCatalog.

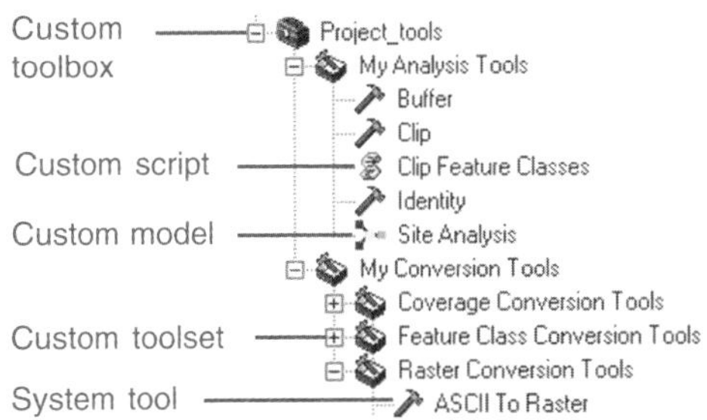

Using dialog boxes and the command line

All tools (system tools or custom tools) can be run from a dialog box or from a command line.

Running tools via dialog boxes

Dialog boxes guide you through the process of running a tool by giving you a form where you specify the data and other necessary parameter values.

All tool dialog boxes have a Help panel that can be opened to display Help topics for the tool being used. A description of what the tool does and a definition of each parameter are just one click away. You can also click the Help button at the top of the Help panel to display more help for the tool including an illustration, a section explaining how the tool works, and examples of how to run the tool at the command line or within a script. Once you understand the tool and its parameters, you can close the Help panel to save space on your desktop.

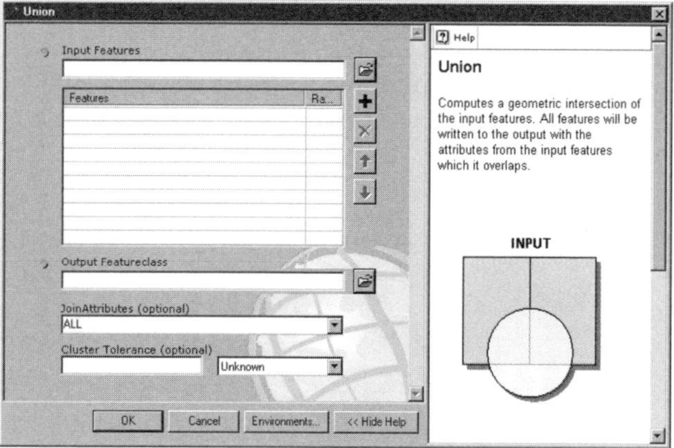

When should you use a dialog box?

Run a tool via its dialog box to become familiar with the tool and its parameters. The dialog box helps you provide valid parameter values and provides options where appropriate.

See Chapter 5, 'Working with toolsets and tools', for more information on running a tool via its dialog box.

Running tools at the command line

Regardless of the ArcGIS product (ArcInfo, ArcEditor, or ArcView) you have installed, you can access a command line in any ArcGIS Desktop application, such as ArcMap, ArcCatalog, or ArcScene™. The command line is similar to the ArcInfo Workstation command line. You type a tool name, set appropriate parameter values, then press Enter to execute the tool.

Tools can be run at the command line, provided that the toolbox containing the tool you want to run is added to the ArcToolbox window. After typing the tool name, the usage is displayed for the tool. It helps you supply values for the tool's parameters.

The command line honors the domain set for each parameter, so valid parameter values will be presented in a dropdown list. This includes any valid layers or keywords.

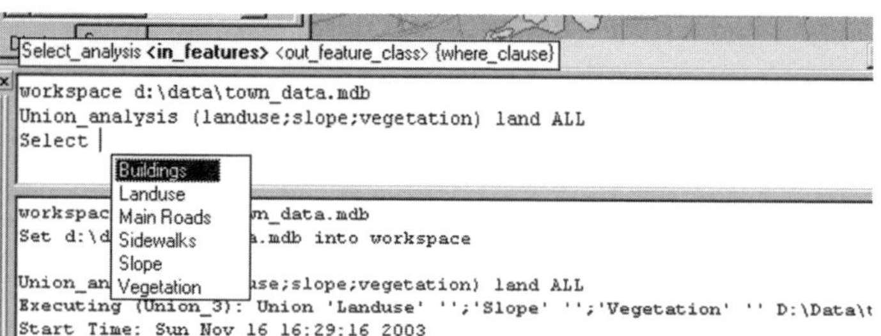

The Command Line window showing commands typed at the command line. Layers can be used as input parameter values, and multiple commands can be entered using Ctrl + Enter after each line.

When should you use the command line?

- If you are familiar with the tool you want to use and the parameter values that need to be supplied, you may find the command line quicker and more convenient.
- You can create *variables* to replace more complex parameter values, such as a variable for the table that reclassifies a raster, to help you to quickly perform your geoprocessing tasks.

See Chapter 7, 'Using the Command Line window', for more details on using the command line to geoprocess your data.

Building a model

For more complex geoprocessing tasks that involve multiple tools, you can create a new model by linking processes together in a graphical environment (a ModelBuilder window). This enables you to create a visual diagram of the steps needed to complete a geoprocessing task. The diagram you build represents a model.

To open a ModelBuilder window, you first create a new model inside a toolbox or toolset. A ModelBuilder window opens automatically, so you can start building your model.

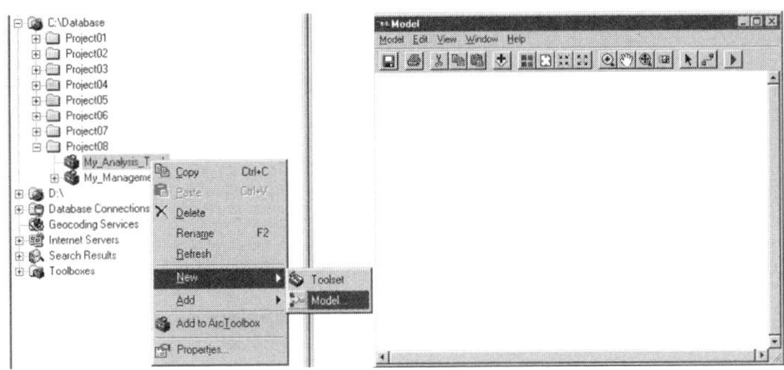

The creation of a model and the ModelBuilder window

In the model diagram, components, called elements, reference input data, a tool that operates on the input data, and the resulting output data. Elements are connected together into processes. Connector arrows indicate the sequence of processing. You may have several processes in a model, and they can be chained together so that the output data from one process becomes the input data for another process.

To build a model, you can drag tools from toolboxes in the ArcToolbox window or the ArcCatalog tree and data from the ArcCatalog tree or the table of contents of any other ArcGIS Desktop application, such as ArcMap. Alternatively, use the Add Data and Tools button in the ModelBuilder window to add tools and data, which places their representative elements on the model diagram. When you fill in necessary parameter values for each tool and connect processes, your model will become ready to run.

Any parameter within the model can be set as a variable that can be shared between processes. This means you don't have to make updates to every tool that uses a particular parameter value; you can just update the value in one dialog box.

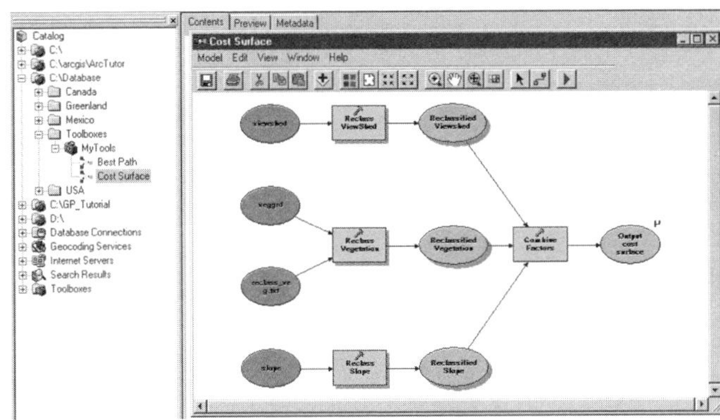

The ModelBuilder window with a typical model

Running a model

When the model is run, the project data is processed in the order specified, and output data is created. You can save, modify, and rerun the model. It's easy to redo the same procedure multiple times or alter data and other parameter values and rerun the model.

You can run the model from within the ModelBuilder window or from its dialog box. Variables in the model can be set as model parameters, so values for these parameters can be specified within your model's dialog box.

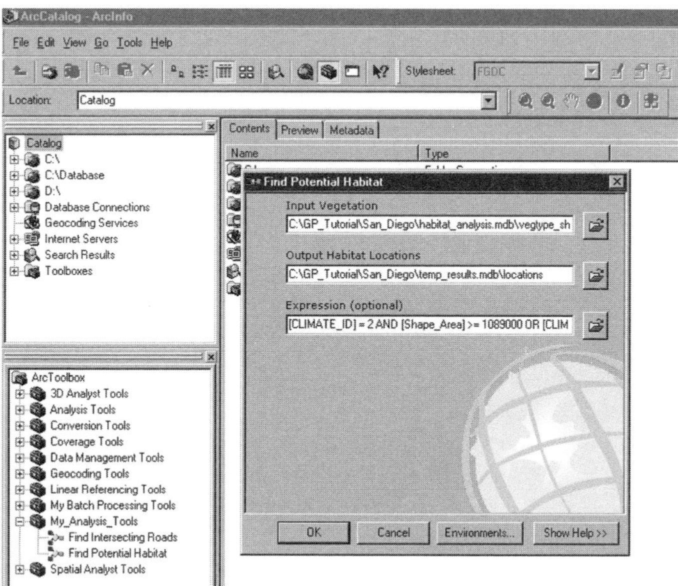

A model's dialog box with parameters displayed

When should you build models?

- Models automate the process of performing a sequence of geoprocessing tasks on your data, so build models when you want to perform multiple geoprocessing tasks, such as clipping your data, unioning the results together, then selecting areas from the unioned result that meet a criteria.
- You can quickly change parameter values for tools in your model, then rerun the model to experiment with different outcomes.
- Build models when you need to present a work flow to decision makers or the public.

See Chapter 8 for more conceptual information on building models and Chapter 9 for more information on using the ModelBuilder window.

Running a script

In many instances, the geoprocessing work that must be done is repetitive. Whether it involves geoprocessing a large number of datasets or large datasets with numerous records, these geoprocessing tasks require automation. Any COM-compliant scripting language can be used to write scripts that execute ArcGIS geoprocessing tools, providing an efficient and effective way to automate your geoprocessing tasks.

If you are new to scripting languages, you do not need to be an expert programmer to create and use scripts. You can build a model in a ModelBuilder window and export the model to a script that you can then run or modify.

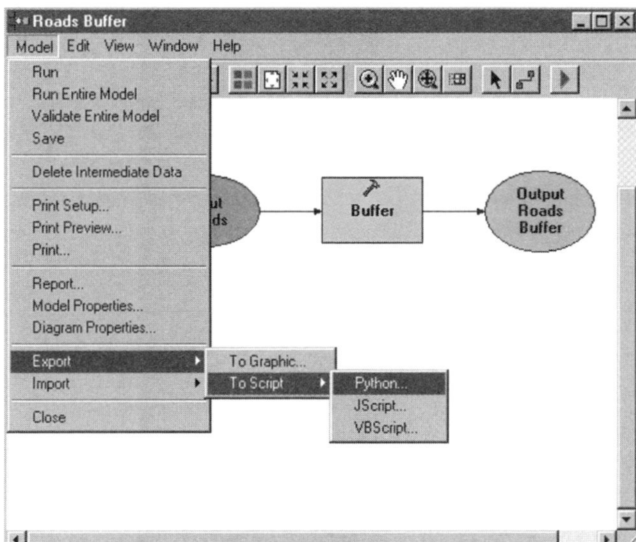

A model consisting of one process (a buffer tool with its parameter values defined). The model can be exported to a script.

The exported script is shown in the graphic on the following page.

```
# ---------------------------------------------------------------------------
# buffer.py
# Created on: Tue Dec 02 2003 10:25:18 AM
#   (generated by ArcGIS/ModelBuilder)
# ---------------------------------------------------------------------------

# Import system modules
import sys, string, os, win32com.client

# Create the Geoprocessor object
gp = win32com.client.Dispatch("esriGeoprocessing.GpDispatch.1")

# Load required toolboxes...
gp.AddToolbox("C:/Program Files/ArcGIS/ArcToolbox/Toolboxes/Analysis Tools.tbx")

# Local variables...
majorrds_shp = "C:/DATA/majorrds.shp"
Output_Featureclass = "C:/DATA/majorrds_Buffer.shp"

# Process: Buffer...
gp.Buffer_analysis(majorrds_shp, Output_Featureclass, "Rd_width", "FULL", "ROUND", "LIST", "'Rd_width'")
```

The above script was created by exporting a model containing one buffer process.

By adding lines of code and wrapping the buffer process in a loop, all feature classes inside the specified workspace will be buffered.

```
# ---------------------------------------------------------------------------
# bufferAll.py
# Created on: Tue Dec 02 2003 10:25:18 AM
#   (generated by ArcGIS/ModelBuilder)
# ---------------------------------------------------------------------------

# Import system modules
import sys, string, os, win32com.client

# Create the Geoprocessor object
gp = win32com.client.Dispatch("esriGeoprocessing.GpDispatch.1")

# Set the input workspace
gp.workspace = "C:/DATA/Roads/workspace.mdb"
# Set the output workspace
out_workspace = "C:/DATA/Roads/results.mdb"

try:
    # List all feature classes in workspace
    fcs = gp.listfeatureclasses()
    #Loop through list of feature classes
    fcs.Reset()
    fc = fcs.Next()
    while fc != "":
       #GDB's don't support "." in the fc name, so replace these with "_".
       outFeatureClass = out_workspace + "\\" + fc.replace(".","_")
       # Process: Buffer...
       gp.Buffer_analysis(fc, outFeatureClass, "Rd_width", "FULL", "ROUND", "LIST", "'Rd_width'")
       fc = fcs.Next()
except:
    gp.AddMessage(gp.GetMessages(2))
    print gp.GetMessages(2)
```

You can run a script as a standalone operation, or you can add a script to a toolbox and run the script from its dialog box. To do this you would add the script to a toolbox, then define the parameters, which you set as system arguments within the script, for which the user of the script must specify the values. See Exercise 3 of the quick-start tutorial or Chapter 5, 'Working with toolsets and tools', for information on adding a script to a toolbox and setting parameters within the Properties dialog box of the script.

When should you write a script?

- Scripts let you execute simple processes that include single tools, complex processes linked together, or looping to perform batch processing on a set of input datasets.
- Scripts are recyclable, meaning they can be data nonspecific and, therefore, used over and over again. They can increase efficiency because they can be executed at any time.

See *Writing Geoprocessing Scripts With ArcGIS* for more information on writing scripts that will perform geoprocessing tasks.

Working with data

Files, such as spreadsheets and documents, are organized into folder hierarchies on your desktop computer. A GIS manages the *data sources* that geoprocessing tools work with in a similar hierarchy of folders, files, and geodatabases. You can use ArcCatalog to explore, access, and manage data sources, as well as items such as folders and geodatabases. For more information, see *Using ArcCatalog.*

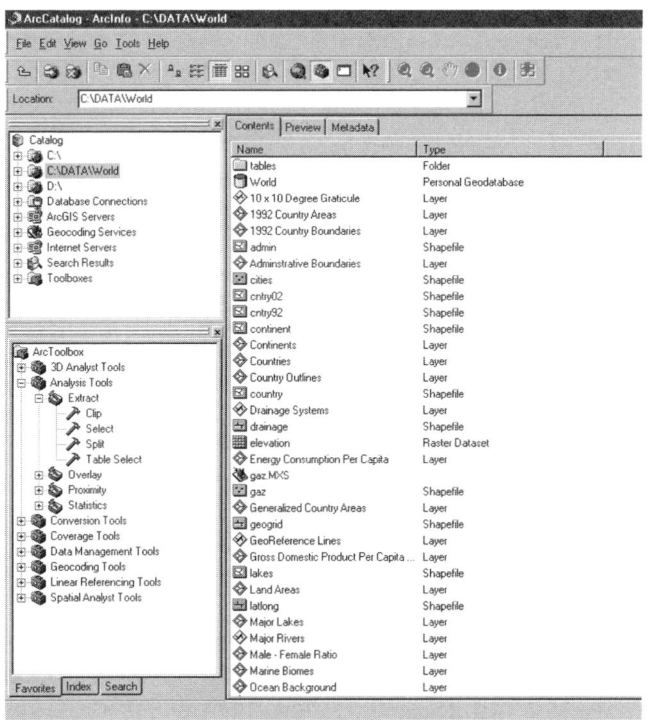

ArcCatalog and the ArcToolbox window

About workspaces

A workspace is a container of geographic data that will be used by geoprocessing tools. There are three main workspaces supported—folders, personal geodatabases, and ArcSDE® geodatabases—but you can also set a feature dataset as a workspace if you want to work with its feature classes.

By setting a workspace prior to running tools, you can simply type the name of a dataset—or a feature class if the workspace is a feature dataset—within the workspace as the parameter value when running tools, rather than the full path to the data. Workspaces can be set in the Environment Settings dialog box. See 'Working with workspaces' in Chapter 6 for more information.

Folders

A *folder* can store other folders, geodatabases, data sources, and toolboxes. Folders in the ArcCatalog tree represent folders on disk. Coverages, shapefiles, TIN datasets, layers, and layer files can only reside inside folders. Other data sources, such as raster data, feature classes, and tables, can be exported from a folder to a geodatabase, personal or ArcSDE. Feature datasets within a folder, such as coverage, Smart Data Compression [SDC] or Vector Product Format [VPF] datasets, can be set as a workspace so just the names of the feature classes within the dataset can be entered as inputs when running tools. However, results cannot be written to such datasets. A path to a valid workspace should be entered for the values of output data parameters.

Folders themselves can be used as inputs and outputs to certain geoprocessing tools, such as the Create Folder tool or the Create Personal GDB tool. In the Search tab of the ArcToolbox window, type "folder" to locate tools that work with folders.

Geodatabases

A *geodatabase* is a relational database that contains geographic information. Geodatabases can contain feature classes, feature datasets, tables, and toolboxes. Feature classes can be organized into a feature dataset, or they can exist independently in the geodatabase.

For more information on geodatabases, see *Using ArcCatalog*, *Modeling Our World*, and *Building a Geodatabase*.

Personal geodatabases

To manage your own spatial database, you can create a *personal geodatabase* that is stored in ArcCatalog in a Microsoft Access database. With personal geodatabases, many readers are supported, but only a single editor is supported. You can create and work with personal geodatabases in ArcGIS without any other software.

By setting a personal geodatabase as a workspace prior to running a tool, you can simply type the name of a feature dataset or a feature class contained within a geodatabase as the input data when running a tool, and it will be taken from the location set for the workspace. Feature datasets within a personal geodatabase can also be set as a workspace, so you can work with the feature classes it contains.

A personal geodatabase can be a source of input and a destination for output of data sources from geoprocessing operations. Personal geodatabases can be managed using the system tools.

You can easily create a personal geodatabase using the Create Personal GDB tool located in the Workspace toolset in the Data Management Tools toolbox.

In the Search tab of the ArcToolbox window, type “geodatabase” to locate tools that accept personal geodatabases as input or output.

 ArcSDE geodatabases

To allow multiple users to simultaneously update data, use an *ArcSDE geodatabase*, where data is stored in a centrally located relational *database management system*, such as IBM® DB2®, Informix®, Oracle®, or Microsoft® SQL Server™. ArcSDE geodatabases can be used with any ArcGIS product (ArcView, ArcInfo, or ArcEditor) but require ArcSDE for editing and schema management.

There are two main advantages to using ArcSDE geodatabases: many people in the organization can update the geodatabase’s contents at the same time through the use of versioning, and long transactions and large datasets can be handled efficiently.

By setting an ArcSDE geodatabase as a workspace prior to running a tool, you can simply type the name of a feature dataset or a feature class contained within a geodatabase as the input data when running a tool, and it will be taken from the location set for the workspace. Feature datasets within an ArcSDE geodatabase can also be set as a workspace, so you can work with the feature classes it contains.

An ArcSDE geodatabase can be a source of input and a destination for output of data sources from geoprocessing operations. It can also be used as input to and output from certain geoprocessing tools. In the Search tab of the ArcToolbox window, type "geodatabase" to locate tools that accept ArcSDE geodatabases as input or output.

About data sources

A data source is any geographic data used as input to or output from a geoprocessing tool. Supported data sources include geodatabase feature datasets and feature classes, shapefile datasets, coverage datasets and feature classes, computer-aided design (CAD) datasets and feature classes, SDC feature datasets and feature classes, VPF datasets and feature classes, raster datasets and raster dataset bands, TIN datasets, layers, layer files, tables, and table views. For more information on the structure of data, see *Modeling Our World*.

The *feature class is* one of the most commonly used data sources. It is composed of a collection of geographic features with the same type of geometry—**point, line, or polygon—and the same set of attributes**. Feature classes are discussed in the sections that follow. They can be contained within a geodatabase as well as within the following datasets: geodatabase feature, shapefile, coverage, CAD, SDC, and VPF.

Feature data

Geodatabase feature datasets

Geodatabase feature datasets reside inside geodatabases, personal or ArcSDE. They contain a collection of geodatabase feature classes that share the same extent and coordinate system.

The following geoprocessing tools accept geodatabase feature datasets as input or output: Create Feature Dataset, Copy, Delete, Rename, and Project.

Geodatabase feature classes

You can perform geoprocessing tasks on the feature classes contained within a geodatabase feature dataset. Geodatabase feature classes store geographic features represented as points, lines, polygons, annotation, dimensions, and multipatches and their attributes. They store simple features, so they can be organized inside or outside a feature dataset, but always inside a geodatabase, personal or ArcSDE. Simple feature classes that are outside a feature dataset are called standalone feature classes. Feature classes that store topological features must be contained within a feature dataset to ensure a common coordinate system.

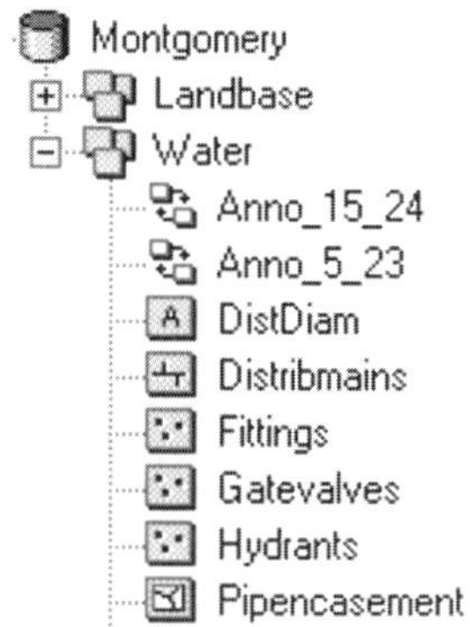

In the Search tab of the ArcToolbox window, type "feature class" to locate tools that accept geodatabase feature classes as input or output.

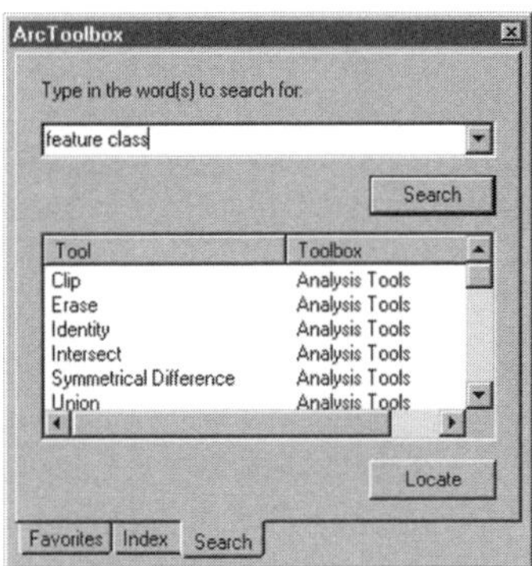

To locate tools that work with geodatabase feature classes, type "feature class" as the word to search for.

Shapefile datasets

A *shapefile* dataset (.shp) is stored in a folder, is composed of geographic features and their attributes, and contains one feature class. Geographic features in a shapefile can be represented with points, lines, or polygons (areas).

Because shapefile datasets contain one feature class, you can use them like any other feature class.

In the Search tab of the ArcToolbox window, **type "feature class"** to locate tools that accept shapefiles as input or output.

Coverage datasets

A *coverage* is stored in a workspace that is a folder in your file system. It contains an integrated set of feature classes that represent geographic features.

In the Search tab of the ArcToolbox window, **type "coverage"** to locate tools that accept coverages as input or output.

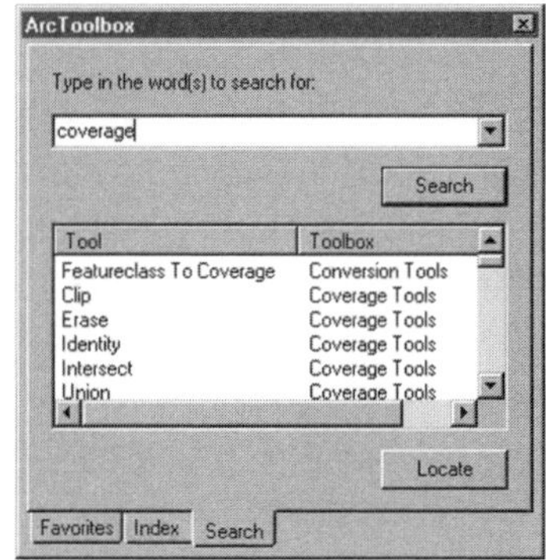

To locate tools that work with coverages, type "coverage" as the word to search for.

Coverage feature classes

Coverage feature classes can store a set of points, lines (arcs), polygons (areas), regions, routes, tics, links, and annotation (text). They can also have *topology*, which determines the relationships between features that are within or between feature classes.

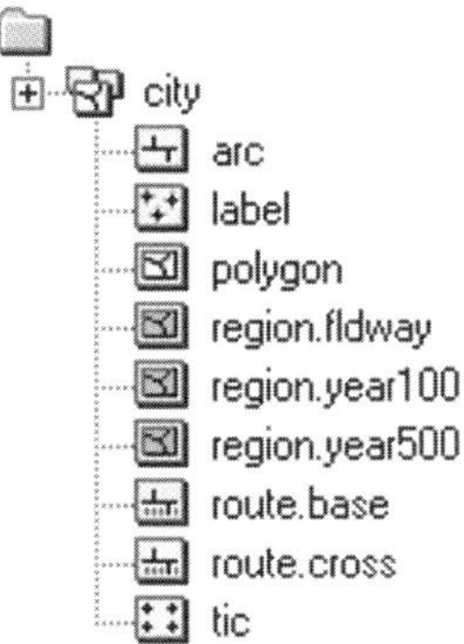

A coverage feature class can be used as input to tools that produce a distinct output or to tools that update the attributes of the input feature class. They cannot be used by those tools that update the geometry of the input.

If you want to update the geometry of a coverage feature class, export it to a geodatabase feature class and perform updates to the geometry of the geodatabase feature class.

In the Conversion Tools toolbox, click the To Geodatabase toolset and use the Feature Class To Geodatabase tool to convert coverage feature classes into geodatabase feature classes.

In the Search tab of the ArcToolbox window, type "feature class" to locate tools that work with coverage feature classes.

CAD feature datasets

A *CAD feature dataset* is the ArcGIS feature representation of a *CAD file*. A CAD feature dataset is composed of five read-only feature classes: points, polylines, polygons, multipatches, and annotation.

In the Search tab of the ArcToolbox window, type "CAD" to locate all tools that accept CAD feature datasets as input or output.

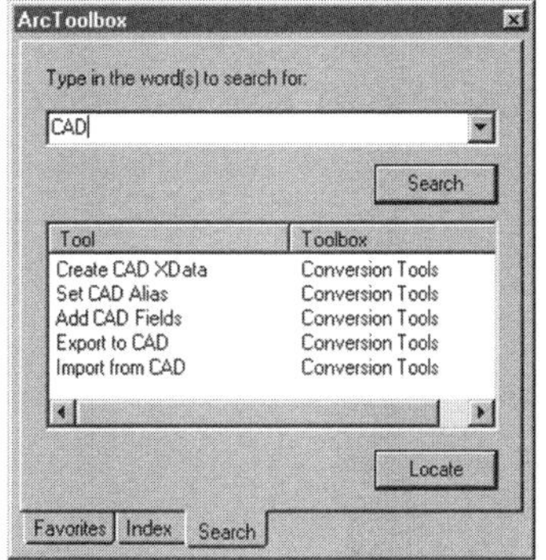

To locate tools that work with CAD feature datasets, type "CAD" as the word to search for.

Supported CAD file formats include DWG (AutoCAD®), AutoDesk® Drawing Exchange Format (DXF), and DGN (the default MicroStation file format). If the extension of your CAD file is not .dwg, .dxf, or .dgn, you will not see its feature dataset representation in the ArcCatalog tree. Use the File types tab in the Options dialog box (accessed via the Tools menu) to add the file extension of your CAD file so its feature dataset representation is displayed in the ArcCatalog tree.

Another representation of a CAD file in ArcGIS is the *CAD drawing dataset*, which is a pictorial representation of an entire CAD file. The symbology in a CAD drawing dataset mimics the symbology of the originating CAD file. It is useful to display the CAD drawing dataset in any ArcGIS application with a display, but for geoprocessing purposes you'll generally use only the CAD dataset or feature classes as input to geoprocessing tools.

CAD feature classes

A *CAD feature class* is a read-only, in-memory member of a CAD feature dataset. CAD feature classes include points, polylines, polygons, multipatches, or annotation.

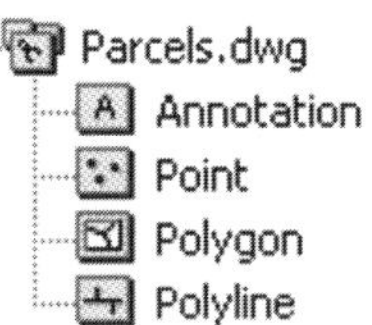

A CAD feature class is the result of a direct file read into memory of the objects contained within a single CAD drawing file of a specific geometric type. The feature attribute table of a CAD feature class is a virtual table composed of select read-only CAD graphic properties and any existing block or cell attribute values.

Because they are read-only, CAD feature classes can only be used by geoprocessing tools that produce a distinct output. They cannot be used by tools that alter the geometry or attributes of the input feature class. To alter the geometry or attributes of a CAD feature class, export the feature class to a geodatabase feature class and perform the alteration on the geodatabase feature class.

In the Search tab of the ArcToolbox window, type "feature class" to locate tools that accept CAD feature classes as input or output.

SDC datasets

A *Smart Data Compression dataset* contains read-only feature classes, all of which have the same attribute information but with different levels of generality in the shapes. SDC data is encrypted and highly compressed. This format is used by ESRI to provide StreetMap™ data and also by commercial data vendors who distribute street data for geocoding and routing with various ESRI® software products.

The following geoprocessing tools accept SDC datasets as input or output: Copy, Delete, and Rename.

SDC feature classes

SDC feature classes can store a set of points, lines, and polygons. They can also have topology, which determines the relationships between features that are within or between feature classes.

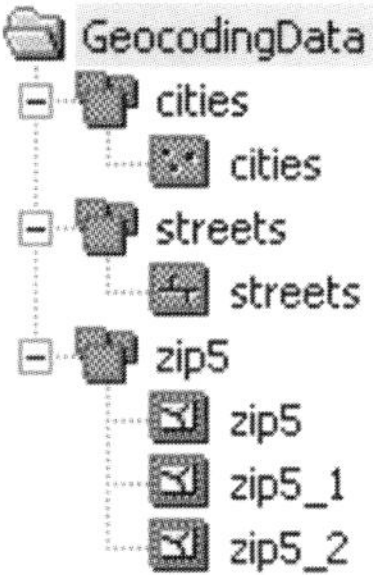

SDC feature classes differ in respect to other ArcGIS feature classes because they support multiple geometries for a single record. This design feature allows the storage of generalized versions of detailed geometry with only one set of attribute information. They support a licensing mechanism that can be used to bind access to a specific ArcGIS application, such as ArcReader™, or to a specific extension, such as ArcGIS Business Analyst.

SDC feature classes are created using ESRI's Data Development Kit Professional (DDKP). They are composed of an SDC file that stores both geometry and data, a spatial index (.sdi) file that represents the spatial index, an attribute index descriptor (.idi) file that is a pointer to attribute indexes, and any number of attribute indexes that are defined in the IDI file. Filename extensions for individual attribute indexes are open as long as they are correctly defined in the IDI file.

SDC feature classes can only be used as input to geoprocessing tools that produce a distinct output or tools that update attributes. They cannot be used by those tools that update the geometry of the input.

If you want to update the geometry of an SDC feature class, export it to a geodatabase feature class and perform updates to the geometry of the geodatabase feature class. In the Conversion Tools toolbox, click the To Geodatabase toolset and use the Feature Class To Feature Class tool or the Feature Class to Geodatabase tool (for multiple inputs) to convert SDC feature classes into geodatabase feature classes.

In the Search tab of the ArcToolbox window, type "feature class" to locate tools that work with SDC feature classes.

VPF datasets

The *VPF dataset* is a U.S. Department of Defense military standard that defines a standard format, structure, and organization for large geographic databases. A VPF dataset contains read-only feature classes.

You can export from VPF to coverage and vice versa. In the Search tab of the ArcToolbox window, type "VPF" to locate all tools that accept VPF datasets as input or output.

VPF feature classes

A *VPF feature class* is a collection of features (primitives) that have the same attributes. Each feature class contains point (node), line (edge), polygon (face), or annotation features and has an associated feature attribute table. VPF feature classes are read-only. They can only be used by geoprocessing tools that produce a distinct output. They cannot be used by tools that alter the geometry or attributes of the input feature class.

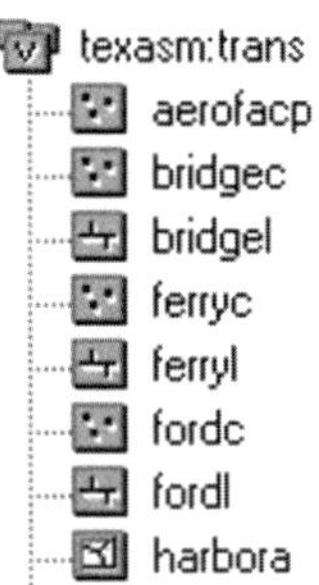

In the Search tab of the ArcToolbox window, type "feature class" to locate tools that accept VPF feature classes as input or output.

If you want to alter the schema of a VPF feature class, you must export the feature class to a geodatabase feature class. In the ArcToolbox window, under Conversion Tools, click the To Geodatabase toolset and use the Feature Class To Feature Class tool or the Feature Class to Geodatabase tool (for multiple inputs) to convert VPF feature classes into geodatabase feature classes.

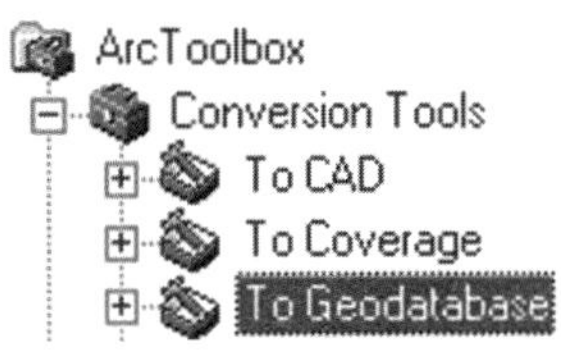

Raster data

Raster datasets

A *raster* dataset is any valid raster format that is organized into bands. A raster dataset can contain one or more bands. It can be stored in a folder in a file system, a geodatabase (personal or ArcSDE), or a raster catalog. The following raster datasets are supported: ESRI GRID, ERDAS IMAGINE, TIFF, MrSID, JFIF (JPEG), ESRI BIL, ESRI BIP, ESRI BSQ, Windows Bitmap, GIF, ERDAS 7.5 LAN, ERDAS 7.5 GIS, ER Mapper, ERDAS Raw, ESRI GRID Stack File, DTED Level 1 & 2, ADRG, PNG, NITF, CIB, and CADRG

When you use a raster dataset as input to a geoprocessing tool, you can simply click the raster dataset in the Browse dialog box and click Add. Band 1 of the raster dataset is used, unless it is a tool that operates on multiband data, such as those in the Multivariate toolset (of the ArcGIS Spatial Analyst Tools toolbox), or some of the raster data management tools such as Copy and Rotate.

In the Search tab of the ArcToolbox window, type "raster" to locate all tools that accept raster datasets as input or output.

Raster dataset bands

A *raster dataset band* consists of a rectangular matrix of cells that describe characteristics of an area and their relative positions in space. Each cell has a value, indicating what is represented in the cell. All cells in a raster band have the same height and width and represent a portion of an area.

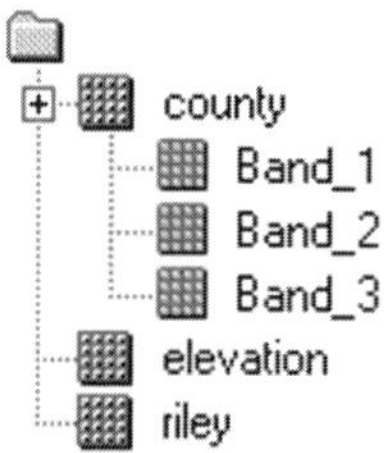

If you know you want to use a specific band of a raster dataset as input, double-click the raster dataset in the Browse dialog box for the input parameter and click the band you want to use. If you use the dataset, Band 1 will automatically be used.

Type "raster" in the Search tab of the ArcToolbox window to locate tools that accept raster bands as input or output.

Raster catalogs

A *raster catalog* is a collection of raster datasets organized in a table, in which the records define the individual raster datasets that are included in the catalog. A raster catalog is used to display adjacent or overlapping raster datasets without having to mosaic them into a large file. Geodatabase raster catalogs can be created using the Create Raster Catalog tool. The raster datasets within a raster catalog can be used as input to geoprocessing tools that accept raster datasets as input.

Type "raster" in the Search tab of the ArcToolbox window to locate tools that accept raster catalogs or raster datasets—that could be stored inside a raster catalog—as input or output.

TIN data

TIN datasets

TIN datasets can be used to display and analyze surfaces. They contain irregularly spaced points that have x,y coordinates describing their location and a z-value that describes the surface at that point. The surface could represent elevation, precipitation, or temperature. A series of edges join the points to form triangles. The resulting triangular mosaic forms a continuous faceted surface, where each triangle face has a specific slope and aspect.

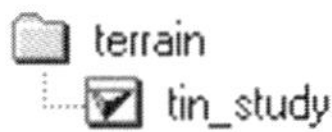

In the Search tab of the ArcToolbox window, **type** "TIN" to locate tools that accept TIN datasets as input or output.

Note: You will only see the ArcGIS 3D Analyst™ Tools toolbox if you have installed the 3D Analyst extension. If the tools are disabled (a lock icon appears next to the tools), the extension is not enabled. You need to enable the extension in the Extensions dialog box of the application you're working in (click the Tools menu, then click Extensions) to be able to work with the extension's tools.

Layer data

 Layer files

Layer files reference geographic data stored on disk. They can reference most data sources supported in ArcCatalog, such as feature classes, CAD datasets, CAD drawing datasets, coverage datasets, shapefile datasets, raster datasets and bands, TIN datasets, and so on. Think of them as a cartographic view of your geographic data. They are separate files on disk and have a .lyr extension.

Many geoprocessing tools will handle layer files. Some tools only accept layer files or layers in memory as input or output. If this is the case with the tool you are using, the display name for the parameter will contain the word "layer", and all other data sources will be filtered out of the parameter's Browse dialog box. For more information on layer files, see *Modeling Our World.*

Geostatistical layer files

Geostatistical layer files are created by the ArcGIS Geostatistical Analyst extension.

You can export geostatistical layer files to ESRI GRID using the Geostatistical Layer To Raster tool located inside the

Geostatistical Analyst Tools toolbox. Converting the geostatistical layer file to ESRI GRID will enable you to perform further geoprocessing.

Layers

A layer is an in-memory file that references data stored on disk. It is the same as any layer created when you add data to the display of ArcMap. The Make Layer tools (such as Make Feature Layer) create a layer from input data. This layer is temporarily stored in memory; it is not saved on disk. It only remains available within the current session. If the session is closed, the layer is deleted. Layers created in ArcCatalog cannot be used in ArcMap and vice versa. ArcCatalog does not display created layers, but they can be used as inputs to other geoprocessing tools in the session in which you are working.

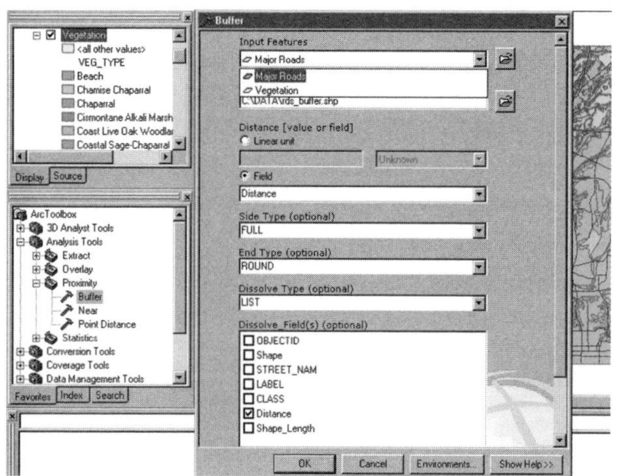

Using ArcMap layers as input data: Major Roads is a layer in the table of contents of ArcMap and is used as the input data for the Buffer tool. The layer is added by clicking the dropdown list for the Input Features parameter and selecting the layer. Layers created in ArcCatalog can be used as inputs to tools in the same way.

One of the main reasons for making and using layers is performing attribute or locational selections. By making a layer in memory first, you can perform selections on the layer without affecting the original data source.

If desired, you can save a layer to a layer file (.lyr) using the Save To Layer File tool.

Table data

Tables

A *table* can be stored in a folder or a geodatabase. It contains a set of data elements arranged in rows and columns. Each row represents an individual entity, record, or feature, and each column represents a single field or attribute value. Tables can contain attributes that can be joined to datasets to provide additional information about geographic data.

Folder-based tables include INFO and dBASE tables. Both these tables can be joined to any feature class, so the attributes from the table can be used in future processing.

Tables stored in a geodatabase, for example, Access, FoxPro®, Oracle, or SQL Server, may contain additional attributes for a particular feature class, or they might contain geographic information, such as addresses or x, y, and z coordinates.

The graphic below shows an INFO™ table inside a folder and an Access table inside a geodatabase.

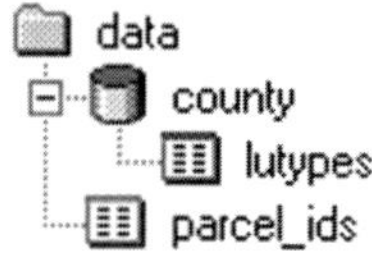

Type "table" in the Search tab of the ArcToolbox window to locate tools that accept tables as input or output.

Table views

Table views are the table equivalent of a layer. They are tables stored in memory and are the same as the table view created when a table is added to ArcMap. The Make Table View tool can be used to create a table view. ArcCatalog does not display these tables, but they can be used as inputs to other geoprocessing tools in the session in which you are working. Once you exit the application, the tables in memory are removed.

Type "table" in the Search tab of the ArcToolbox window to locate tools that accept table views as input or output.

Results from running tools

When you run a tool from its dialog box or the command line, a result is produced. By default, results are not overwritten. If you are working in an application with a display, such as ArcMap, results are permanently created and added to the display by default. Such settings can be changed on the Geoprocessing tab of the Options dialog box, accessed via the Tools menu of the ArcGIS Desktop application in which you are working.

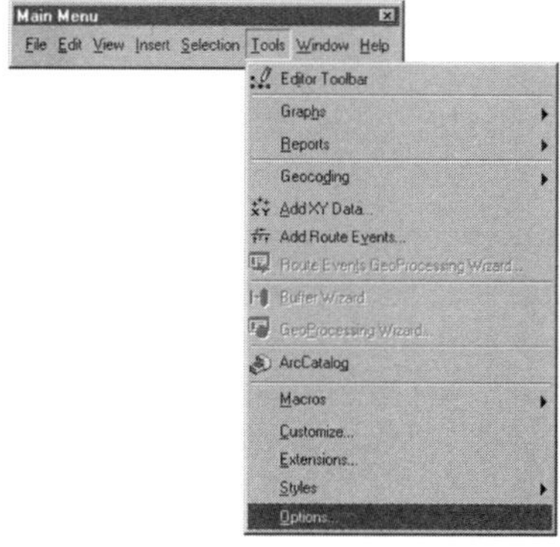

Accessing the options that can be set for results from running tools

Adding results to the display

The Add results of geoprocessing operations to the display check box is only available if you are working in an ArcGIS Desktop application with a display, such as ArcMap. It is not available in ArcCatalog. This option is checked on by default, so results produced from running tools via their dialog box or the command line will be added to your display. For custom tools, only data referenced by derived data variables that are set as parameters in the model or the script's dialog box will be added to the display. The Add to Display option in the ModelBuilder window only applies when a model is run from the ModelBuilder window, not from its dialog box. For more information, see 'Running a model' in Chapter 9.

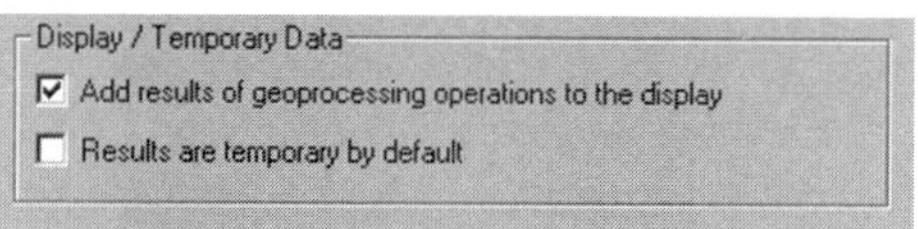

Uncheck the Add results of geoprocessing operations to the display check box if you don't want results added to the display. This option is not available in ArcCatalog.

Creating temporary results

The Results are temporary by default check box is only available if you are working in an ArcGIS Desktop application with a display, such as ArcMap. It is not available in ArcCatalog. By default, all results are permanent. Check this check box if you want all the results that are created from running tools (and added to the display) to be temporary by default. This can be useful if you are testing scenarios and don't want to worry about cleaning up data you don't want to keep. Once you exit the session all results will be deleted; so before you exit the session, you must make permanent those temporary results you want to keep.

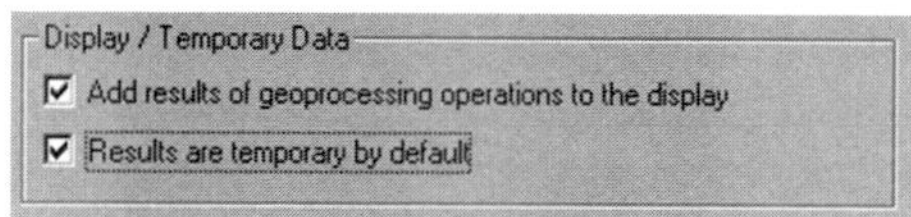

Check to create temporary results from running tools. This option is not available in ArcCatalog. This option is only available when Add results of geoprocessing operations to the display is checked.

To make a temporary result permanent, right-click the layer in the display table of contents and click Make Permanent. A quick way to make multiple layers permanent is to save the application's document. When you do this, all temporary results added to the display will become permanent.

If you have unchecked the Add results of geoprocessing operations to the display check box, the Results are temporary by default check box will become unchecked (if already checked) and will be disabled. You can only create temporary results if the Add results of geoprocessing operations to the display check box is checked.

Overwriting results

The results of geoprocessing operations in any application can be overwritten. This is a useful option if you are running operations over and over to achieve a certain result and you don't want to produce multiple undesirable outputs that take up disk space and will have to be deleted later.

If this check box is not checked (the default), each time you run an operation and name the result the same as an existing output, you will receive an error indicating that the output exists. It is useful to retain the default state if you want to make sure you do not overwrite existing data.

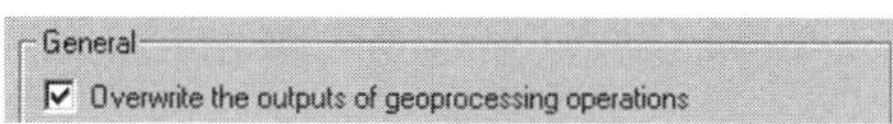

Check to overwrite the results from running tools. Care must be taken when this option is checked. After running a tool, then rerunning the tool (using the same output name), the result from the first run is overwritten by the result from the second run of the tool.

Information on other options

For information on the other options on the Geoprocessing tab of the Options dialog box, see the following sections:

For information on logging geoprocessing operations to a history model, see 'Keeping track of geoprocessing operations' in this chapter.

See Chapter 4 for information on specifying the location of the My Toolboxes folder.

For information about changing the current environment settings, see Chapter 6.

See Chapter 9 for information on displaying valid parameters when connecting a variable to a tool and more than one parameter is valid.

Geoprocessing settings

Geoprocessing settings include the state of the ArcToolbox window, the state of the Environment Settings dialog box, and any variables that have been created at the command line.

Geoprocessing settings are saved automatically between ArcCatalog sessions and when you save a map document in ArcMap. Also, when working on different projects you might use different settings based on the nature of each project. For instance, you may have removed unnecessary toolboxes or changed default environment settings. You can save the geoprocessing settings you've specified as the default settings for all applications. Alternatively, you can save settings to a file so you can quickly load them the next time you need them in any application.

Each of these settings is discussed in the sections that follow.

The state of the ArcToolbox window

Any changes you make to the ArcToolbox window will be saved when you save the settings. This includes the addition or removal of toolboxes and the reorganization of tools, such as those in the following graphic.

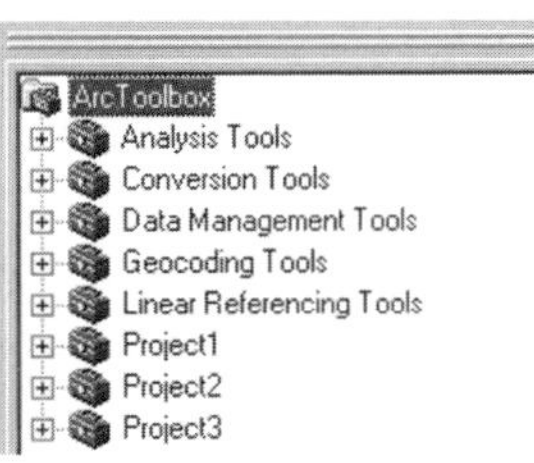

For more information on working with toolboxes see Chapter 4, 'Working with toolboxes'.

The state of the Environment Settings dialog box

Any changes you make to the default environment settings, such as those in the following graphic, will be saved when you save the settings.

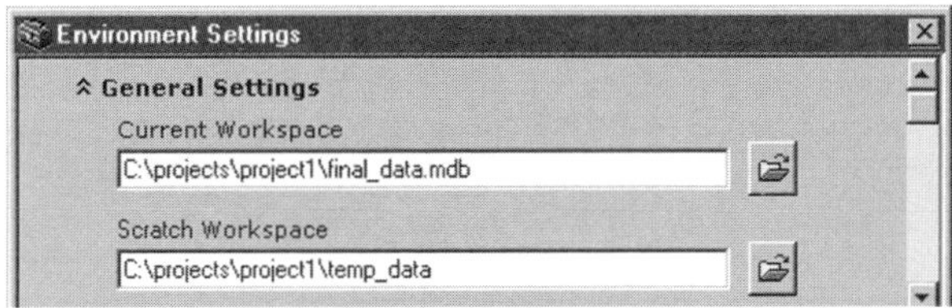

For more information on setting environments see Chapter 6, 'Specifying environment settings'.

Variables that have been created at the command line

If you've created a variable for a complex parameter value that you want to use for certain projects or between applications, saving the settings will save your created variables, so you don't have to re-create them each time they are needed.

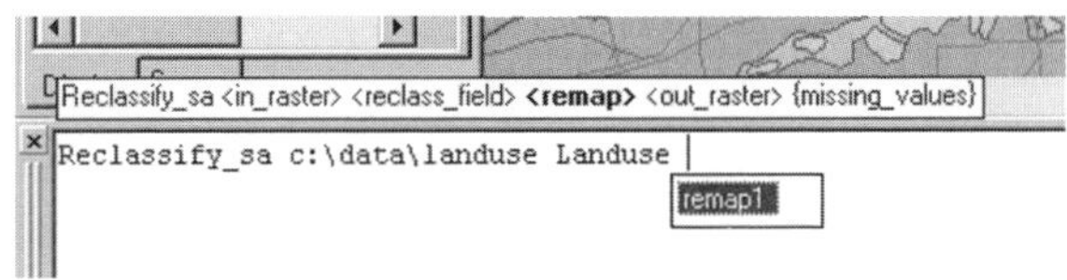

See 'About variables' in Chapter 7 for more information on variables and their creation.

See 'Saving and loading geoprocessing settings' on the following pages for information on how to save and load geoprocessing settings.

Saving and loading geoprocessing settings

Geoprocessing settings include the state of the ArcToolbox window, the state of the Environment Settings dialog box, and any variables that have been created at the command line.

Settings are persisted between ArcCatalog sessions, and if you are working in ArcMap and you save the map document, all currently set geoprocessing settings will be saved with the map document.

When you are switching between different applications or working on a certain project, you may wish to save the geoprocessing settings you have specified to a file so you can quickly load them the next time you need them in any application. Alternatively, you may save the settings as the default for all applications. See 'Geoprocessing settings' earlier in this chapter to gain a better understanding of the settings that are saved.

Saving settings

1. Right-click the ArcToolbox window, point to Save Settings and click either To File or To Default, depending on whether you want to save settings to a file that can be loaded later, or you want to set the current settings as the default for all applications.
2. If you choose To File, click the Save in dropdown arrow and navigate to the location in which you wish to save the settings file.
3. Click the File name text box and type a name for the file.
4. Click Save.

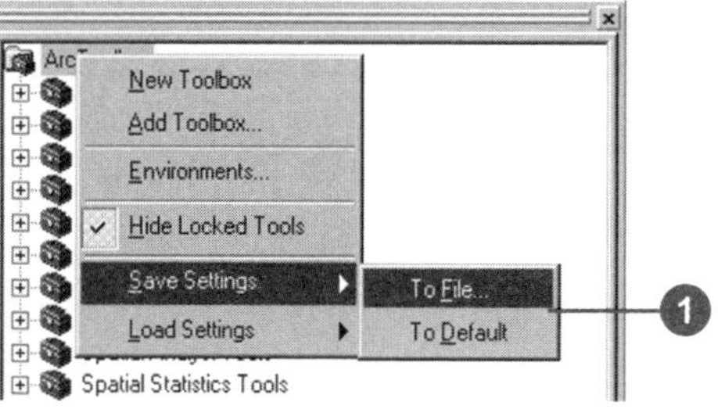

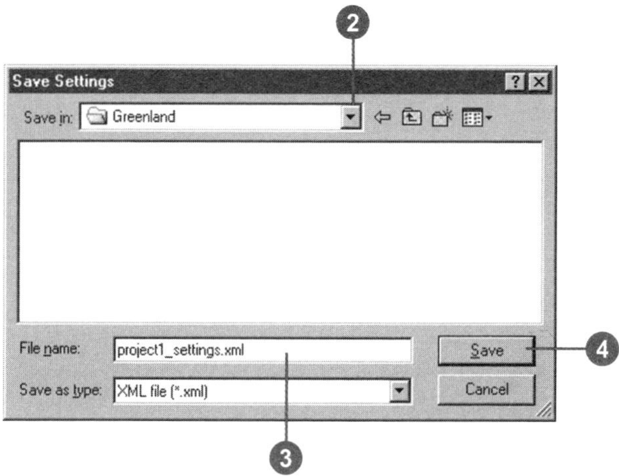

Loading settings

1. Right-click the ArcToolbox window, point to Load Settings and click either From File or From Default, depending on whether you want to load settings from a file, or you want to load the default settings for all applications.
2. If you choose From File, click the Look in dropdown arrow and navigate to the location of your saved settings XML file.
3. Click the XML file and click Open.

   Your saved geoprocessing settings are loaded.

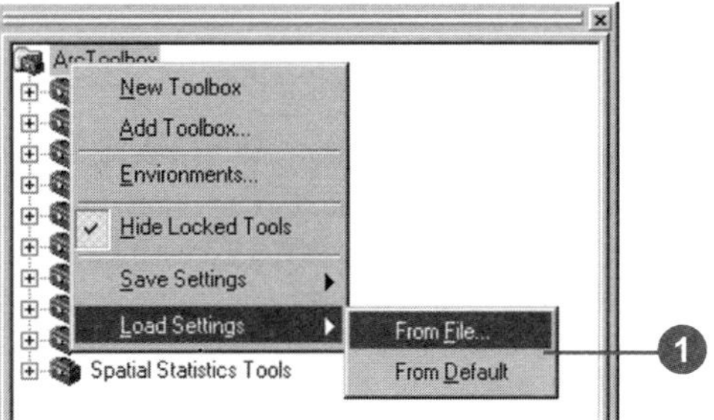

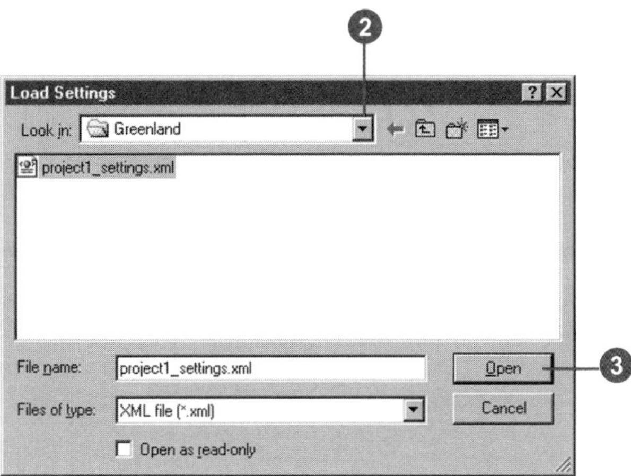

Keeping track of geoprocessing operations

When you perform geoprocessing tasks, regardless of the method, whether from a dialog box, the command line, a model, or a script, there are different ways you can keep track of your geoprocessing operations:

- View the metadata for results from running geoprocessing tools.
- View messages in the Command Line window.
- View the tools run in any session.
- Create a report containing information about the tools run in a session.

Each of these options is discussed in the sections that follow.

Viewing metadata for geoprocessing results

Whenever a result is produced by running a geoprocessing tool, you can find out how the result was produced by viewing its metadata in ArcCatalog.

Viewing messages in the current session

The message section of the Command Line window records messages for all geoprocessing operations, regardless of where they are performed. Tools that have been executed can be edited and rerun from a dialog box.

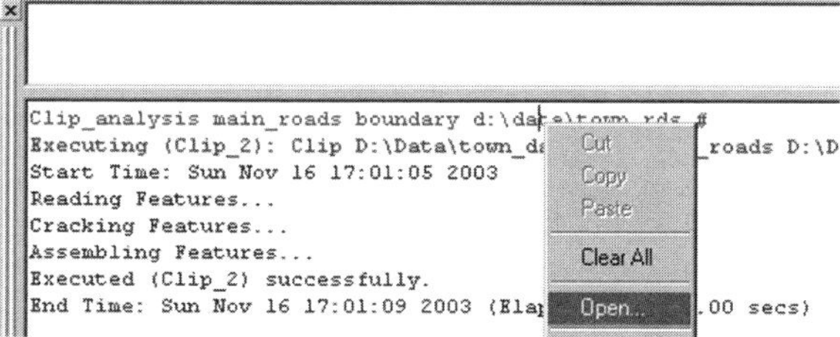

Right-click the execution string and click Open to edit the tool's parameter values and rerun the tool.

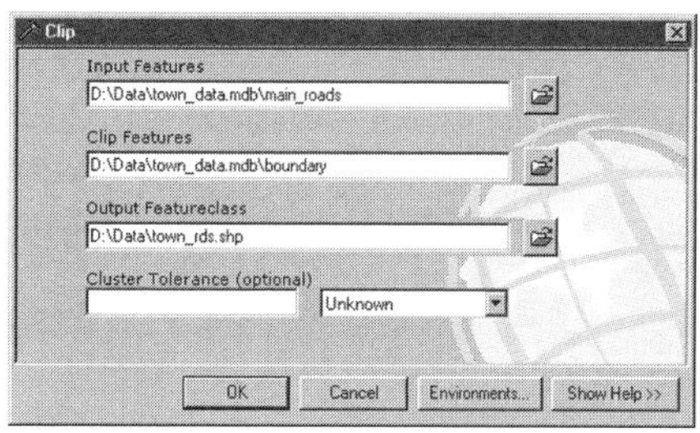

Alter parameter values, then click OK to rerun the tool.

For more information on using the Command Line window, see Chapter 7, 'Using the Command Line window'.

Viewing the tools run in any session

After running a geoprocessing tool using any ArcGIS Desktop application, the tool is added to the session's history model, located in the History toolbox in your My Toolboxes folder.

The history model documents which tools were run and which parameter values were set in the last session. You can view this information the next time you open ArcCatalog by editing the history model and double-clicking each tool within it to view the parameter values that were set.

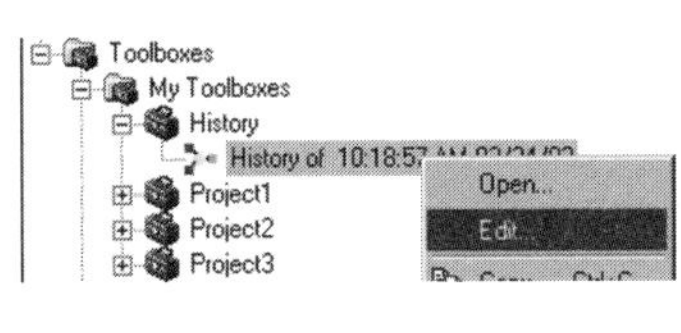

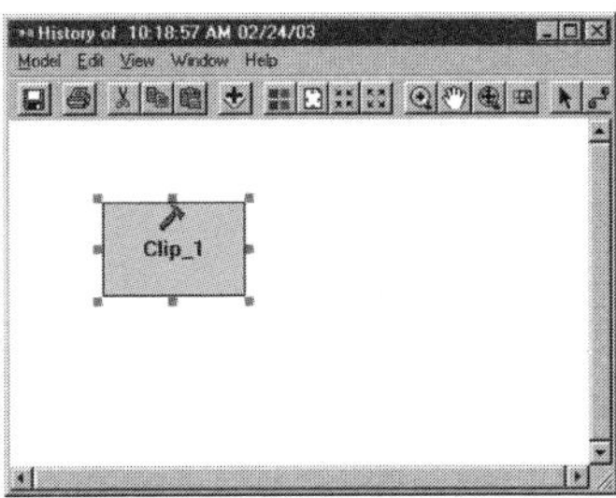

In the history model above, there is one tool: Clip_1. This was the only tool run in this particular session. You can open the tool to view the parameter values that were set. There could be multiple tools inside a history model, depending on the number of tools run in a particular session.

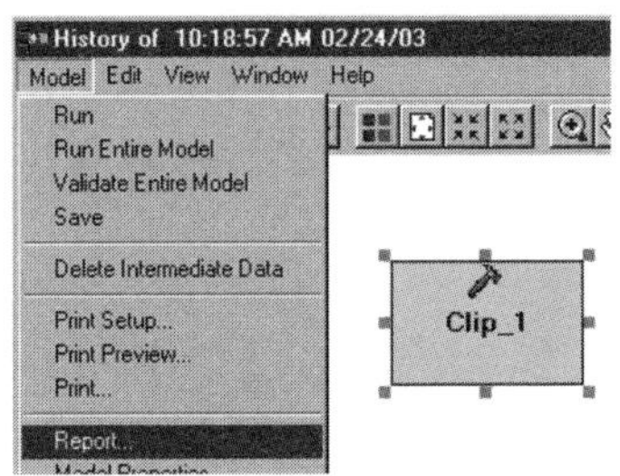

To generate a report, click the Model menu and click Report. A report can be displayed in a window, or you can save it to disk.

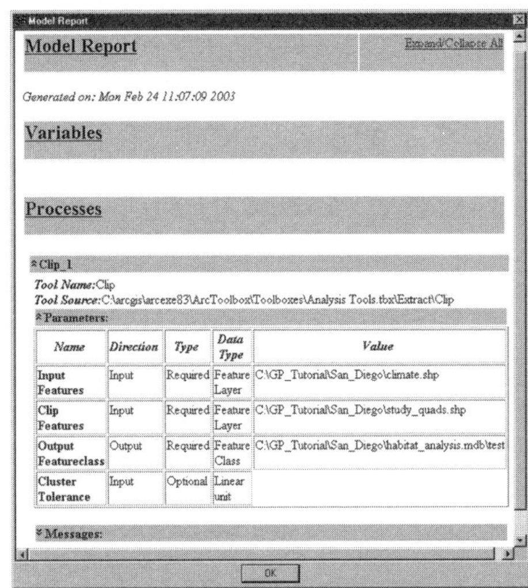
Model Report

Model Report — Expand/Collapse All

Generated on: Mon Feb 24 11:07:09 2003

Variables

Processes

Clip_1

Tool Name: Clip
Tool Source: C:\arcgis\arcexe83\ArcToolbox\Toolboxes\Analysis Tools.tbx\Extract\Clip

Parameters:

Name	Direction	Type	Data Type	Value
Input Features	Input	Required	Feature Layer	C:\GP_Tutorial\San_Diego\climate.shp
Clip Features	Input	Required	Feature Layer	C:\GP_Tutorial\San_Diego\study_quads.shp
Output Featureclass	Output	Required	Feature Class	C:\GP_Tutorial\San_Diego\habitat_analysis.mdb\test
Cluster Tolerance	Input	Optional	Linear unit	

Messages:

OK

In the sample report above, the Clip tool and the parameter values that were set for it are documented for future reference.

You can also view the contents of the history model as it is being generated to view the tools that have been run and parameter values that have been set to date in the current session. After running a tool, right-click the execution string in the message section of the Command Line window and click Show History. For more information on viewing history during a session, see Chapter 7, 'Using the Command Line window'.

If you don't want a history model to be generated for each session, you can turn the option off. Click the Tools menu in the application you are running and click Options. Click the Geoprocessing tab and uncheck the Log geoprocessing operations to a history model check box.

Generating a report of tools run in a session

When you generate a report from the history model generated for a session, you have a detailed account of the tools that were run during that session.

For more information on generating a report, see 'Generating a report' in Chapter 9.

Sharing your geoprocessing work

When sharing your geoprocessing work, you should ensure that you send all relevant sources of information. A toolbox may contain custom tools that require other files to execute. Scripts added to a toolbox are stored outside the toolbox, and system tools added to a toolbox may reference dynamic link libraries (DLLs) not registered on the desktop. Models may contain system tools and scripts, so a toolbox may contain a model that cannot execute because one or more tools may be missing a component. If a toolbox contains a script or a system tool, the script, DLL, or both must also be sent with the toolbox. The receiving user can then register any DLLs and update the properties of the tool to ensure that the connections between the tool and its dependencies are correct, such as the path to the script if relative paths have not been set for the tool.

Setting relative paths

When relative paths are set, all pathnames within the tool will be stored relative to the toolbox containing the tool. By ensuring that the paths to all information sources are set as relative and by keeping your data relative to the location of the toolbox, when the user of your tools opens them, paths to sources of information will not have to be repaired.

Examine the following directory structure as an example.

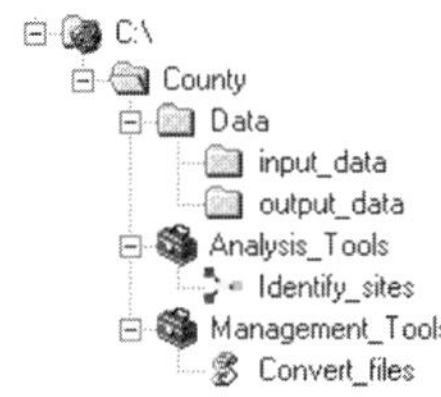

As long as the directory structure between the toolboxes and the data does not change and relative paths are set for the tools—the model and script—the data folders and the toolboxes can be placed in any folder, on any drive, as the graphic that follows shows.

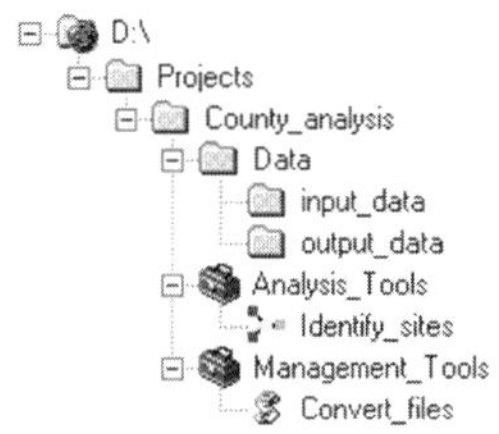

The paths to information sources referenced in the model, for example, input and derived data, and in the script, for instance, the path to the script, will be modified automatically.

To set relative paths for sources of information referenced by a tool, right-click the tool and click properties. Click the General tab and check the Store relative path names (instead of absolute paths) check box.

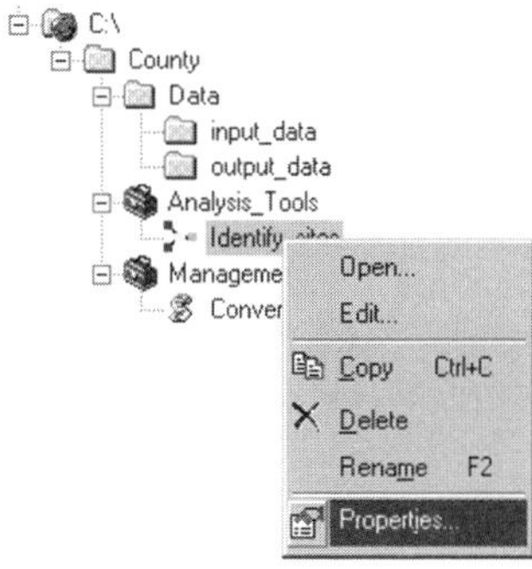

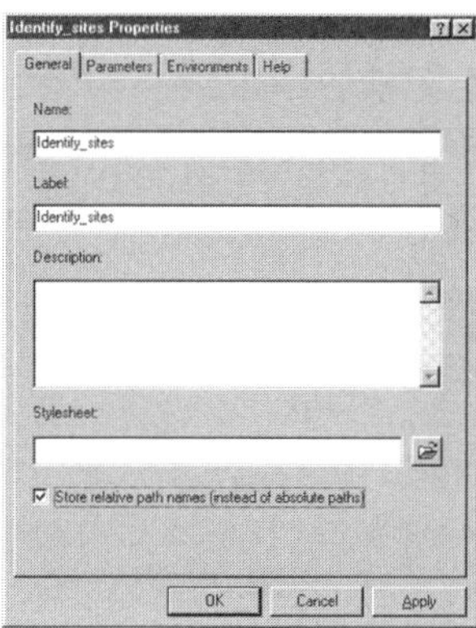

See 'Storing relative pathnames' in Chapter 5 for more information on relative paths.

Setting permissions on a toolbox

A toolbox may be a file in a folder or a table in a geodatabase, so depending on the physical location of the toolbox, certain permissions may be set on the toolbox and its contents. When sharing your toolboxes over a network or within an ArcSDE geodatabase, you should make your toolboxes read-only. Many users can read the same toolbox, but problems can result from allowing multiple users to make changes to the same toolbox. The content of the toolbox is determined by the last application to be closed. For more information, see 'Rules for working with toolboxes' in Chapter 4.

A toolbox in a folder

A toolbox file's (.tbx) permissions are set like any other file on disk. Attributes such as read and write access may be set. In Windows 2000 or Windows XP, you can grant specific users and user groups a wide range of permissions. These permissions are for the entire TBX file, not for the specific tools it contains.

A toolbox in a geodatabase

A toolbox is a table in the database. In geodatabase terms, the toolbox is an object class, and the type of permissions and how they are granted are determined by the database supporting the geodatabase. For example, in Oracle, an object class (table) has an owner who has read and write access. The owner may grant a number of different database access permissions to other users. These permissions are for the entire toolbox table, not for the specific tools it contains. Tools are simply rows in the table, so permissions cannot be set on a tool-by-tool basis.

The geodatabase facilitates a multiuser work flow with the use of versions that allow multiple users to edit an object class in the geodatabase without data replication. Toolboxes do not support multiuser editing in a geodatabase, only multiuser reading.

Distributing your work

Once all relevant sources of information are in the same folder, relative paths are set for your tools, and permissions are set on your toolboxes, you are ready to distribute your work. The safest and quickest way to distribute your work is to e-mail it or place it on an FTP site that can be accessed by the person receiving your work. Alternatively, or as a backup, you could copy the files to a CD and send it through the mail.

Archiving your work

Before you send your work, it is efficient to use an archiving program, such as WinZip®, to combine all the necessary files into one compressed archive. Once the files are compressed, you then distribute the archive. The most commonly used archive format is the ZIP file, but you could use others, such as TAR, gzip, or CAB files.

There are two benefits to using archives to distribute your geoprocessing work. First, only one file transfer operation is required to obtain all related files, and second, file transfer time is minimized because the files in an archive are compressed.

When the user receives an archive file, it needs to be unzipped to a folder. The paths to all information sources are automatically set relative to the position of the toolbox, provided that you set relative paths for your tools.

Sharing your work via a network

If you are working in a networked environment, you can simply share the folder containing the necessary files, such as toolboxes, data, script files, and XML files for stylesheets, geoprocessing settings, or documentation files, so others can access and copy them locally. Remember to set relative paths for sources of information referenced by the tools, and be sure the same directory structure is set between the toolbox and the data on the machine your work is copied to.

You can use the standard Universal Naming Convention (UNC) format for paths, such as \\yourmachinename\folder name\etc, so others in your local area network can access and run your tools from, and place results on, your machine. However, this can introduce issues with overwriting your toolboxes' content. Remember that a toolbox's contents are determined by the last application to be closed. See 'Rules for working with toolboxes' in Chapter 4 for more information.

Working with toolboxes

4

IN THIS CHAPTER

- **Opening and docking the ArcToolbox window**
- **Locating the system toolboxes**
- **Creating new toolboxes**
- **Managing toolboxes**
- **Adding documentation to toolboxes**
- **Viewing documentation for toolboxes**
- **Rules for working with toolboxes**

A toolbox is a persistent entity that can contain toolsets and geoprocessing tools. It takes the form of a TBX file on disk or a table in a geodatabase. The collection of system toolboxes contains more than 400 system tools categorized into toolsets.

A toolbox is the mechanism for sharing tools. Toolboxes are accessed through the ArcCatalog tree or the ArcToolbox window. They may be added to or removed from the ArcToolbox window to create a list of shortcuts to often used, or favorite, toolboxes stored on disk.

The behavior of a toolbox is similar to a dataset. A toolbox may be:

- Created inside a folder or a geodatabase
- Opened to see its contents, such as tools and toolsets
- Copied and pasted from one location to another
- Altered by adding, deleting, or renaming the tools and toolsets it contains
- Documented by creating or editing documentation

Opening and docking the ArcToolbox window

When the ArcToolbox window is initially opened, a list of available toolboxes is displayed in the window. Contained within the toolboxes are system tools that can be run. The toolboxes listed are shortcuts that point to toolboxes stored on disk.

You can customize the ArcToolbox window to centralize the location of the toolboxes you use most often. In addition to working with the system toolboxes, you can create your own toolboxes in folders or geodatabases that might be spread out over your system or on other machines in your network.

The ArcToolbox window is a separate window inside the ArcGIS Desktop application you are running. It can be docked anywhere within the application, or it can be placed on your desktop.

Opening the ArcToolbox window

1. Click the Show/Hide ArcToolbox Window button on the Standard toolbar of the ArcGIS Desktop application you are working in to open the ArcToolbox window.

   A list of toolboxes that contain toolsets and system tools is displayed in the window.

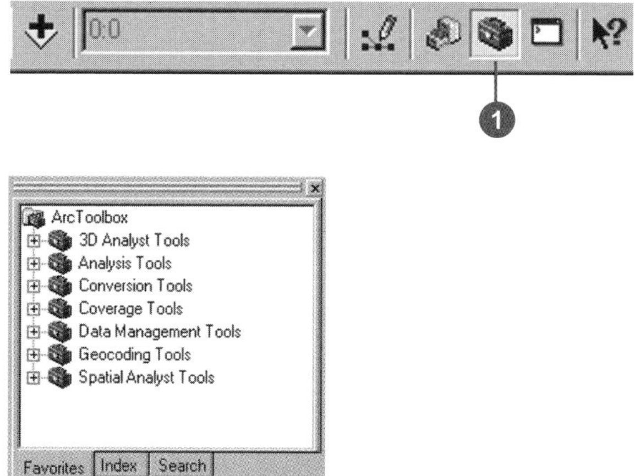

Docking the ArcToolbox window

1. Click the bar at the top of the ArcToolbox window and drag the window to your preferred location within the application or outside it.
2. Drop the panel by releasing the mouse.

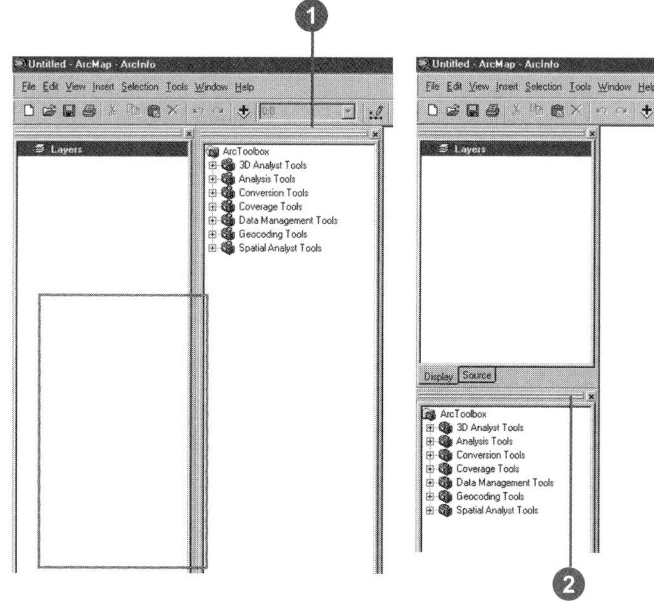

Locating the system toolboxes

The system toolboxes can be accessed from the Toolboxes folder in the ArcCatalog tree. By expanding the Toolboxes contained within the System Toolboxes folder, you will see the toolsets stored within them. Inside each toolset is a collection of tools that can be used.

See Also

See Chapter 5, 'Working with toolsets and tools', for more information on working with system tools.

See Also

See 'Opening and docking the ArcToolbox window' earlier in this chapter for information on using the ArcToolbox window to create shortcuts to the toolboxes you use most frequently.

See Also

Type "Quick Reference Guide" in the Search tab of the online Help system and double-click the link to the Geoprocessing Commands Quick Reference Guide for a list of available tools depending on the ArcGIS product (ArcView, ArcEditor, or ArcInfo) and extension products you have installed.

1. Click the Tools menu and click Options.
2. Click the General tab.
3. Check Toolboxes, then click OK on the Options dialog box.

   A Toolboxes folder is added to the ArcCatalog tree so you can access the toolboxes installed on your system.
4. Navigate in the ArcCatalog tree to the Toolboxes folder.
5. Expand the Toolboxes folder to view its contents.
6. Expand the System Toolboxes folder to view the toolboxes you have stored on your system.

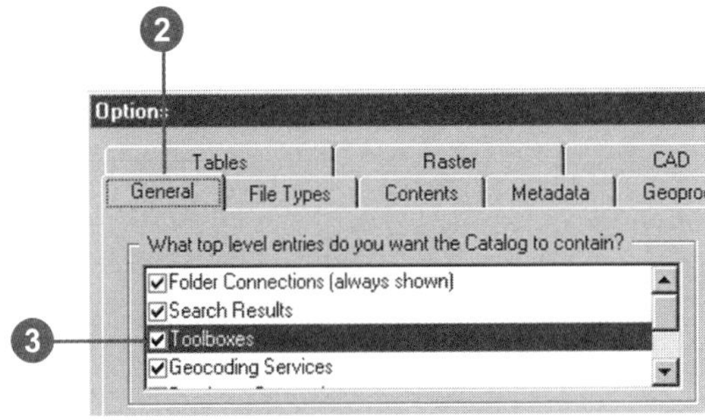

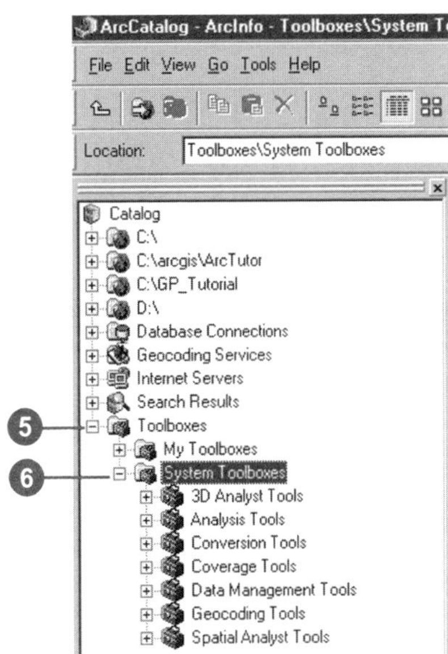

Creating new toolboxes

You can create new toolboxes in the ArcToolbox window or the ArcCatalog tree to hold the tools you want to use.

In the ArcToolbox window, new toolboxes point to a default location on disk: the location of the My Toolboxes folder in the Toolboxes folder in the ArcCatalog tree.

Toolboxes can also be created inside existing folders or geodatabases directly within the ArcCatalog tree. Once you have created a new toolbox, add it to the ArcToolbox window as a shortcut and use it from there. This saves having to navigate to the location of your toolbox in the ArcCatalog tree each time you want to use it. ▸

Tip

Locating toolboxes

If you are unsure about the location of a toolbox, right-click the toolbox and click Properties to see the location on disk where the toolbox was created.

See Also

See Chapter 5, 'Working with toolsets and tools', for information on adding tools to toolboxes and creating new tools inside toolboxes.

Creating a new toolbox in the ArcToolbox window

1. Right-click the ArcToolbox folder in the ArcToolbox window and click New Toolbox.

   A toolbox shortcut appears in the ArcToolbox window.

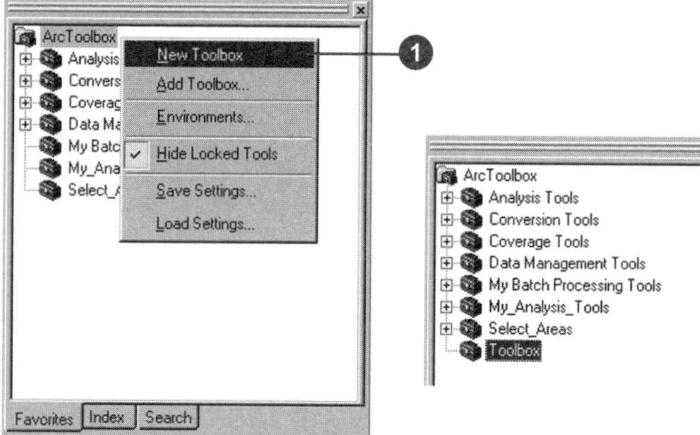

Creating a new toolbox in the ArcCatalog tree

1. Right-click the folder or geodatabase in which you want to create a new toolbox, point to New, and click Toolbox.
2. Type a new name for the toolbox.
3. Press Enter.

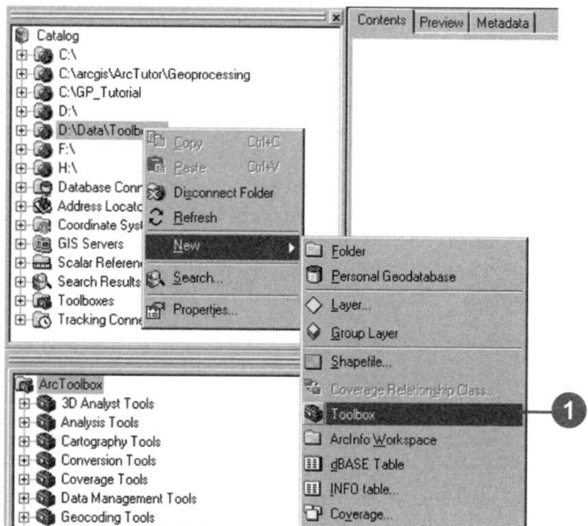

The My Toolboxes folder is located inside the Toolboxes folder, which is at the root level in the ArcCatalog tree. New tools can be created directly in this folder. Alternatively, they can be created inside the ArcToolbox window. Tools created inside the ArcToolbox window are shortcuts that point to this folder.

You can change the default location on disk for the My Toolboxes folder. Any new toolboxes you create directly inside the My Toolboxes folder or the ArcToolbox window will be stored on disk in the location specified.

Tip

Resetting the location of the My Toolboxes folder

To reset the location on disk of the My Toolboxes folder back to the default, click Reset on the Geoprocessing tab of the Options dialog box, accessed via the Tools menu in the application you are working in.

Changing the default location of the My Toolboxes folder

1. Click the Tools menu and click Options.
2. Click the Geoprocessing tab.
3. Click the Browse button.
4. Browse to the location where you would prefer new toolboxes, which are created inside the ArcToolbox window or directly inside the My Toolboxes folder in the ArcCatalog tree, to be stored.
5. Click OK on the Browse for Folder dialog box.
6. Click OK.

   All new toolboxes that you create inside the ArcToolbox window or the My Toolboxes folder will be stored in this location on disk.

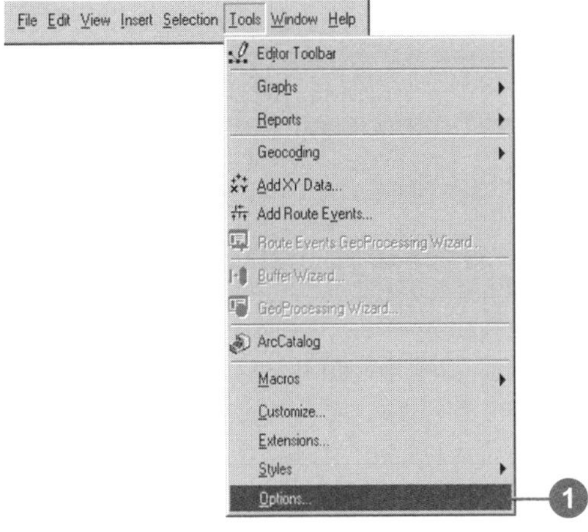

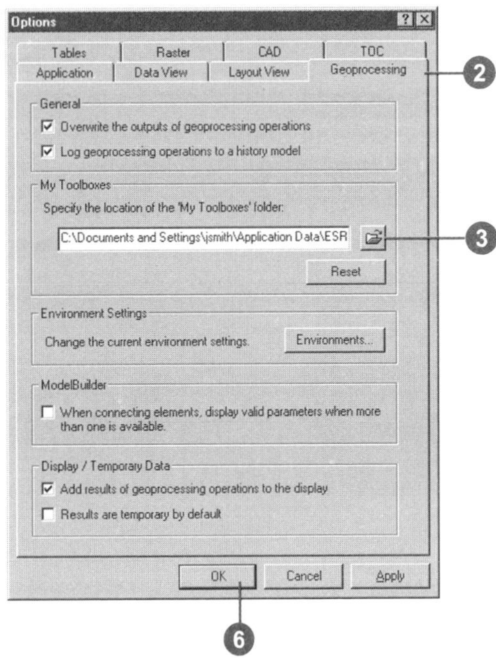

Managing toolboxes

You manage toolboxes in a similar way to how you manage data.

Toolboxes can be copied and pasted between folders, between geodatabases, or between geodatabases and folders.

Toolbox shortcuts can be copied from the ArcToolbox window and pasted into a folder in the ArcCatalog tree. They can be copied and pasted into a geodatabase, provided the copied toolbox resides inside a geodatabase.

You cannot copy and paste toolbox shortcuts within the ArcToolbox window. Instead, create a new toolbox and copy tools or toolsets into it from existing toolboxes in the ArcToolbox window. ▸

Tip

About read-only toolboxes

The tools within toolboxes that are set as read-only can be run and copied but cannot be altered. Set toolboxes as read-only if you want others to access and run your tools, but you don't want their contents to be altered.

Copying and pasting a toolbox within the ArcCatalog tree

1. Right-click the toolbox in the ArcCatalog tree and click Copy.
2. Right-click the folder or geodatabase into which you want to place a copy of the toolbox and click Paste.

   A copy of the toolbox is placed in the destination location.

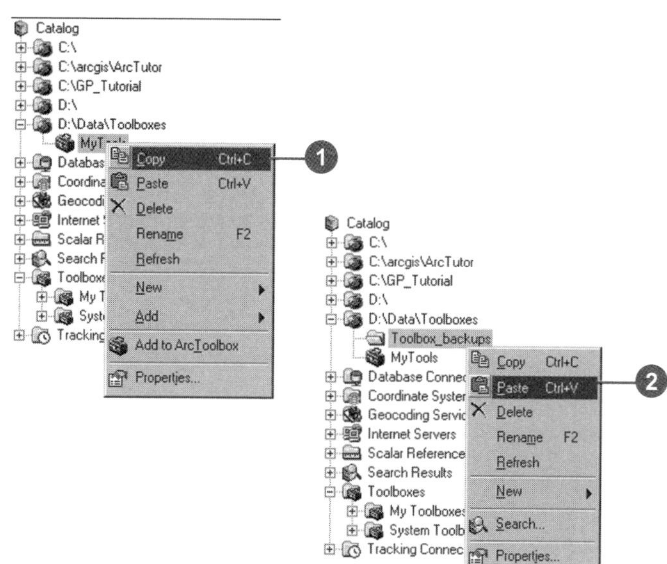

Copying and pasting a toolbox from the ArcToolbox window to the ArcCatalog tree

1. Right-click the toolbox shortcut in the ArcToolbox window and click Copy.
2. Navigate in the ArcCatalog tree to the location into which you want to paste the toolbox.
3. Right-click the folder or geodatabase and click Paste.

   A copy of the toolbox is placed in the destination location.

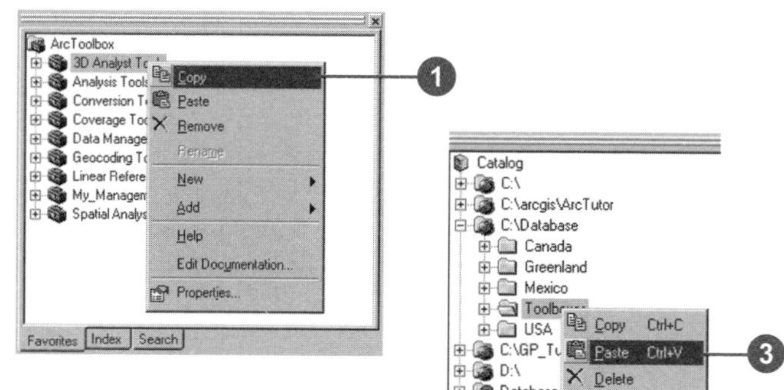

Toolboxes can be dragged and dropped in the ArcCatalog tree in the same way you drag and drop data. When a toolbox is dragged into another folder or geodatabase, it is moved to the new location if the new location is on the same disk. If the new location is on a different disk, a copy of the toolbox is made.

Toolboxes that you have created in folders and geodatabases can be added or dragged to the ArcToolbox window, providing quick access to often used tools. ►

Tip

About toolboxes in the ArcToolbox window

Toolboxes in the ArcToolbox window are shortcuts that point to toolboxes in folders or geodatabases on disk. When working with toolboxes in the ArcToolbox window, you are always accessing and updating the actual toolbox on disk, not a copy of the toolbox.

Dragging and dropping toolboxes

1. Click the toolbox you want to move, then drag and place it in another location.

   The location you place it into can be a folder or a geodatabase.

   Note that if you drag the toolbox into the ArcToolbox window, it is not moved. A shortcut to the toolbox on disk is created.

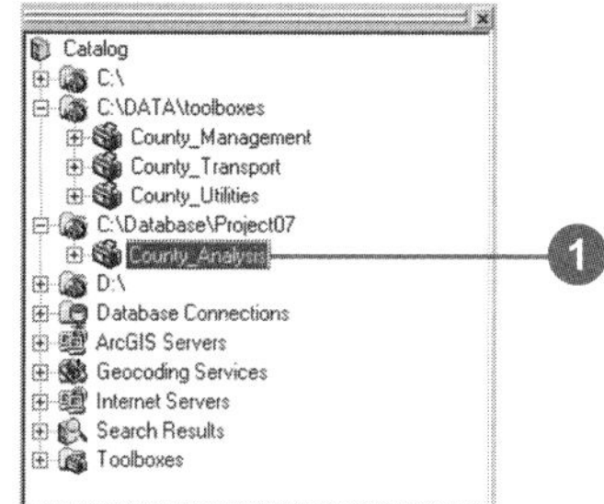

Adding toolboxes to the ArcToolbox window from the ArcCatalog tree

1. Right-click the toolbox in the ArcCatalog tree and click Add to ArcToolbox.

   The toolbox is added to the ArcToolbox window as a shortcut to the toolbox stored on disk.

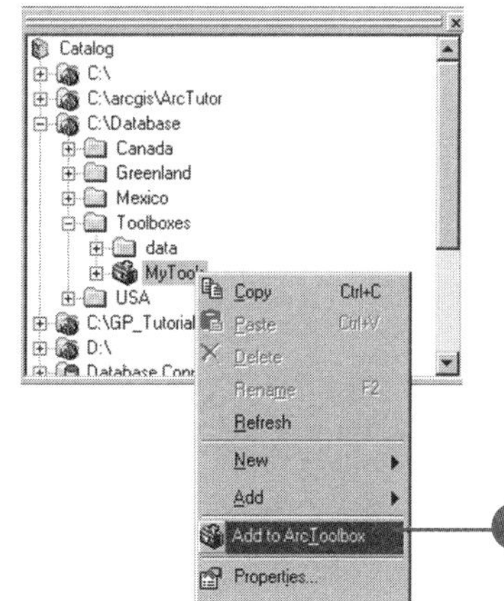

If you are working in a desktop application, such as ArcMap, where you do not have access to the ArcCatalog tree, you'll always use the ArcToolbox window to access toolboxes. Toolboxes can be added as shortcuts to the ArcToolbox window by using the Add Toolbox option. ▸

See Also

If you have set up your ArcToolbox window with appropriate toolboxes for a particular project, you can save the geoprocessing settings by right-clicking the ArcToolbox window and clicking Save Settings. Settings can be loaded in the ArcToolbox window in any ArcGIS application by right-clicking the ArcToolbox window and clicking Load Settings. For more information, see 'Geoprocessing settings' in Chapter 3.

Adding toolboxes to the ArcToolbox window via the ArcToolbox context menu

1. Right-click the ArcToolbox folder inside the ArcToolbox window and click Add Toolbox.
2. Click the Look in dropdown arrow and navigate to the location of the toolbox on disk that you want to add as a shortcut.
3. Click the toolbox.
4. Click Open.

   The toolbox is added to the ArcToolbox window.

   The toolbox that is added is a shortcut that points to the toolbox stored on disk. Use the shortcut as a fast way to access the toolbox on disk.

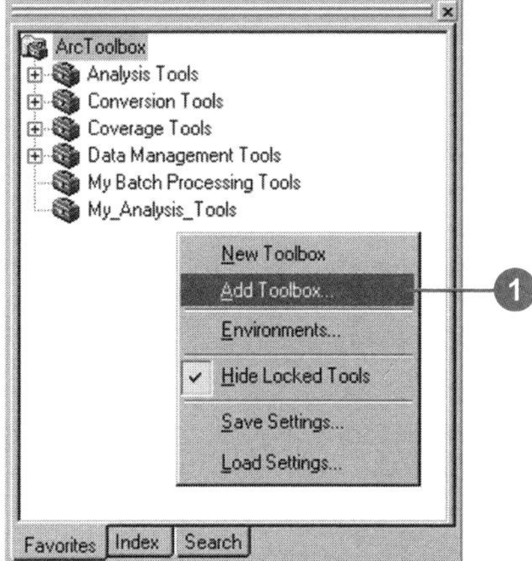

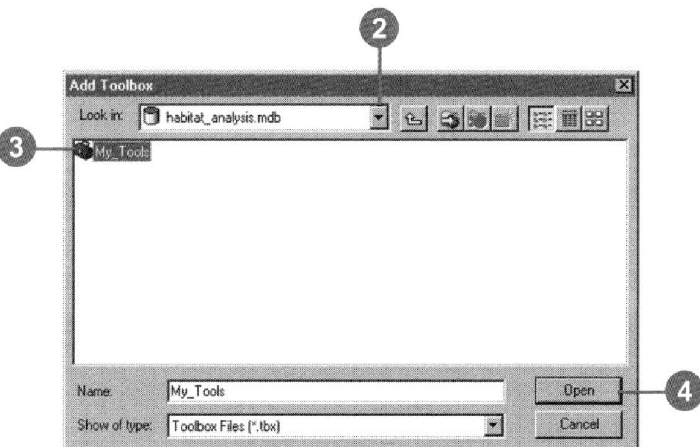

Toolboxes can be deleted from the ArcCatalog tree similarly to datasets.

If the toolbox you delete refers to a script, the script added to the toolbox is not also deleted from disk. You must delete the script from disk using, for instance, Windows Explorer, if it is also no longer needed.

Toolboxes you add as shortcuts to the ArcToolbox window cannot be deleted, but they can be removed from the ArcToolbox window. The toolbox will still exist on disk. This provides you with an easy way to manage the toolboxes you frequently use in the ArcToolbox window and remove those you don't use often. You can always add toolboxes back to the ArcToolbox window.

To protect toolboxes from being deleted, make them read-only. For toolboxes inside a folder, right-click the TBX file on disk and click Properties to set the read-only property on your toolbox. For toolboxes inside a personal geodatabase, right-click the geodatabase on disk and click Properties to set the read-only property. For toolboxes inside an ArcSDE geodatabase, set privileges for the toolbox. Right-click the toolbox in ArcCatalog and click Privileges. ►

Deleting toolboxes

1. Right-click the toolbox in the ArcCatalog tree and click delete.

   The toolbox is permanently deleted from disk.

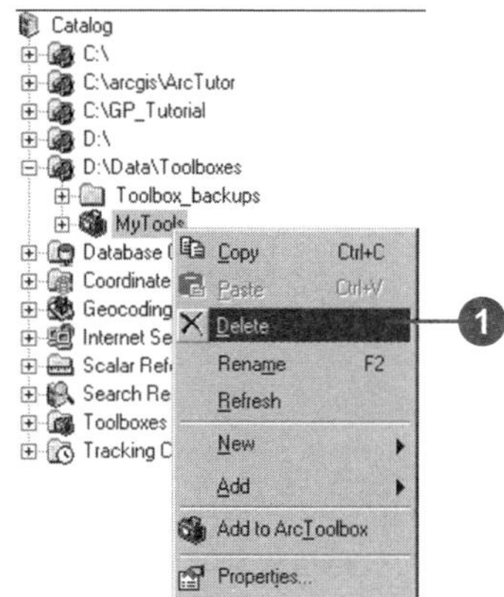

Removing toolboxes from the ArcToolbox window

1. Right-click the toolbox and click Remove.

   The toolbox shortcut will be removed from the ArcToolbox window, but the toolbox that is stored on disk will not be deleted.

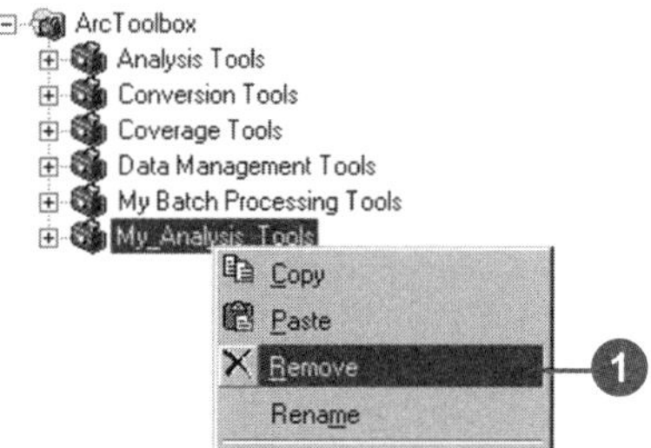

Toolboxes you create can be renamed to something more meaningful in the ArcCatalog tree.

If you have already added a toolbox as a shortcut to the ArcToolbox window within ArcCatalog and you rename the toolbox in the ArcCatalog tree, the shortcut in the ArcToolbox window will be updated automatically. If you rename the shortcut, the toolbox on disk will be automatically renamed in the ArcCatalog tree.

However, if you add the toolbox to the ArcToolbox window in ArcMap, then rename the toolbox in ArcCatalog, the toolbox name will not be updated in the ArcToolbox window in ArcMap because it has already been read into memory. You have to remove the toolbox and add it back to the ArcToolbox window to obtain the name change made to the toolbox.

Toolboxes inside a geodatabase act as any other object. Certain characters are not allowed in the name of the toolbox. ▸

Tip

Renaming toolboxes

For a quick way to rename a toolbox, click the toolbox to select it, then click again to rename it.

Renaming toolboxes

1. Click the toolbox you want to rename.
2. Click File and click Rename.
3. Type the new name.
4. Press Enter.

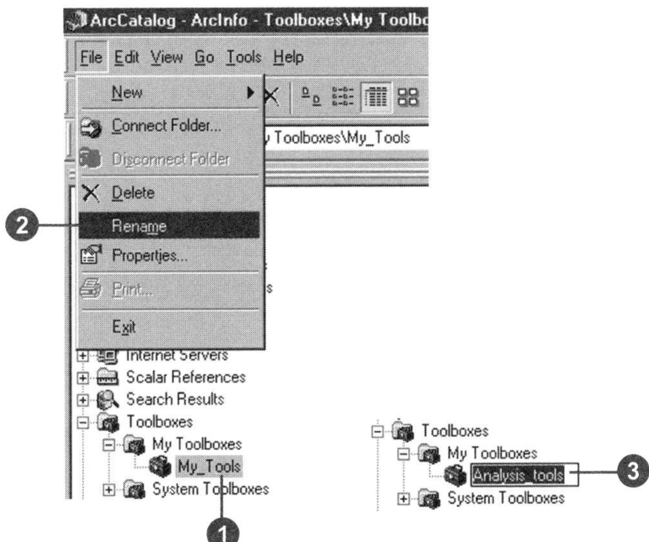

Sometimes the ArcCatalog tree's list of a toolbox's contents will not match the toolbox's actual contents. One reason this could happen is if you had two ArcCatalog applications open at once and you deleted a tool from a toolbox in one application. The other application will not know about this change until you exit the application and open another application. To avoid having to close the application, you can just refresh the toolbox. Note that to avoid such conflicts it is always safest to work with one application open at a time. ▸

See Also

See 'Rules for working with toolboxes' in this chapter for more information on conflicts that can occur when working with the same toolbox in multiple applications.

Refreshing toolboxes

1. Right-click the toolbox in the ArcCatalog tree and click Refresh.

   The toolbox should now contain its actual contents.

By right-clicking a toolbox and clicking Properties, you can find out the location of the toolbox on disk and the alias used.

Obtaining the location of a toolbox on disk is particularly useful if you have added a toolbox as a shortcut to the ArcToolbox window and wish to know where it is located on disk.

The toolbox *alias* is an alternative name for the toolbox. It can be used to avoid confusion when running tools at the command line or within a script, when tools have the same name but are stored in different toolboxes in the ArcToolbox window. For example, typing "clip_analysis", then a space, at the command line will produce usage for the system tool Clip in the Analysis Tools toolbox. ►

See Also

See Chapter 7, 'Using the Command Line window', for more information on using aliases at the command line.

Viewing the location and alias of a toolbox

1. Right-click the toolbox and click Properties.
2. Click the General tab.

   The name of the toolbox, the location in which it is stored, and the alias name for the toolbox are displayed.

   The alias can be used when adding or setting the toolbox containing the tool you wish to use at the command line or inside a script.
3. Click OK.

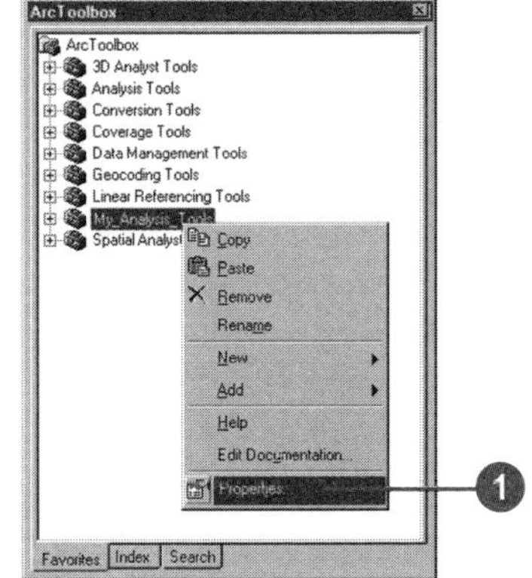

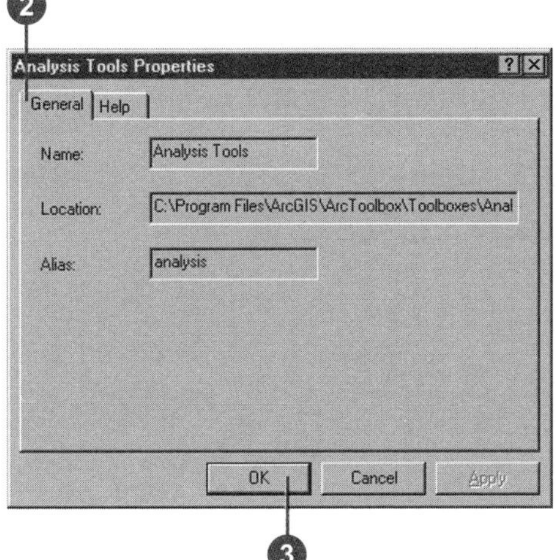

The ArcToolbox window is a separate window in each application. It is saved in its current state when you save a map document. This is useful if you are working on different projects and will use different tools for each project.

See Also

See 'Geoprocessing settings' in Chapter 3 for information on saving geoprocessing settings, including the state of the ArcToolbox window, so settings can be loaded at a later date in any ArcGIS Desktop application.

Saving toolboxes within a map document

1. In ArcMap, set up your toolboxes in the ArcToolbox window with the tools you want to use within the map document.
2. Click File on the Main Menu, then click Save As.
3. Click the Save in dropdown arrow, and navigate to a directory for which you have write access.
4. In the File name box, type a name for the map document.
5. Click Save.

   Your changes to the ArcToolbox window will be saved with the map document.
6. Click File, then Exit to close the ArcMap session.

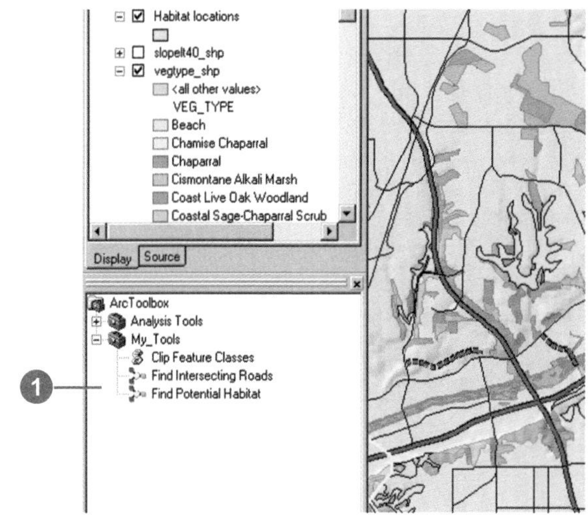

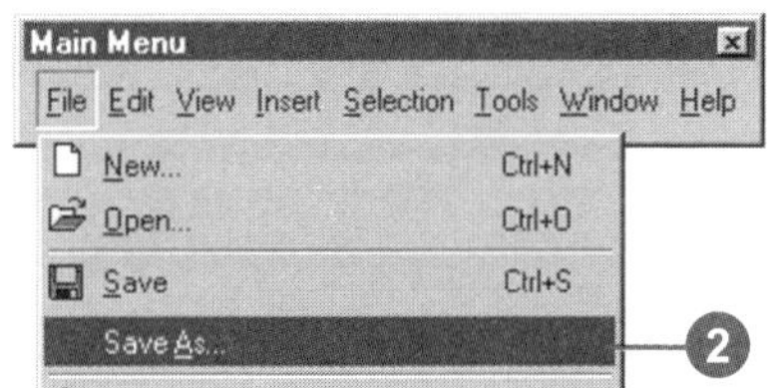

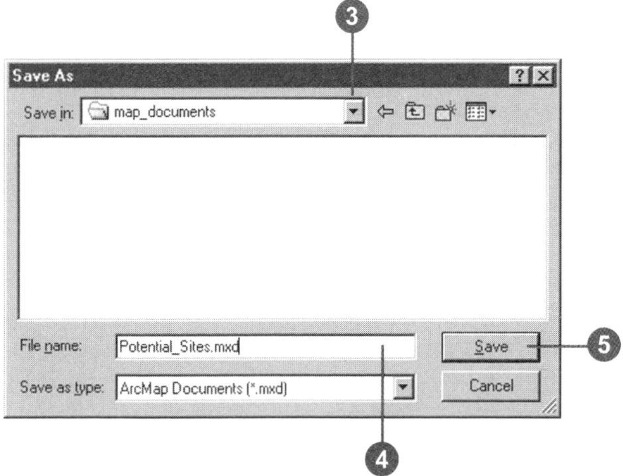

Adding documentation to toolboxes

You can add documentation to toolboxes you have created using the *Documentation Editor*.

The entries in the General Information section of the Documentation Editor will appear in the metadata tab for the toolbox. The abstract provides a paragraph of text describing the toolbox's contents. Keywords enable a toolbox to be found easily when searching for it in ArcCatalog. Author information is fundamental if you want users of your toolbox to be able to contact you, and you should document any constraints as to the use of the toolbox.

Entries in the Help section of the Documentation Editor appear in the Help page accessed via the ArcToolbox window. You can add summary information for the toolbox and also for its toolsets. ▸

See Also

See 'Adding documentation to tools' in Chapter 5 for information on documenting tools.

Adding documentation to a toolbox via the ArcToolbox window

1. Right-click the toolbox and click Edit Documentation.

   The Documentation Editor opens so you can add documentation to your toolbox.

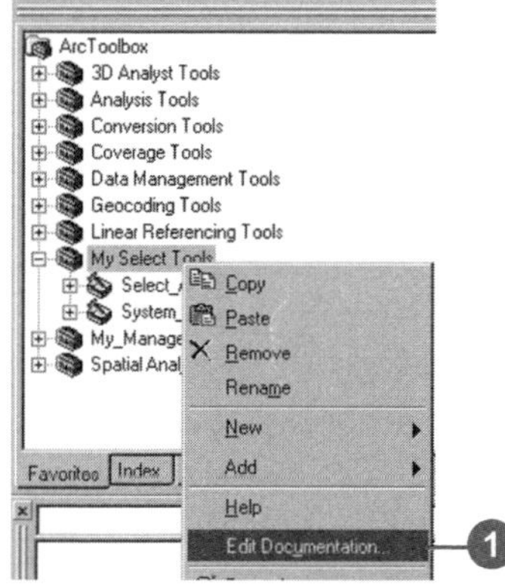

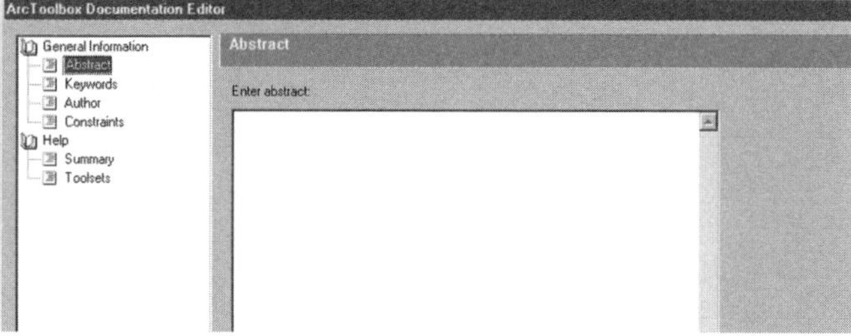

Adding documentation to a toolbox via the ArcCatalog tree

1. Click the toolbox in the ArcCatalog tree, then click the Metadata tab.
2. Click the Edit Metadata button on the Metadata toolbar.

   The Documentation Editor opens so you can add documentation to your toolbox.

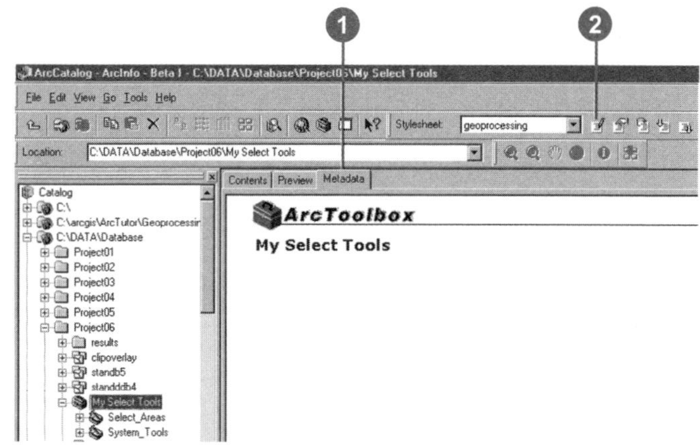

The abstract is a brief description of the contents of the toolbox. It displays only in the Metadata tab, not in the Help page for the toolbox, that is accessed via the ArcToolbox window.

Keywords can make it easy to find toolboxes in the ArcCatalog tree. If your toolboxes are spread out in different folders, you can simply search the ArcCatalog tree for all toolboxes with the same keyword. A list of toolbox shortcuts with the same keyword is provided. ►

Tip

Writing documentation for toolboxes

Use the documentation for the system toolboxes as a guide to help you write documentation for your own toolboxes.

Tip

Previewing your Help file

You can view the Help page as you are creating it. Click the Preview button to view your documentation in an HTML file.

Adding an abstract

1. Open the Documentation Editor.
2. Click Abstract in the General Information section.
3. Enter a paragraph of text for the abstract.
4. Click Finish.

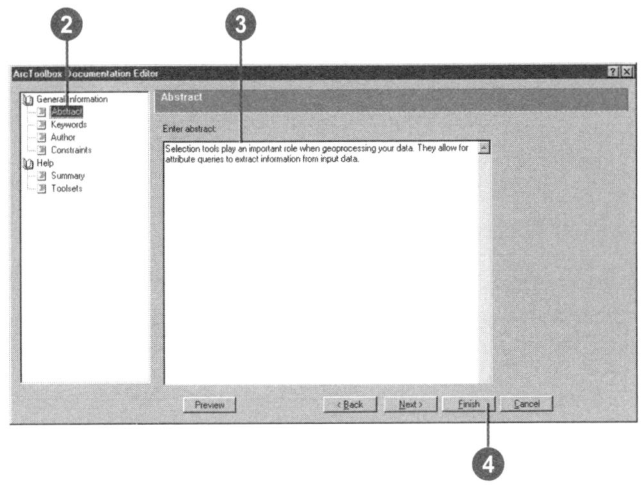

Adding keywords

1. Open the Documentation Editor.
2. Click Keywords.
3. Click the first row and enter the name for the first keyword.
4. Continue adding keywords as appropriate, then click Finish.

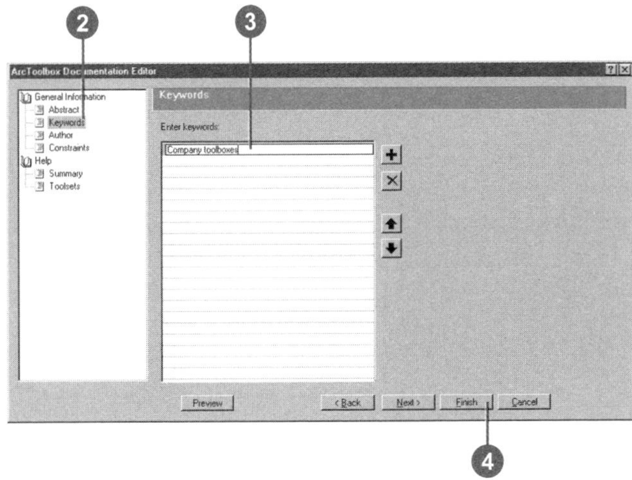

If you are distributing your toolbox, it is important to document your contact details. Users might, for instance, want to credit your work in the work they do using your toolbox and will need your contact information for this purpose. Alternatively, a user might want to contribute information that might increase the usefulness of your toolbox.

You can also list the constraints of the toolbox's contents. Constraints might be related to the proper use or distribution of the toolbox and could also include a liability statement. ►

Adding information about the author

1. Open the Documentation Editor.
2. Click Author in the General Information section.
3. Enter your name, organization, and position.
4. Enter your address details.
5. Click Finish.

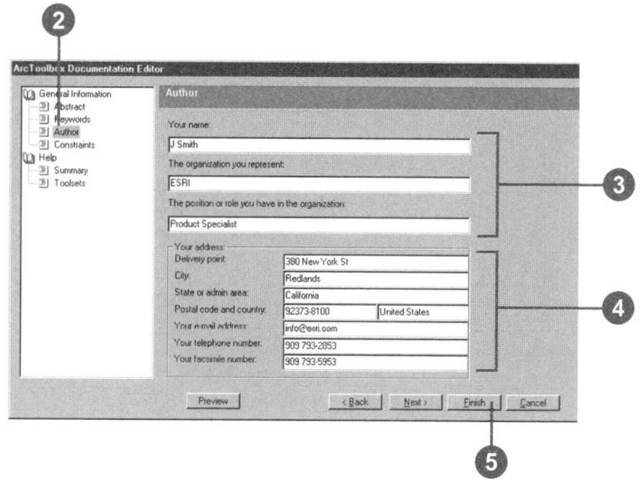

Adding information about constraints

1. Open the Documentation Editor.
2. Click Constraints in the General Information section.
3. Type information regarding the limitations and constraints of the toolbox.
4. Click Finish.

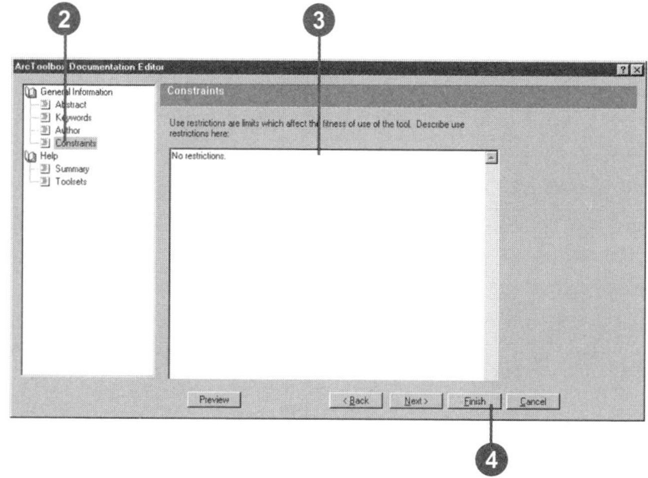

In the Help section of the Documentation Editor, you can add a summary for the entire toolbox and summaries for each toolset. Summary information can take the form of paragraphs, bullet points, illustrations, hyperlinks, subsections, and indented text. This information is displayed in the Help page accessed via the toolbox's context menu in the ArcToolbox window. The next few pages explain how to add this information to your Help pages.

The order of the Contents list in the Documentation Editor is reflected in the Help page. You may want to move the paragraph text for a summary so that, for instance, a link to related information is displayed above the paragraph in the Help page. Click Paragraph in the Contents list and click the down arrow or the up arrow to reposition the paragraph text in the desired location. ►

Adding a summary paragraph for a toolbox

1. Open the Documentation Editor.
2. Click Summary.
2. Click the Paragraph button to insert a paragraph of text about the toolbox.
3. Type the paragraph into the text box.
4. Click Finish.

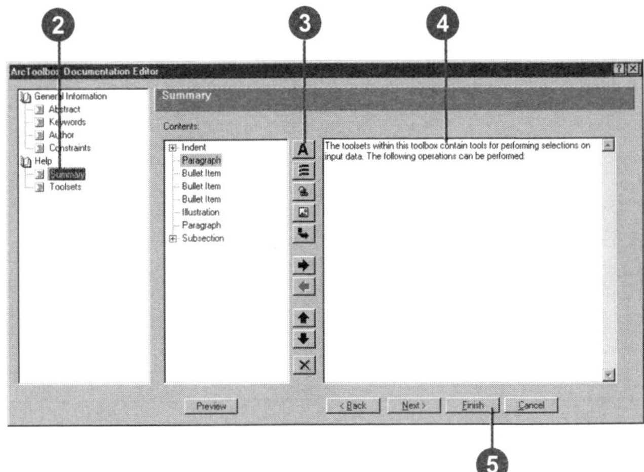

Adding a summary paragraph for a toolset

1. Open the Documentation Editor.
2. Click Toolsets, expand a toolset, and click Summary. Click the Paragraph button to enter text for the toolset.
3. Type a summary paragraph for the toolset into the text box.
4. Click Finish.

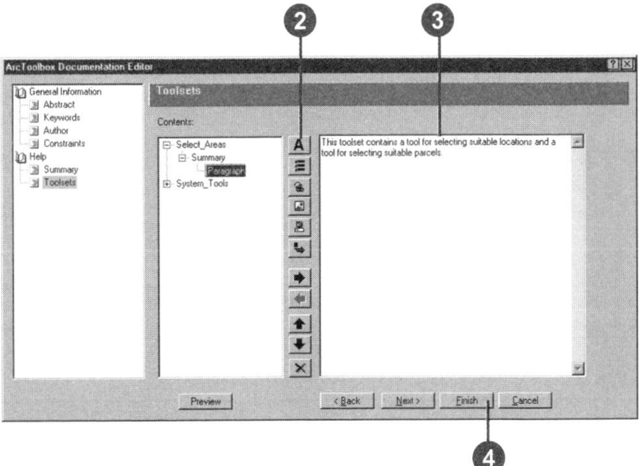

Add bullet items if you need to add a list of bullet points to your toolbox or toolset summaries. ►

Adding bullet items

1. Open the Documentation Editor.
2. Depending on whether you want to add bullet points to the summary of a toolbox or to a specific toolset, click Summary; or click Toolsets, click a toolset in the Contents tab, then click Summary.
3. Click the Bullet Item button to insert a bullet item.
4. Type the text for the bullet item into the input box.
5. Continue adding bullet items and text as appropriate.
6. Click Finish.

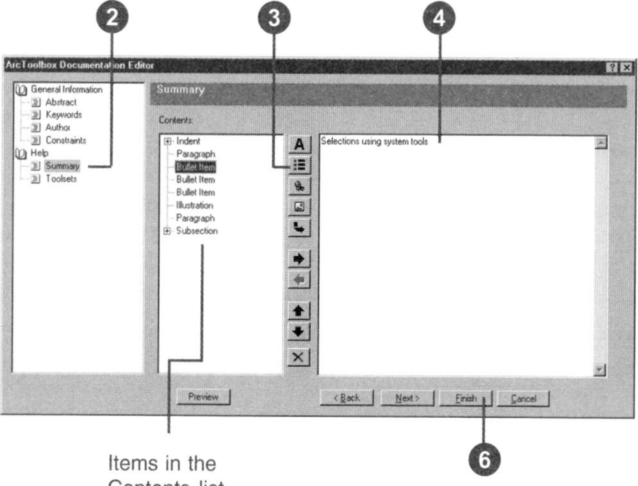

Hyperlinks enable related information to be accessible from a Help page. The user of your Help page simply clicks the link to open another file. It could be, for instance, an HTML file, a graphic, a text document or a Web page. ▶

Tip

Inserting links one after another

If you want to add links in a vertical list, click the Hyperlink button, then click the Paragraph button, then the Hyperlink button. Adding the paragraph prevents the links from being added next to each other horizontally.

Inserting hyperlinks

1. Open the Documentation Editor.
2. Depending on whether you want to add a hyperlink to the summary of a toolbox or to a specific toolset, click Summary; or click Toolsets, click a toolset in the Contents tab, then click Summary.
3. Click the Hyperlink button to add a link to a file or Web site.
4. Type the Path to the filename or Web page.

   The filename could be a graphic, such as a BMP or a JPEG file, or it could be an XML file. When linking to Web pages, type the full address of the page (http://...).
5. Type a Name that will display as the text for the link.
6. Click Finish.

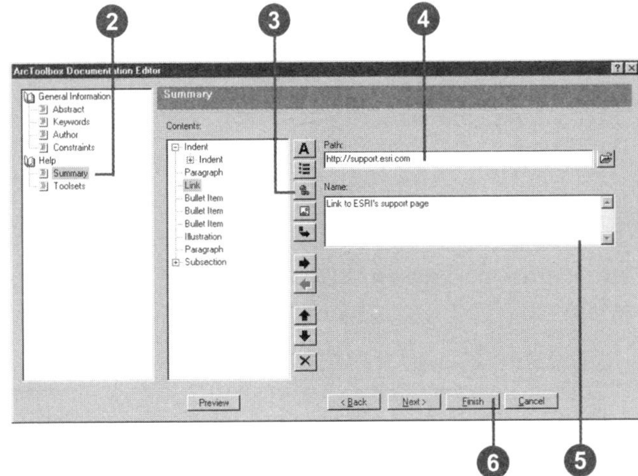

You can add illustrations directly into your toolbox or toolset summaries. Illustrations help to clarify the information given in the text of a summary. ►

Tip

Alternative ways to add graphics

To save space in your Help page or to hide illustrations that the user is not required to see immediately, you can add an illustration as a hyperlink or to a subsection so it only appears when the subsection is expanded.

Adding an illustration

1. Open the Documentation Editor.
2. Depending on whether you want to add an illustration to the summary of a toolbox or to a specific toolset, click Summary; or click Toolsets, click a toolset in the Contents tab, then click Summary.
3. Click the Illustration button to add a path to the illustration that will display directly in the Help page.
4. Type the Path to the illustration or click the Browse button, click the Look in dropdown arrow in the Open dialog box, and navigate to the location of the illustration on disk. Click the illustration and click Open.
5. Click the Name text box and type a name for the illustration.

   The name will display when the mouse pointer is held over the graphic.
6. Click Finish.

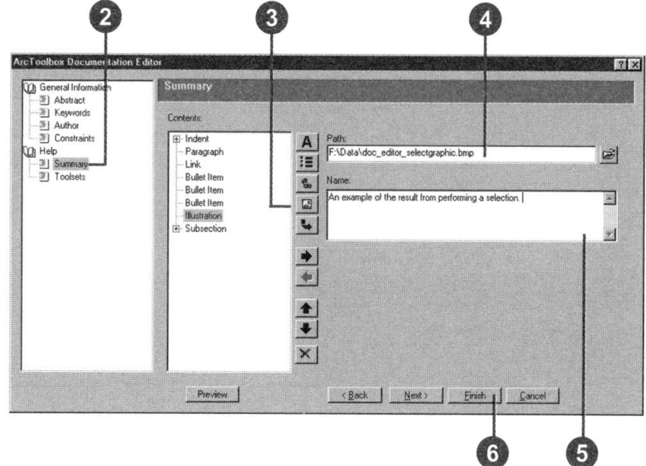

Subsections are expandable sections of text that can include any items, such as illustrations, bullet points, paragraphs, hyperlinks, or additional subsections. ▶

Adding subsections

1. Open the Documentation Editor.
2. Depending on whether you want to add a subsection to the summary of a toolbox or to a specific toolset, click Summary; or click Toolsets, click a toolset in the Contents tab, then click Summary.
3. Click the Subsection button to add an expandable section.
4. Type a name for the title of the subsection.
5. Click any of the buttons to add items to the subsection—for example, an illustration or a paragraph.
6. Click Finish.

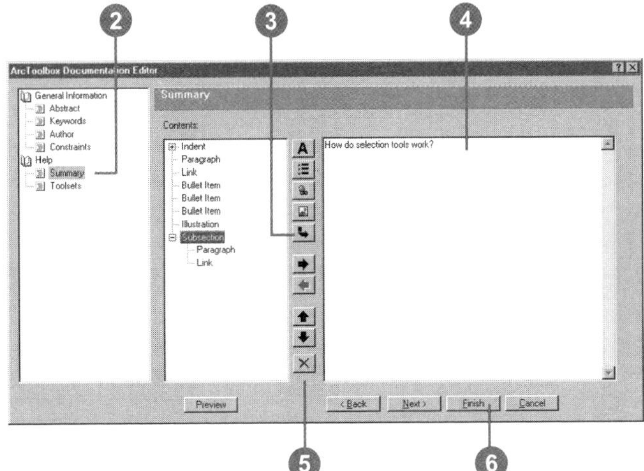

Any item in your Help pages can be indented. Indents are useful for repositioning text or graphics in the Help page. They can be nested inside other indents in order to nudge text or graphics away from the left side of the Help page. ►

Tip

Deleting items

To delete an item from the Contents list, click the item and click the Delete button.

Tip

Placing items in the Contents list

You can control the placement of items in the Summary section of the Contents list. Click Summary, then click a button, such as the illustration button, to place an item at the end of the list of items in the Summary section of the Contents list. To place an item in a specific location, click an existing item added to the Contents list, then click a button to add an item underneath the selected item. Items can also be ordered later using the up and down arrows.

Adding indented text

1. Open the Documentation Editor.
2. Depending on whether you want to add indents to the summary of a toolbox or to a specific toolset, click Summary; or click Toolsets, click a toolset in the Contents tab, then click Summary.
3. Click the Indent button to add indented text.
4. Click the Indent button again to add another indent. Any items added while the second indent is selected will be indented twice.

   Add as many indents as necessary to place text, graphics, hyperlinks, bullet points, or subsections in the desired position on the page.
5. Click the item you want indented, for example, a paragraph or an illustration.
6. Enter the information for the item.
7. Click Finish.

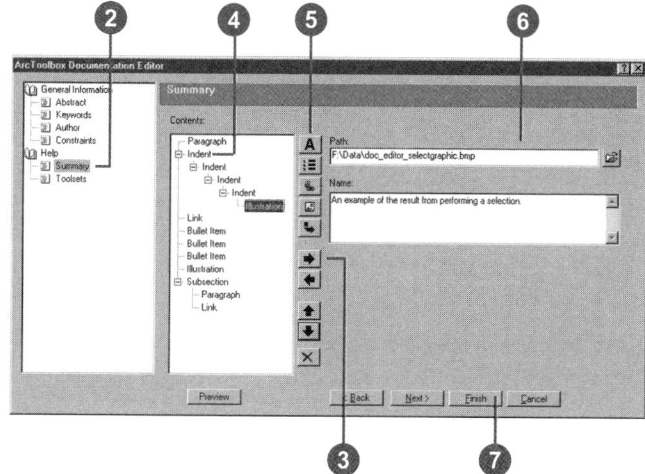

The documentation you add to a toolbox using the Documentation Editor can be exported to an HTML file, providing a static view of your Help page. Note that future changes made in the Documentation Editor will not be reflected in this HTML file. ▸

See Also

After exporting the documentation for all your toolboxes to HTML files and compiling them into a Help system (.chm), you can set up each toolbox to reference an appropriate Help page in your CHM file. See 'Referencing a compiled Help file' in this chapter for more information.

Exporting toolbox documentation to an HTML file

1. Right-click the toolbox and click Properties.
2. Click the Help tab.
3. Click Export.
4. Click the Save in dropdown arrow and navigate to the location into which you want to save the HTML file.
5. Type a File name and click Save.
6. Click OK on the toolbox's properties dialog box.

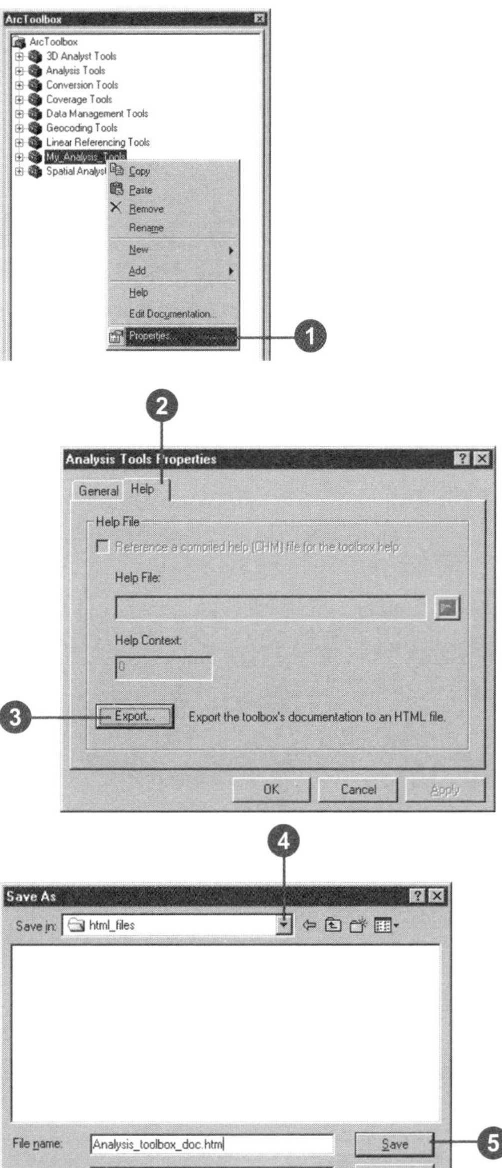

Your toolboxes can reference pages within a compiled Help file (.chm). A CHM file consists of compiled HTML files that are snapshots of the help written for toolboxes.

By providing the Help Context (the HTML topic ID), when the user of your toolbox clicks Help on the toolbox's context menu, the help topic associated with the Help Context ID in the CHM file will display instead of the Help file written in the Documentation Editor. If you don't supply a Help Context, the CHM file will display with its default page.

If modifications are made to the Help file written in the Documentation Editor, this Help file is saved with the tool but is not displayed when the toolbox's Help file is viewed as long as a CHM file is referenced in the toolbox's Help tab.

There are many sources of information on creating and compiling HTML Help, or CHM, files. A good starting point is the 'Microsoft HTML Help 1.4 SDK' topic available from the Microsoft library. Type "http://msdn.microsoft.com" in your Web browser and click the link to the Library. In the Library, you'll find information on creating and compiling CHM files listed under 'Visual Tools and Languages, HTML Help'.

Referencing a compiled Help file

1. Right-click the toolbox and click Properties.
2. Click the Help tab.
3. Check Reference a compiled help (CHM) file for the toolbox help.
4. Browse to the location of the CHM file or type the path.
5. Type a Help Context that corresponds to a topic in the CHM file if you want a certain topic to be displayed by default when the help for the toolbox is opened.
6. Click OK.

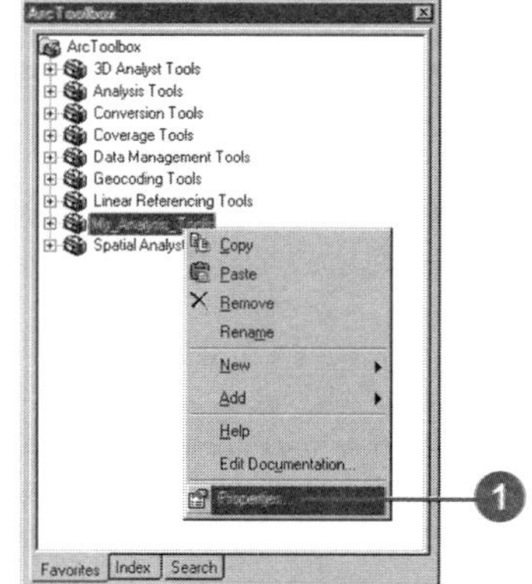

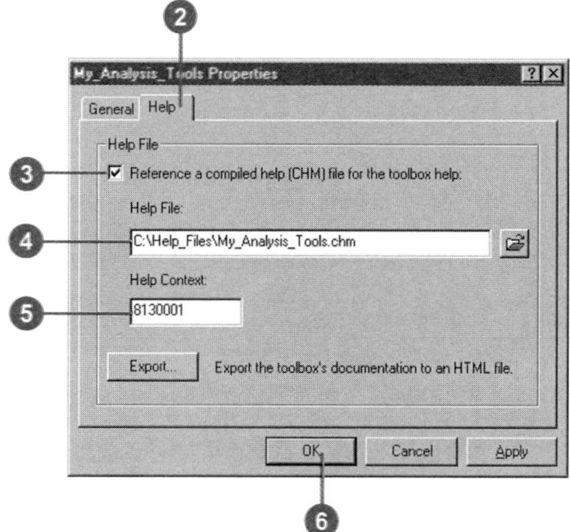

Viewing documentation for toolboxes

You can view documentation for all system toolboxes and for custom toolboxes for which you have added documentation.

All documentation added to the Help section of the Documentation Editor will display when the Help for a toolbox is accessed through the ArcToolbox window. ►

Viewing Help for a toolbox in the ArcToolbox window

1. Right-click the toolbox and click Help.

   An HTML document opens, displaying documentation for the toolbox.

2. Click Close to close the Help window.

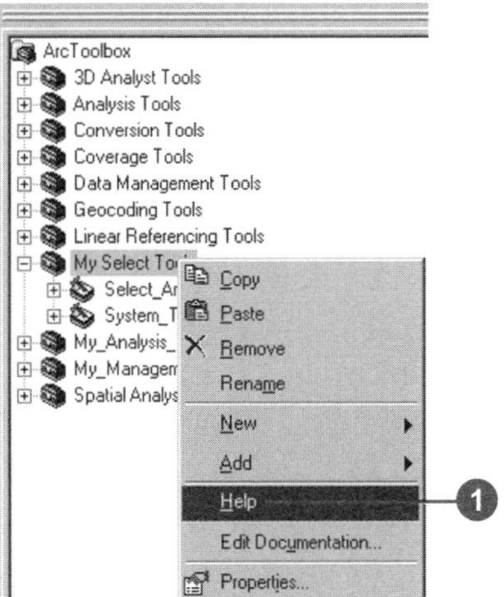

My Select Tools

collapse all

The toolsets within this toolbox contain tools for performing selections on input data. The following operations can be performed:

- Selections using system tools
- Selections of suitable locations
- Selections of suitable parcel types

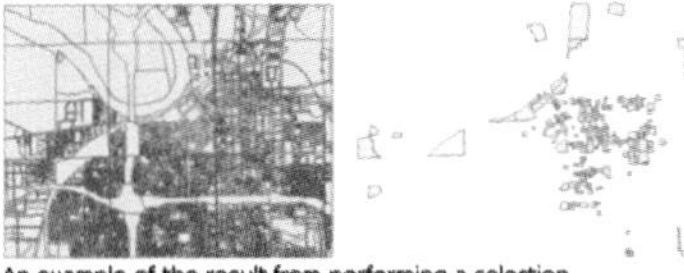

An example of the result from performing a selection.

▼ How do selection tools work?

Selection tools extract selected features from an Input feature class or Input feature layer and store them in the Output feature class. The Output feature class may be created with a subset of fields, a selection set, and an area of interest.

Illustration of a selection tool
Link to ESRI's support page

▼ Toolsets

▼ Select_Areas

This toolset contains a tool for selecting suitable locations and a tool for selecting suitable parcels.

Metadata for a toolbox can be viewed in the toolbox's Metadata tab via the ArcCatalog tree. The information added to the General Information section of the Documentation Editor is displayed, as well as a list of toolsets if any have been created.

See Also

See 'Adding documentation to toolboxes' in this chapter for information on documenting your toolboxes and toolsets.

Viewing metadata for a toolbox

1. Click the toolbox in the ArcCatalog tree, then click the Metadata tab.

   An abstract, keywords, author, constraints information, and a list of toolsets can be displayed in the Metadata tab, depending on what is documented in the Documentation Editor.

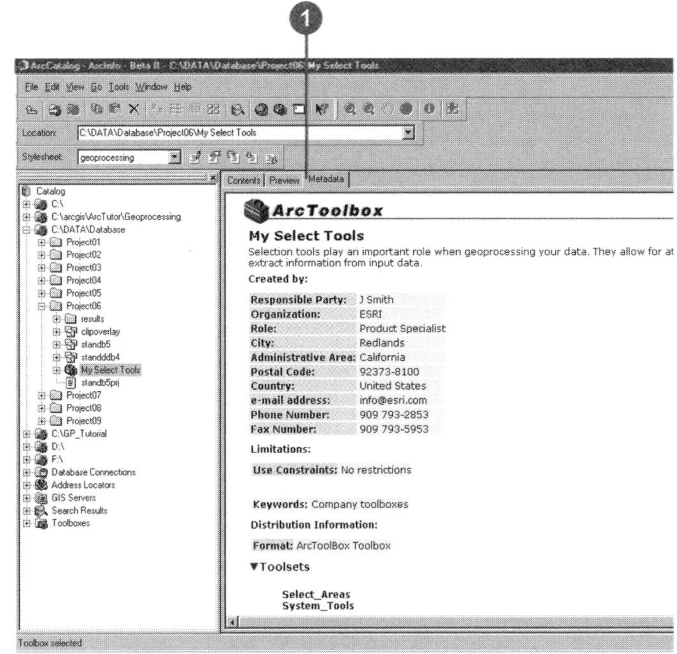

Rules for working with toolboxes

If a toolbox is altered in one application, for instance, in ArcMap, while another application is accessing it, such as ArcCatalog, there is the possibility that changes will be lost. When the last application is closed, it will simply overwrite any changes that were made earlier. When making changes to a toolbox, only one application should be open at a time so this type of conflict is avoided. Errors will not occur when a toolbox is accessed by multiple applications, but the resulting toolbox may not be what you expect.

Rules for accessing a toolbox:

- A toolbox is rewritten when a change has been completed. Refresh the toolbox in other applications that are accessing the same toolbox to refresh the contents of the toolbox.
- If multiple applications access a toolbox, the content of the toolbox is determined by the last application to be closed. Changes made to the toolbox in applications closed earlier will be lost.
- If two applications access a toolbox at the same time, changes made in one application will not be seen by the toolbox in the other application.

  For example, if a toolbox is open in ArcCatalog and ArcMap and a new tool is added in ArcCatalog, ArcMap will not see the tool unless it reopens the toolbox. Removing the toolbox from the ArcToolbox window in ArcMap, then adding it back will refresh the toolbox's contents.
- If a toolbox is accessed in two applications at the same time, deleting the toolbox in one application will not cause it to be deleted in the other.

  For example, if a toolbox is open in ArcCatalog and ArcMap and it is deleted in ArcCatalog, ArcMap will still have access to it because it has already read it into memory. When ArcMap is closed, it will rewrite the toolbox, so instead of being deleted, a newer version is created.

See 'Sharing your geoprocessing work' in Chapter 3 for information on the issues to consider when sharing your toolboxes with others.

Working with toolsets and tools

5

IN THIS CHAPTER

- Managing toolsets
- Working with tools
- Creating models and adding scripts
- Setting parameters
- Changing tool properties
- Adding documentation to tools
- Viewing documentation for tools
- Finding tools
- Understanding tool licensing

Toolsets and tools are contained within toolboxes. Toolsets are used to group collections of tools together into logical groupings. A tool is a geoprocessing operation that performs a geoprocessing task. There are hundreds of system tools available, categorized into toolsets for ease of access. When you run a tool, you simply open the tool's dialog box, supply values for the tool's required parameters and any optional parameters if you don't want to accept the default, then run the tool.

Toolsets can be created inside writable toolboxes, and you can create toolsets inside toolsets to further organize the contents of a toolbox. If you're working with a group of system tools from different toolboxes, it's easy to add the tools you want to work with into your own toolbox for ease of access.

If you need to repeat the same sequence of tools, it is most efficient to create your own model inside a toolbox or write a script and add it to a toolbox. The script can be created in any COM-compliant scripting language, such as Python or VBScript, or it could be an existing AML or EXE file. You can set parameters for models and scripts so users can specify their values when they run your tools from their dialog boxes.

There is documentation available for each system tool, helping you understand how each tool works and what parameter values you need to specify. When you create your own models inside or add scripts to toolboxes, it is equally important to document them so the user understands how they work, so their life cycles are infinite. You can easily add Help using the Documentation Editor.

Managing toolsets

The system tools are organized into logical groupings, called toolsets. You can create your own toolsets inside new toolboxes or inside existing toolsets to further organize the contents of a toolbox.

You may, for instance, have a set of conversion toolsets inside a toolbox, such as a Raster Conversion toolset and a Feature Conversion toolset. The tools inside the Raster Conversion toolset might be further organized into a Raster to Feature toolset and a Feature to Raster toolset.

Toolsets you create can be renamed. If you have the ArcToolbox window open in one ArcGIS Desktop application, such as ArcMap, and you change the name of a toolset inside a toolbox in ArcCatalog, you will have to remove the toolbox containing the toolset in the ArcToolbox window in ArcMap and add it back again to see the change to the toolset name. ►

Tip

Read-only toolboxes

If your toolbox is read-only you cannot create toolsets inside it.

Creating a new toolset

1. Right-click the toolbox or toolset you have created, point to New, and click Toolset.

   A new toolset is created inside the toolbox or toolset.

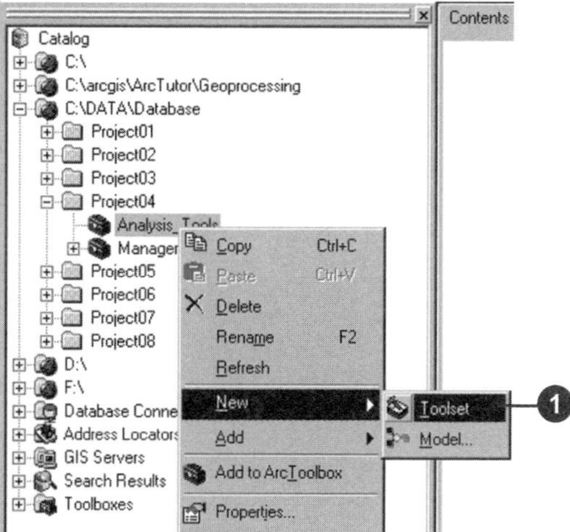

Renaming toolsets

1. Right-click the toolset you want to rename and click Rename.
2. Type the new name.
3. Press Enter.

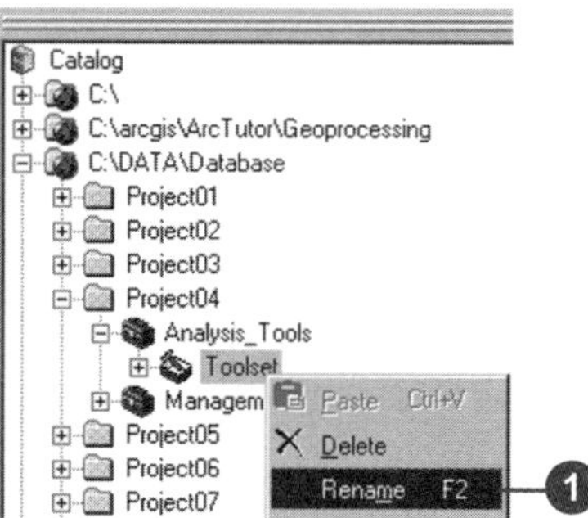

Toolsets you create can be deleted from either the ArcCatalog tree or the ArcToolbox window.

If you are working in the ArcToolbox window with a toolbox you created that contains toolsets you are not working with, create a new toolbox and copy the toolsets you are currently using into it. Then remove the toolbox you don't need from the ArcToolbox window. By doing this, you will retain the original contents of the toolbox, but you'll remove the toolsets you are not currently working with from your ArcToolbox window.

Tip

Creating/Viewing documentation for a toolset

Toolset documentation can be created and viewed at the toolbox level. See Chapter 4, 'Working with toolboxes', for information on creating and viewing documentation for toolsets.

See Also

See 'Creating new toolboxes' in Chapter 4 for information on the creation of new toolboxes.

Deleting toolsets

1. Right-click the toolset that you want to delete and click Delete.

   The toolset is permanently deleted from disk.

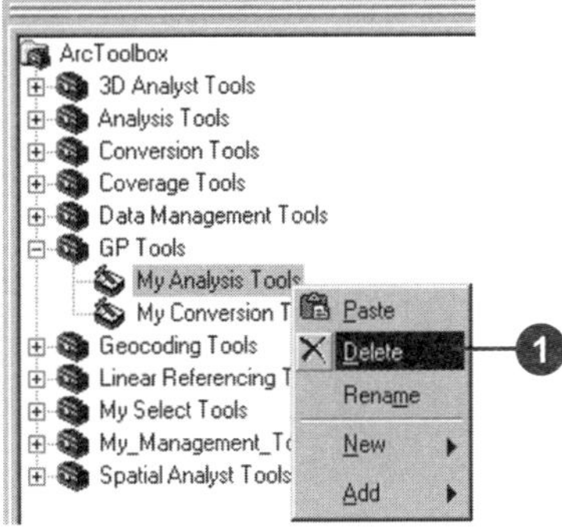

Working with tools

Contained within the System Toolboxes folder in the ArcCatalog tree are toolboxes that contain toolsets. Within each toolset are system tools that you can run. System toolboxes are added to the ArcToolbox window by default, providing a quick alternative to work with your favorite system tools. ►

Tip

Viewing the system tools in the ArcCatalog tree

Click the Tools menu in the application you are using and click Options. Click the General tab and check Toolboxes in the list of top-level entries.

Tip

Environment settings

Click Environments on a tool's dialog box. Environment settings specified will be used for the current run of the tool only.

See Also

You can run your own tools as any system tool by opening the tool's dialog box and running the tool. See 'Creating models and adding scripts' in this chapter for more information.

Opening a tool's dialog box

1. Right-click the tool and click Open.

   Alternatively, double-click the tool.

   The tool's dialog box is opened so you can set values for its parameters and run the tool.

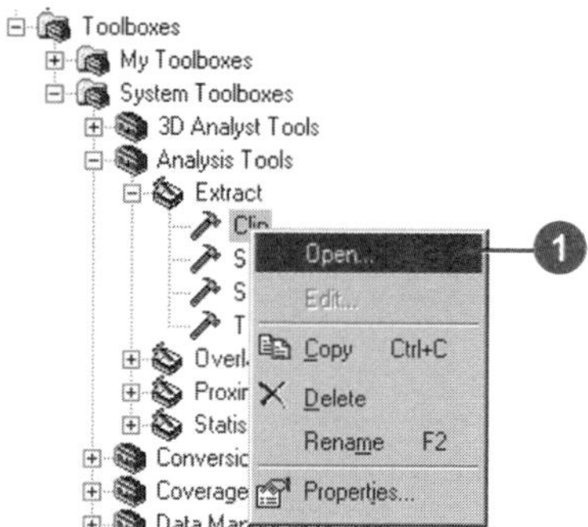

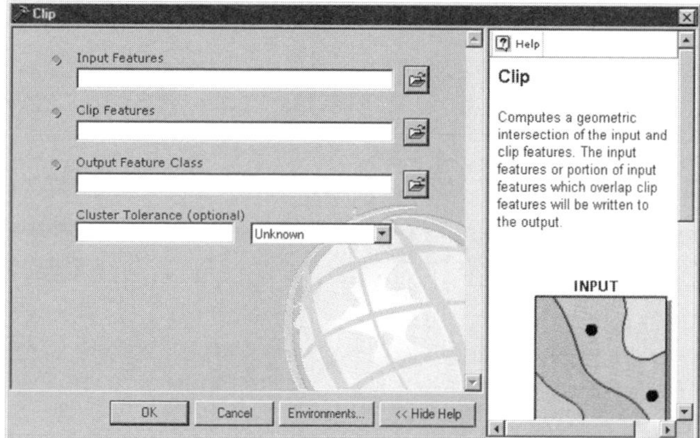

Rules dictate where results will be placed. Specify the path in the tool's dialog box, or set up a workspace in the Environment Settings dialog box, then type a name for the result inside a tool's dialog box—it will be placed in the location set for the workspace. Alternatively, if no workspace is set, results will automatically be placed in the location of the input data to a tool. If there is no workspace set and the input data's location is read-only, the result will be placed in the location of your system's TEMP environment. ▸

Tip

Using a selected set

Select data in the display, then run a tool from the ArcToolbox window. The tool will perform the operation on the selected set.

See Also

For information on changing the settings applied to results from running tools, see 'Results from running tools' in Chapter 3.

See Also

The message section of the Command Line window documents the execution of tools and allows for the reexecution of tools. See Chapter 7, 'Using the Command Line window'.

Running a tool from its dialog box

1. Right-click the tool you want to run and click Open.
2. Fill in the tool's parameter values.

   Required parameters have a green circle to the left of them, indicating that a value is required.

   Those parameters that are optional do not require a parameter value unless you want to change the default value.

   Check that all parameters have valid values. Those that have invalid values will have a red cross to the left of them.
3. Click Environments to alter environment settings specified for the entire application. Changed settings are only applied for this run of the tool. The application's environment settings are not permanently altered.
4. Click OK to run the tool.

   A progress dialog box appears, displaying the status of the processing.
5. Click Details to view the execution messages.
6. Click Close on the Progress dialog box.

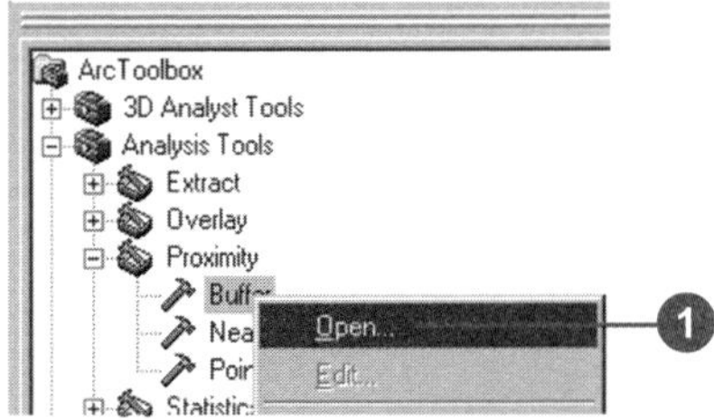

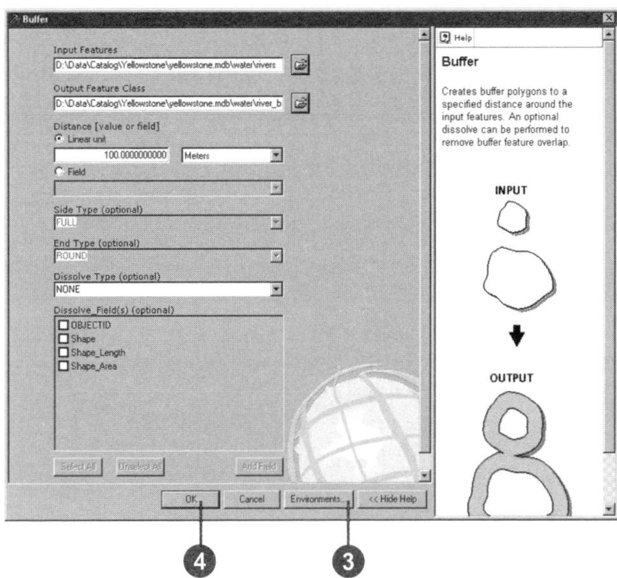

In either the ArcCatalog tree or the ArcToolbox window, you can organize your tools by copying and pasting or dragging and dropping them between toolboxes or toolsets.

When you copy and paste a tool, a copy is placed in the destination toolbox or toolset. When you drag and drop a tool, the tool is moved to the destination location, provided you are working within the same disk. If you drag and drop a tool from a toolbox or toolset on one disk to a toolbox or toolset on another disk, a copy of the tool will be created. ▸

Tip

Dragging or pasting tools into read-only toolboxes

You cannot drag or paste a tool into a read-only toolbox. Make the toolbox writable or create a new toolbox and paste the tool into the new toolbox instead.

Copying and pasting tools

1. Right-click the tool you want to copy and click Copy.
2. Right-click the toolbox or toolset that you want to paste the tool into and click Paste.

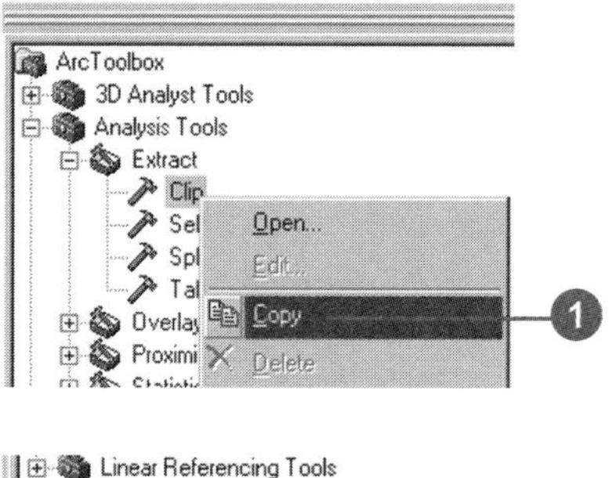

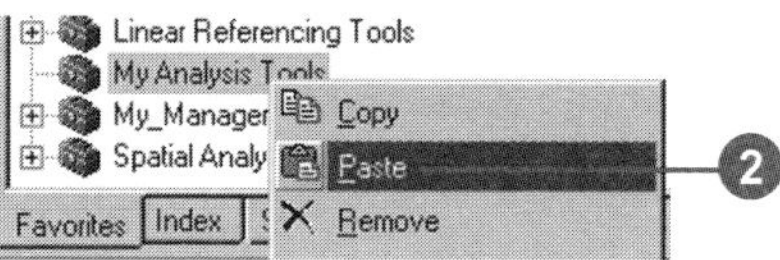

Dragging and dropping tools

1. Click the tool you want to move.
2. Drag the tool and drop it onto another toolset or toolbox.

   The tool is moved to the new location.

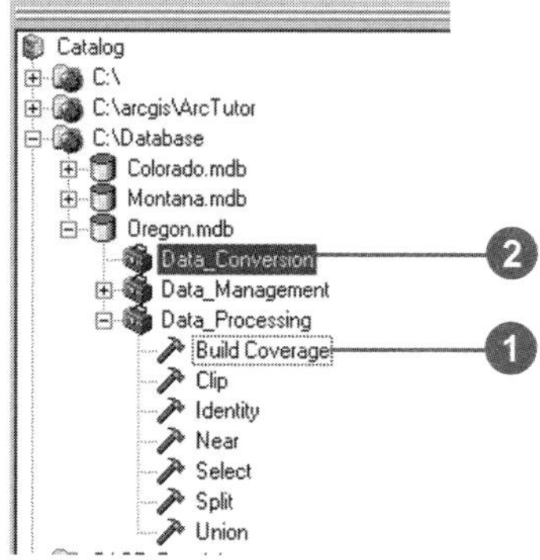

You can add system tools into writable toolboxes or toolsets provided that the DLL has been registered with the system and the tools in the DLL have been registered with the correct component categories. All system tools within the System Toolboxes are, by default, registered correctly. If you have tools that have not been registered correctly, they will not display in the Add Tool dialog box.

By checking a toolbox or toolset in the Add Tool dialog box, all tools contained within the particular toolbox or toolset will be added into your created toolset or toolbox. Using the Add Tool dialog box to add system tools into your own toolbox can be quicker than copying and pasting them. It is also useful for placing system tools that are spread out in ▸

Tip

Read-only toolboxes

You cannot add tools into read-only toolboxes. Either make the toolbox writable, or create a new toolbox to add tools into.

See Also

See 'Creating new toolboxes' in Chapter 4 for information on how to create your own toolbox.

Adding system tools

1. Right-click the toolbox or toolset into which you want to add system tools, point to Add, and click Tool.
2. Expand the toolboxes and toolsets and check the tools you would like to add into your toolbox or toolset.

   If you check the toolbox, all tools within the toolbox will be added. If you check a toolset within a toolbox, all tools within the toolset will be added.
3. Click OK.

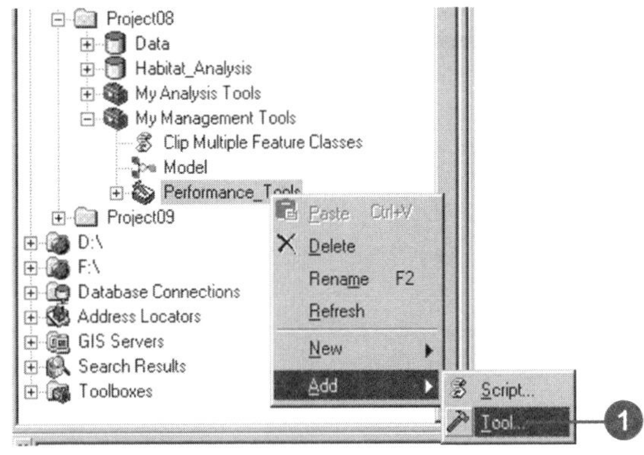

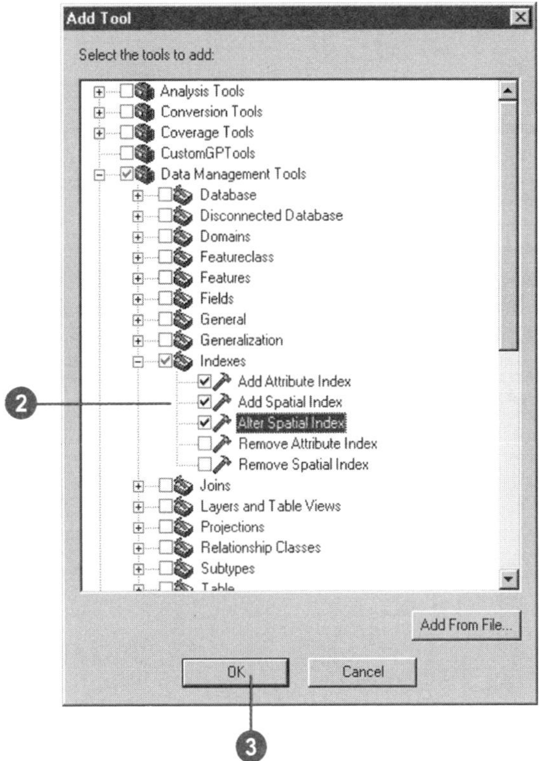

toolsets in different toolboxes into the same toolbox for ease of access.

If you don't see the tools you are expecting in the Add Tool dialog box, the DLL might not be registered with the system and the tools in the DLL might not be registered with the correct component categories. You can click the Add From File button to select the DLL and click Open. Doing so performs the necessary registrations so the tools will display in the Add Tool dialog box.

Adding tools from a DLL

1. Right-click the toolbox or toolset into which you want to add tools, point to Add, and click Tool.
2. Click Add From File. ▸

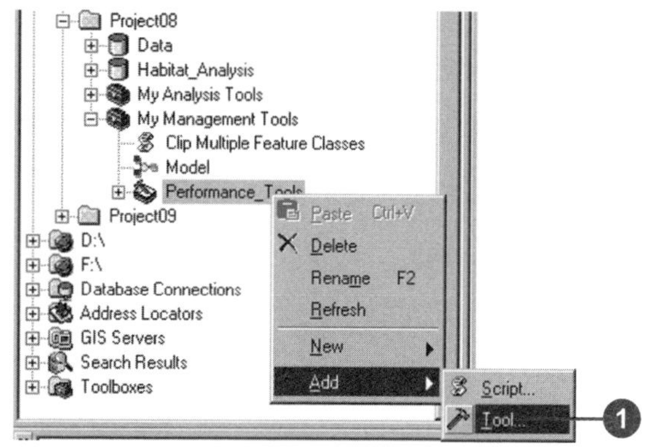

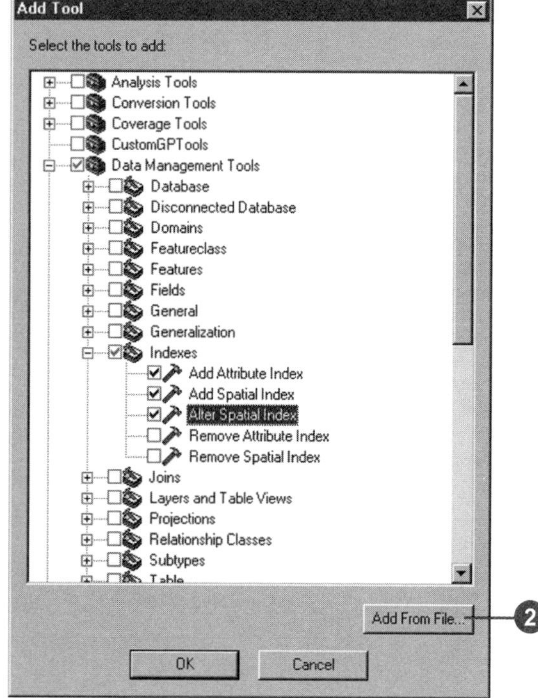

3. Click the Look in dropdown arrow and browse to the location of the DLL file.
4. Click the DLL file to select it.
5. Click Open.

   The DLL is registered with the system, and the tools within the DLL are registered with the correct component categories. This registration enables the toolbox containing the tools to be added to the Add Tool dialog box.
6. Check the added toolbox.
7. Click OK on the Add Tool dialog box.

   All the tools within the checked toolbox are added to your toolbox or toolset.

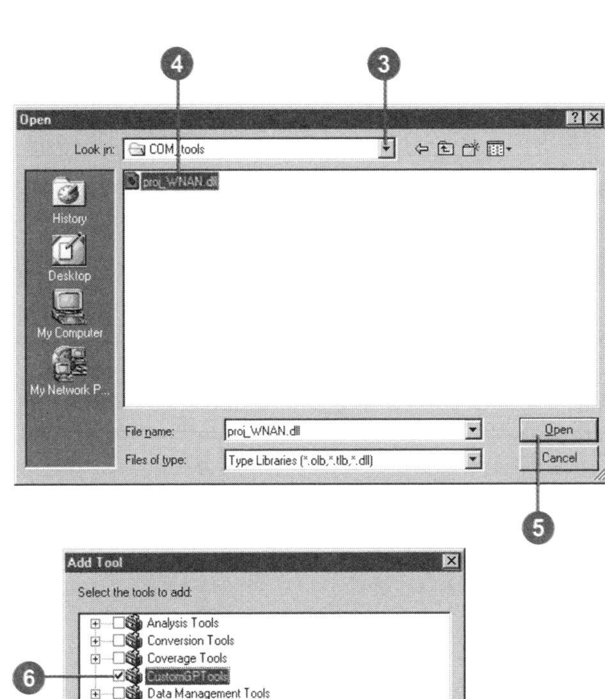

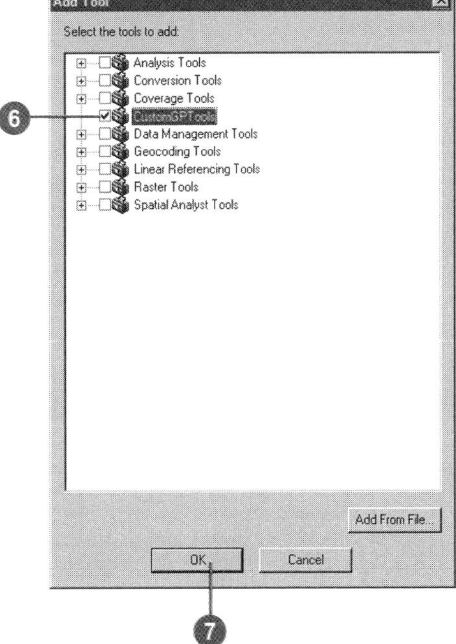

Tip

About tools in read-only toolboxes

You cannot delete or rename tools that are in read-only toolboxes. If you cannot make the toolbox writable, create a new toolbox and add to it the tools you want to use.

Tip

About renaming tools

When you rename a tool by right-clicking and clicking Rename, you are only renaming the tool's label, not the name of the tool. See 'Changing the name, label, and description for your tools' in this chapter for more information on tool names and labels.

Deleting tools

1. Right-click the tool you want to delete and click Delete.

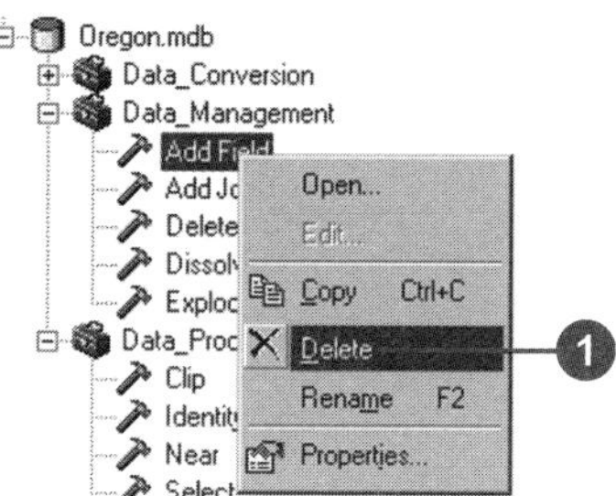

Renaming a tool's label

1. Right-click the tool and click Rename.
2. Type a new name for the tool's label and press Enter.

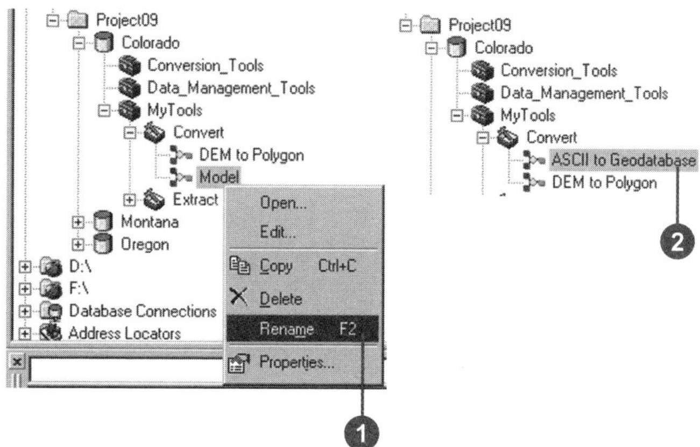

Displaying the properties of a tool

1. Right-click the tool and click Properties.

   The name of the tool (used at the command line or inside a script), the label (used to label the tool in the ArcCatalog tree or the ArcToolbox window), and a description for the tool are displayed.

   The default stylesheets are applied if the Stylesheet text box is empty. The location for the default stylesheets is \Program Files\ ArcGIS\ Desktop\ArcToolbox\ Stylesheets on the drive where you installed ArcGIS.

2. Click OK.

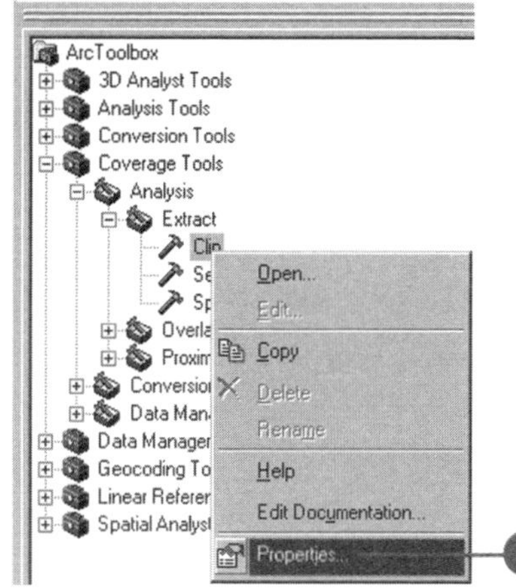

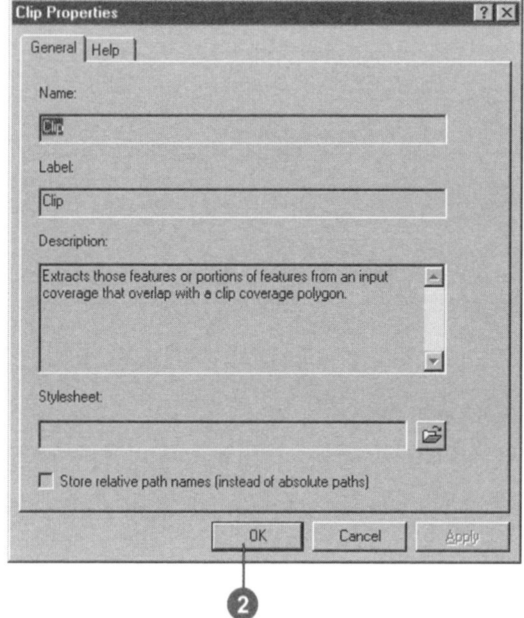

Tip

Changing the properties of tools

You cannot change the properties of system tools, even if they are copied to a writable toolbox. You can change the properties for a tool you have created if the toolbox containing the tool is writable.

See Also

See 'Changing the name, label, and description for your tools' in this chapter for more information on modifying the properties of a tool you have created.

See Also

See 'Setting a modified stylesheet for a tool' in this chapter for information on changing the stylesheet used by a tool you have created.

See Also

For information on relative path names, see 'Storing relative pathnames' in this chapter.

See Also

For information on the Help tab, see 'Referencing a compiled help file' in this chapter.

Creating models and adding scripts

You can use the tools contained within the system toolboxes to perform your geoprocessing tasks. However, when you know you will need to repeat the use of the same tool or sequence of tools many times, either on the same input data and refining parameters or on different input data, you should create a model or add a script. Doing so enables you to perform multiple operations at one time, rather than performing them one by one.

Models run processes that you have built and strung together in a ModelBuilder window. ▶

See Also

For information on building models see Chapter 8, 'Introducing model building', and Chapter 9, 'Using the ModelBuilder window'.

See Also

For information on creating toolboxes, see Chapter 4, 'Working with toolboxes'.

Creating a new model

1. Right-click the toolbox into which you want to create a new model, point to New, and click Model.

   A new model is created in the toolbox with a default name, and a ModelBuilder window is opened so you can build your model.

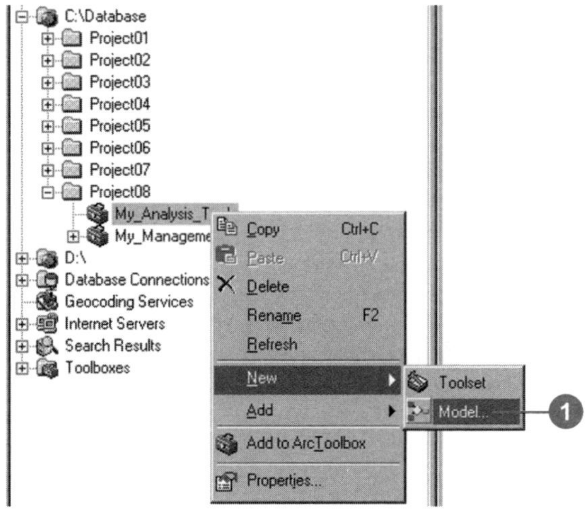

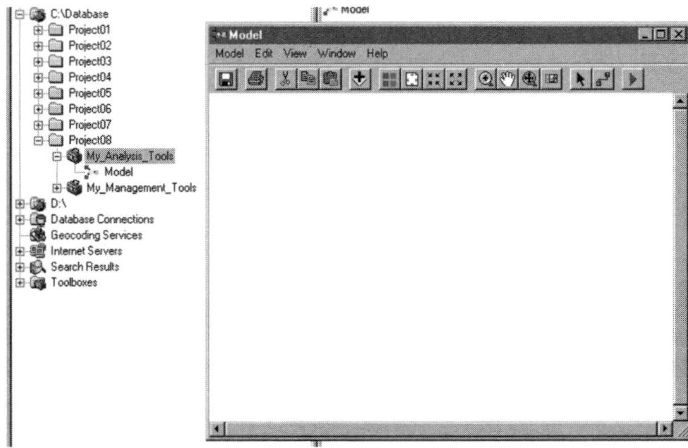

A created model inside a toolbox and its ModelBuilder window

Scripts are useful for performing batch processing and can be run standalone or from inside a toolbox. After adding a script to a toolbox, you can define the parameters for variables that are set as system arguments—for example, sys.argv [ ] in Python—inside the script, so the user of the script will be able to set values for these parameters in the script's dialog box. See 'Setting script parameters' in this chapter for information on the process of defining parameters.

In the first panel of the Script Wizard, the options that must be specified are the name and the label for the script. The name is used when the script is run at the command line or from inside another script. There should be no spaces in the name. The label is the display name for the script and may contain spaces. ▸

See Also

You do not have to create a script from scratch. You can build a model, then export the model to a script. See 'Exporting a model to a script' in Chapter 9.

See also Writing Geoprocessing Scripts With ArcGIS *for more information on writing scripts.*

Adding a script to a toolbox

1. Right-click the toolbox or toolset into which you want to add a script, point to Add, and click Script.
2. Type a Name for the script.
3. Type a Label for the script.
4. Optionally, type a Description for the script.
5. Optionally, click the Browse button to change the default stylesheet used for the tool.

   If the Stylesheet text box is empty, the default stylesheet is used. See 'Setting a modified stylesheet for a tool' in this chapter for more information on stylesheets.
6. Optionally, check Store relative path names—instead of absolute paths—to store relative pathnames. See 'Storing relative pathnames' in this chapter for more information.
7. Click Next. ▸

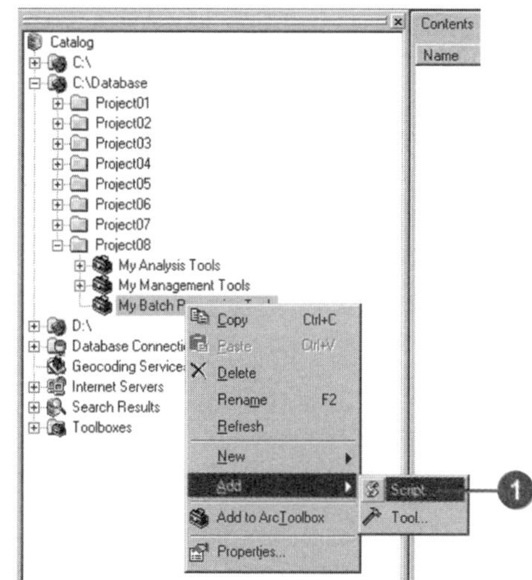

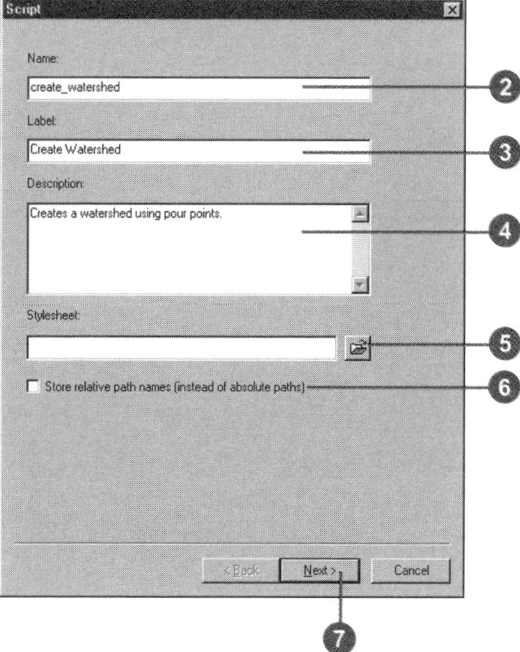

In the second panel of the wizard you add the script that will be added to the toolbox.

A script is a simple text file that can be created in any text editor. It can be written in any COM-compliant scripting language, such as Python, JScript, or VBScript.

Note that the script you add can also be an AML script or an EXE file.

Refer to *Writing Geoprocessing Scripts With ArcGIS* to learn more about writing geoprocessing scripts.

If you are sent a script, you can add it to a toolbox and run it, provided all values for variables are hard coded within the script and paths to input data are correct. ▸

See Also

See 'Editing a custom model or script' in this chapter for information on how to edit the contents of models or scripts.

8. Click the Browse button, then click the Look in dropdown arrow and navigate to the location of your script.
9. Click the script file and click Open.

   Note that you can type the name of a script file that does not exist—with the appropriate extension for the script type. The wizard will ask if you want to create it.
10. Optionally, check Show command window when executing script to view messages that are not piped through the geoprocessor in a command window.

    Messages that are piped through the geoprocessor will display in the message section of the Command Line window.
11. Click Next. ▸

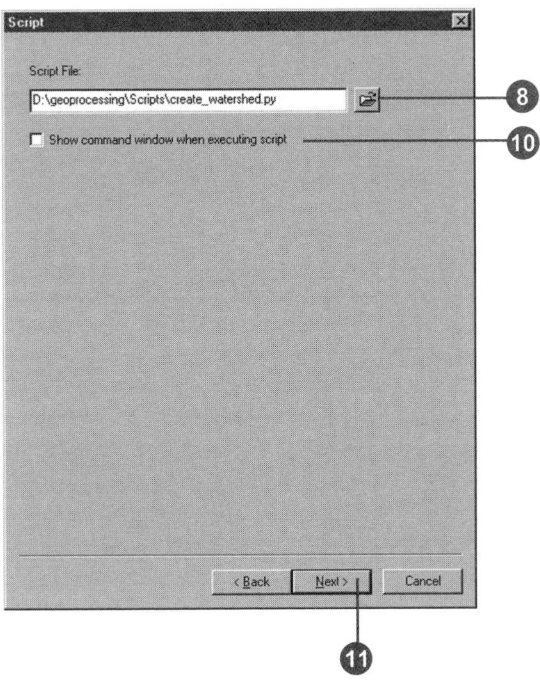

The last panel of the wizard allows for the setting of parameters and their properties.

If all values set for variables within the script are hard coded—for example, a path is specified as the value for an input data parameter—you do not need to set any parameters. If there are variables set as system arguments (sys.argv [ ]) within the script, this means that they need to be defined as parameters within the Properties dialog box of the script so the user of the script can specify the value for these parameters. Without defining the parameters that will appear on the script's dialog box, the script will not run. If parameters need to be defined, see 'Setting script parameters' in this chapter for information on this process.

12. If you need to define parameters, you can do so in this panel of the wizard, then click Finish.

    Your script is added to the toolbox and will run when you open its dialog box, supply any values for parameters, then click OK.

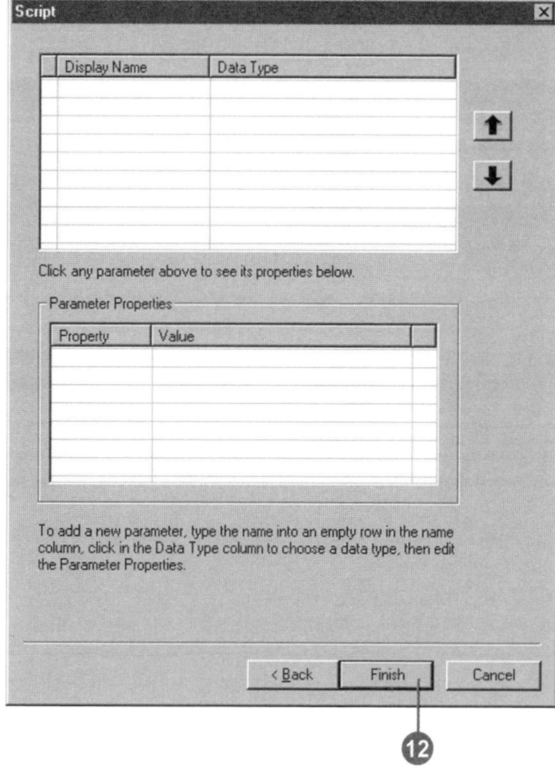

Tip

Editing system tools

Models and scripts installed with the system cannot be edited.

See Also

If your tools are displaying as shown, with a red x over them, there could be a tool (a model or script) referenced in the model that no longer exists in its location on disk, or the DLL of a COM tool could be unregistered. You'll also see a red x if a parameter name in a model or a script that is added to the model has changed. See 'Repairing a model' in Chapter 9 for information on fixing broken models.

Editing a custom model or script

1. Right-click the model or script and click Edit.

   If the tool is a model, the ModelBuilder window will open, so you can edit the contents of the model.

   If the tool is a script, the script will be opened in its native application, so you can edit the code.

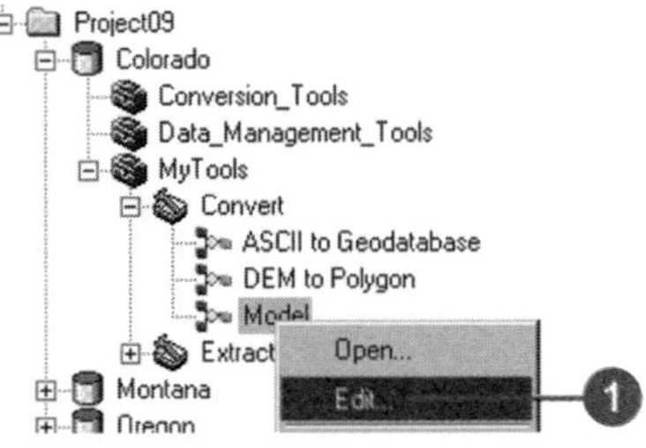

Setting parameters

Once you have created your model, you can set variables as model parameters so the parameters display on your model's dialog box. In the process of adding a script to a toolbox, you can also set parameters. Note that in the process of setting parameters for a script, the parameters and their properties are defined. Parameters are already defined for a model as they inherit the properties of the variable.

The user of your model or script will specify values for set parameters in the model's or the script's dialog box.

For a model, there are two ways to set parameters: via the model's properties dialog box or within the ModelBuilder window after building your model. See 'Working with variables' in Chapter 9 for ▸

See Also

See 'Creating a new model' earlier in this chapter and Chapter 9, 'Using the ModelBuilder window', for information on building models and exposing parameters as variables so they can then be set as model parameters for a model's dialog box.

Setting model parameters

1. Right-click the created model and click Properties.
2. Click the Parameters tab.

   A list of model parameters is displayed if any have already been set.
3. Click Add to add variables as model parameters to the list.
4. Click the variables you want to add as model parameters; press Ctrl to select more than one variable.
5. Click OK.

   The variables are added as model parameters to the list.
6. Click OK on the properties dialog box.
7. Right-click the model and click Open to see the model parameters that have been set on the model's dialog box.

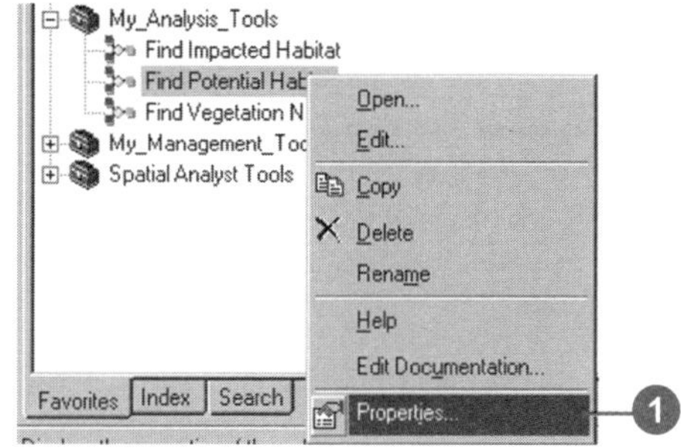

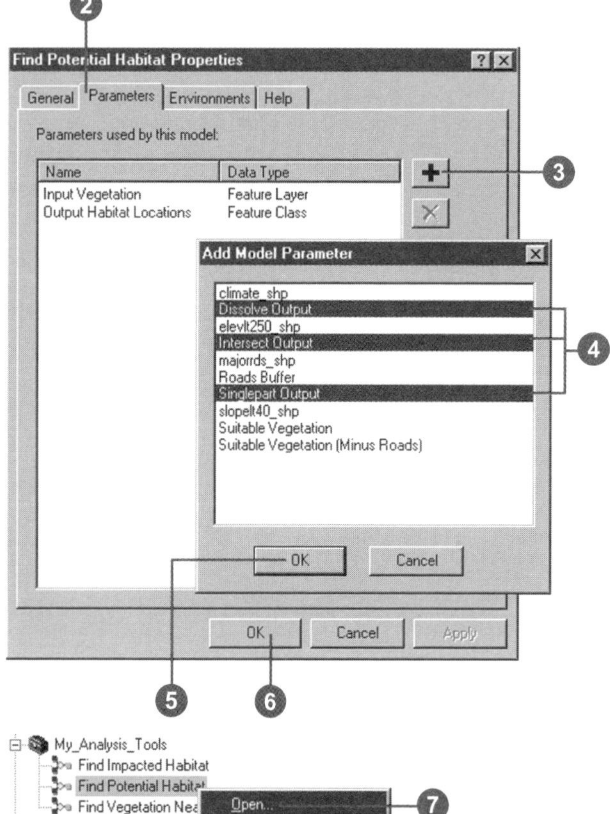

information on setting variables as model parameters within the ModelBuilder window.

If you have received a script and information on its parameters and their properties, it is an easy task to define and set parameter information for a script.

If the values for variables inside your script are set as system arguments (sys.argv [ ]), you must define the parameters that correspond to these system arguments so the user can specify the values for these parameters inside the script's dialog box.

For example, the value for the input workspace may have been set as a system argument in the script as follows:

gp.workspace = sys.argv [1]

rather than a hard-coded path:

gp.workspace = "C:\\Workspace" ►

See Also

See Writing Geoprocessing Scripts With ArcGIS *for information on writing scripts.*

Setting script parameters

1. Right-click the script and click Properties.
2. Click the Parameters tab.
3. Type a display name for the first parameter.

   In this example, Input Folder is the display name for the parameter. This name will appear as the parameter name in the script's dialog box.
4. Click the Data Type cell for the parameter, then click the dropdown list and select the appropriate data type.

   In this example, the appropriate data type is a folder. All other data types will be filtered from the Browse dialog box for the Input Folder parameter. ►

```
#Import standard library modules
import win32com.client, sys, os
#Create the Geoprocessor object
gp = win32com.client.Dispatch("esriGeoprocessing.GpDispatch.1")

#Set the input workspace
gp.workspace = sys.argv[1]
#Set the clip feature class
clipFeatures = sys.argv[2]
#Set the output workspace
out_workspace = sys.argv[3]
#Set the cluster tolerance
clusterTolerance = sys.argv[4]

try:
    #Get a list of the feature classes in the input folder
    fcs = gp.ListFeatureClasses()
    #Loop through the list of feature classes
    fcs.Reset()
    fc = fcs.Next()
    while fc != "":
        #GDB's don't support "." in the fc name, so replace these with "_".
        #For example "climate.shp" will be converted to "climate_shp".
        outFeatureClass = out_workspace + "\\" + fc.replace(".","_")
        #Clip each feature class in the list with the clip feature class
        #Do not clip the clipFeatures, it may be in the same workspace
        if str(fc) != str(os.path.split(clipFeatures)[1]):
            gp.clip_analysis(fc, clipFeatures, outFeatureClass, clusterTolerance)
        fc = fcs.Next()
except:
    gp.AddMessage(gp.GetMessages(2))
    print gp.GetMessages(2)
```

A Python script showing values for variables set as system arguments (sys.argv [ ])

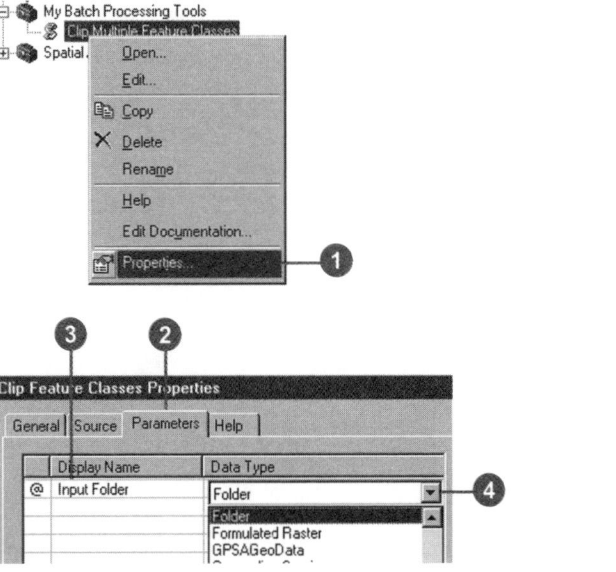

This is for the user to be able to specify the input workspace inside the script's dialog box.

The order you add the parameters must match the order of the arguments in the script. Each parameter has a set of properties associated with it:

Display Name—The name you see for the label of the parameter in the script's dialog box.

Data Type—The data type accepted by the parameter. For instance, you might only want raster datasets to be accepted for an input parameter. The Browse dialog box for the parameter will filter out all other data types when you set the data type property for the parameter to be Raster Dataset.

Type—This property defines whether the parameter is required, optional, or derived. Required parameters require an input value from the user. Optional parameters do not require a value from the user. Set output parameters to be derived if the tool updates the input to the tool or if you ▸

See Also

See Chapter 8 of Writing Geoprocessing Scripts With ArcGIS *for information on setting parameter values.*

5. In the Parameter Properties section, click the cell next to the property Type and select the appropriate type for the parameter, either Required, Optional, or Derived.

   In the example, it is required to specify an input folder.

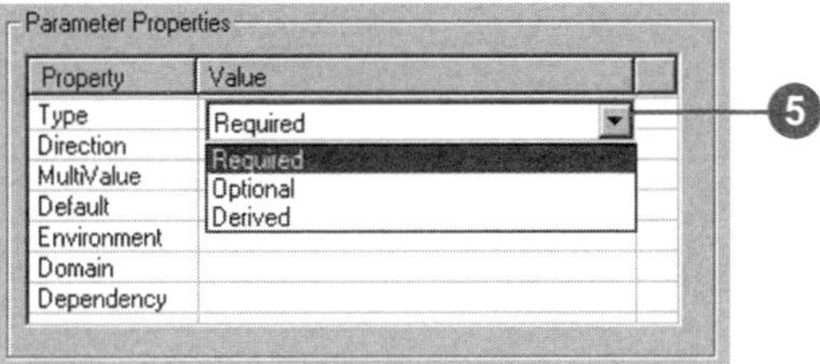

6. Click the cell next to the property Direction and click either Input or Output for the direction of the parameter.

   In the example, the parameter Input Folder is an input to the tool, so its direction property should be set as Input.

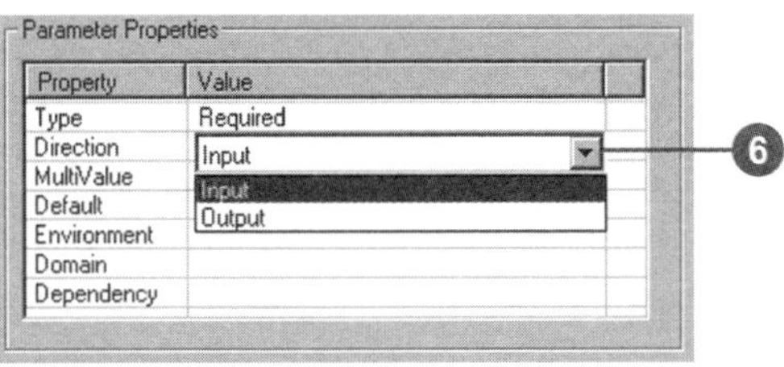

7. Click the cell next to the property MultiValue and click Yes or No, depending on whether the parameter should accept multiple values or not.

   In the example, the parameter Input Folder only accepts one value, a folder, so this property should be set to No.

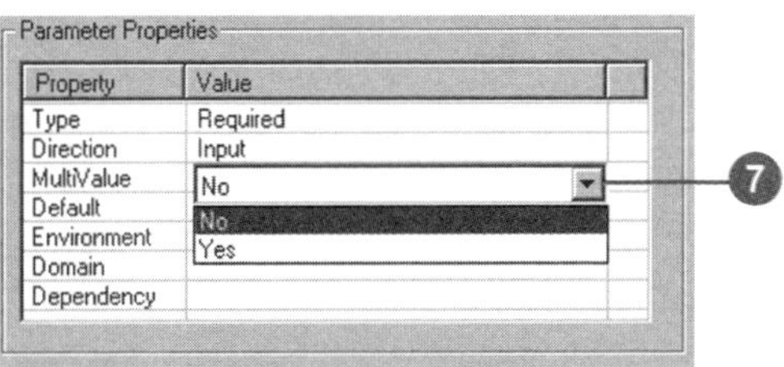

8. Set a Default value if desired.

   In this example, the default value is set to C:\GP_Tutorial\San_Diego. This value will be displayed by default in the dialog box of the tool as the Input Folder parameter's value. ▸

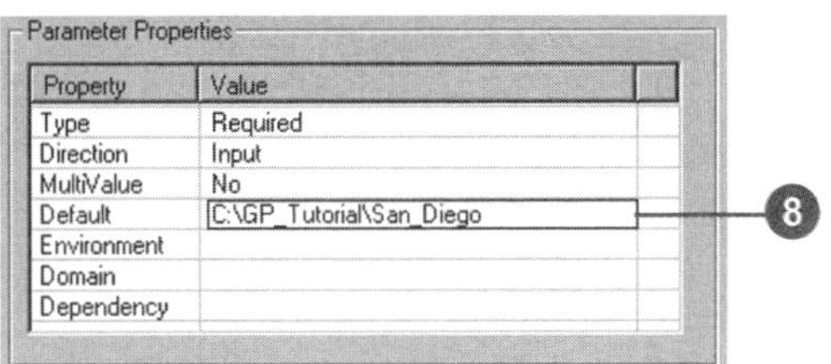

need an output for use in a model but you don't want the output parameter to show in the script's dialog box.

Direction—This property defines whether the parameter is an input or an output.

MultiValue—If you need your parameter to be able to handle an array of values rather than just one value, you should set the MultiValue property to Yes. For example, you might want to clip multiple feature classes. The Data Type would be Feature Class, and by making the parameter support multiple values, you allow the user to specify multiple input feature classes.

Default—You can set up a default value for the parameter that will show up when the script's dialog box is opened, such as the path to the input data for the input data parameter or a value for the cluster tolerance parameter. If you don't specify a value for the Default property, the parameter value will be blank when the script's dialog box is opened. If you specify a value for this property, the Environment property will be disabled. Clear the value set for the Default property to enable the Environment property. ▸

9. If a default value is not set, you can specify to display the value set in the Environment Settings dialog box. Click the cell next to the property Environment and choose the appropriate environment setting from the dropdown list.

10. If only certain parameter values are allowed, you can set a domain.

    Click the cell next to the property Domain and click the Domain button.

11. Check either Coded Value or Range, depending on whether valid values are a range or specific values.

    For a range, specify the start and end values. If the user of the tool specifies a value outside this range, the parameter value will become invalid. For coded values, type the valid values one below the other in the text box.

    In the example, a domain is not necessary.

12. Click OK on the Domain dialog box. ▸

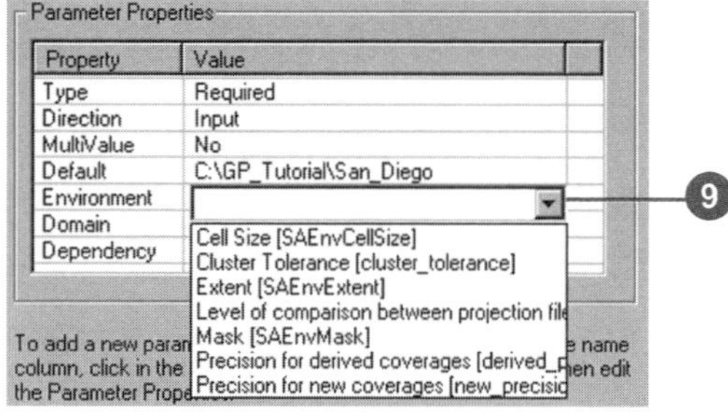

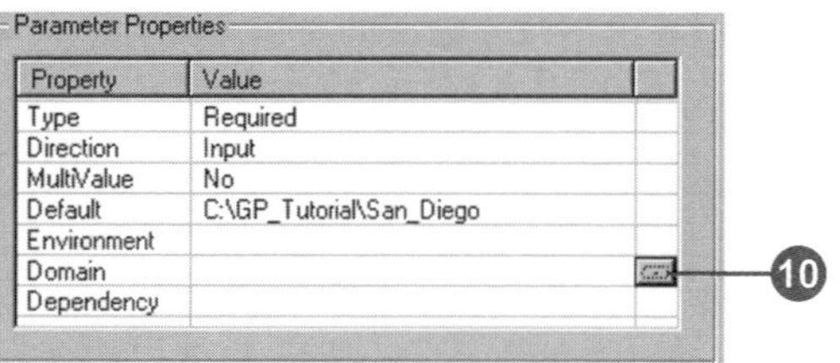

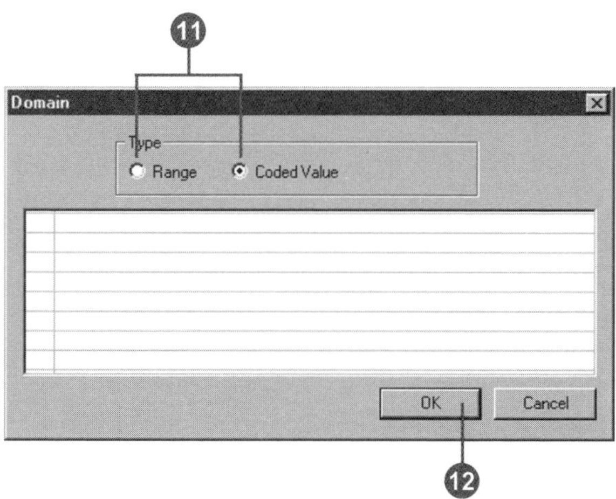

Environment—You can set the value for the parameter to display whatever is set in the Environment Settings dialog box. If a value is set for this property, the Default property is disabled and the Coded Value domain check box is disabled. Setting this property to NONE will enable the Default property and allow a Coded Value domain to be set again.

Domain—Set the domain property to limit the number or range of acceptable parameter values. For instance, acceptable values may be 1 of 3 values, or there may be a range of acceptable values for a parameter.

Dependency—You can specify that a parameter is dependent on a value being specified for another parameter. For example, a field parameter needs to know about the input feature class before it can be populated with the fields from the input. If the output from a script tool is derived, where the output is the same as the input, the output type needs to be set as derived and the output parameter value set to have a dependency on the input parameter value. When a value is entered for the input parameter on the tool's dialog box, the value is automatically also used for the output parameter.

13. If the parameter is dependent on another parameter for information, click the cell next to the Dependency property and select the parameter that it is dependent on.

    In the example, the Input Folder is the first parameter entered, so there are no other parameters available yet in the dropdown list.

14. Continue adding parameters and setting their properties until all necessary parameters have been defined.

15. Click OK.

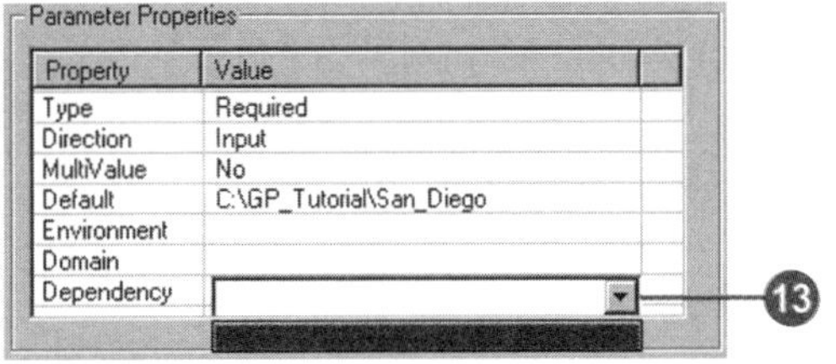

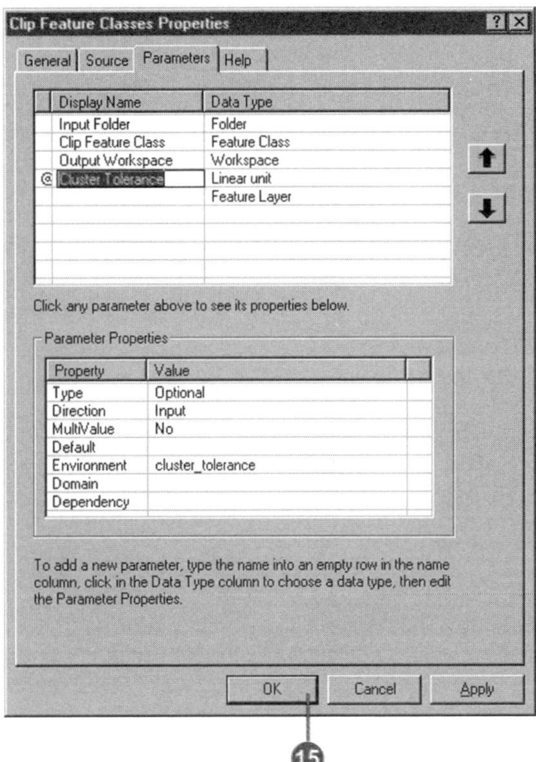

Changing tool properties

System tools are read-only so their properties cannot be modified. You can modify or add additional properties for your own tools, provided they reside in writable toolboxes.

Properties you can modify include the name, label, and description for the tool; the stylesheet used; whether relative pathnames should be stored; the source for a script; and parameters available in the tool's dialog box. ►

Tip

Accessing the properties dialog box for a model

You can also access the properties dialog box for a model inside the ModelBuilder window by clicking Model on the Main menu, then clicking Model Properties.

See Also

See Chapter 6 for information on the Environments tab of a tool's properties dialog box.

See Also

See 'Referencing a compiled Help file' in this chapter for information on the Help tab.

Changing the name, label, and description for your tools

1. Right-click the tool (model or script) and click Properties.

   Note that the toolbox containing the tool must not be read-only. You cannot alter the properties of tools in read-only toolboxes.

2. Click the General tab.
3. Click the Name text box and type a new name for the tool.

   The name is used when the tool is run at the command line or inside a script.

   Note that spaces are not allowed in the name.

4. Click the Label text box and type a new label for the tool.

   The label is the display name for the tool. Spaces in the label are allowed.

5. Click the Description text box and type a description for the tool.

   The description appears in the Help panel in the tool's dialog box.

6. Click OK on the properties dialog box.

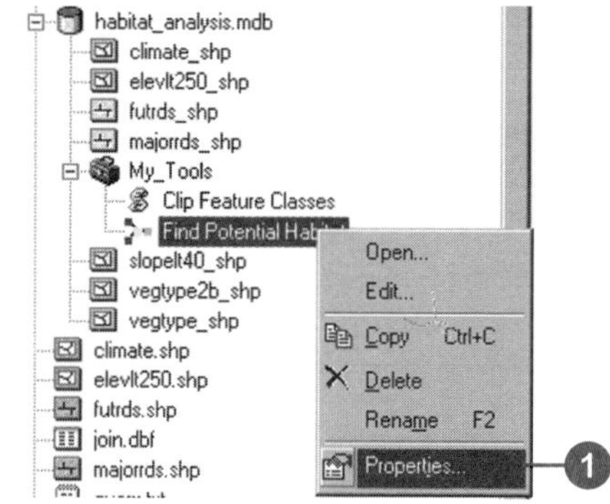

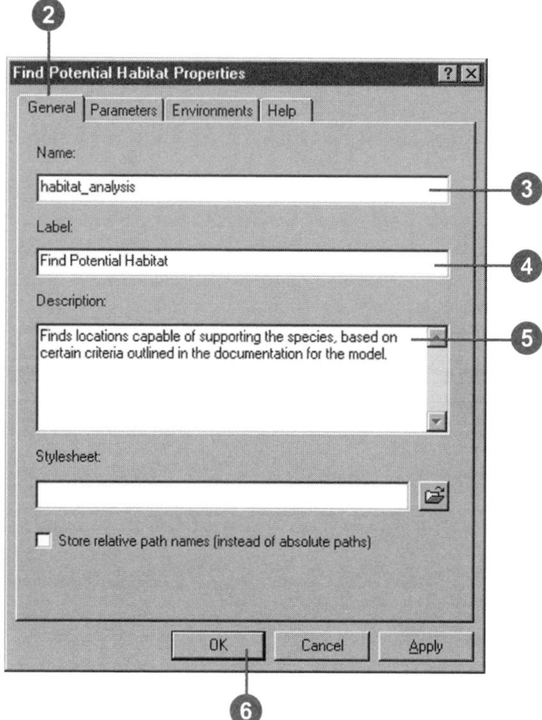

Stylesheets are used to control the properties of items inside tool dialog boxes. The default stylesheets are usually sufficient for most purposes, but you may want to customize some of the properties, for example, by changing the background image used or by adding text.

The MdDlgContent.xsl default stylesheet contains the properties for the contents panel. It can be found in your ArcGIS\ArcToolbox\Stylesheets folder. To override it, first make a copy of MdDlgContent.xsl locally, then proceed to modify the contents of the copied stylesheet. You can simply open the .xsl file in Notepad to modify its contents.

Note that the default stylesheets are used if the Stylesheet text box on the General tab of the Properties dialog box for the tool is empty.

The main stylesheet you'll want to customize is MdDlgContent.xsl. However, you can also customize MdDlgHelp.xsl, which is the stylesheet applied to the Help panel of the tool's dialog box. ▸

Setting a modified stylesheet for a tool

1. Copy the default stylesheet or stylesheets (MdDlgContent.xsl, MdDlgHelp.xsl, or both) from your ArcGIS\ArcToolbox\ Stylesheets folder to the location of the toolbox containing the tool. If the toolbox is located inside a geodatabase, copy the stylesheet to the location of the geodatabase.
2. Open the copied stylesheets in Notepad or any editor and modify the contents.

   In the example, the default background image is replaced by a graphic of the model diagram. Only the name of the graphic is specified because the graphic is located in the default location: \Documents and Settings\username000\ Application Data\ESRI\ ArcToolbox\Dlg (on Windows NT, the username000 folder is located on \WINNT\Profiles). If you only specify the name, users of your tool will have to place the graphic in the Dlg folder on their machine. Alternatively, a hard-coded path can be specified.
3. Right-click the tool and click Properties. ▸

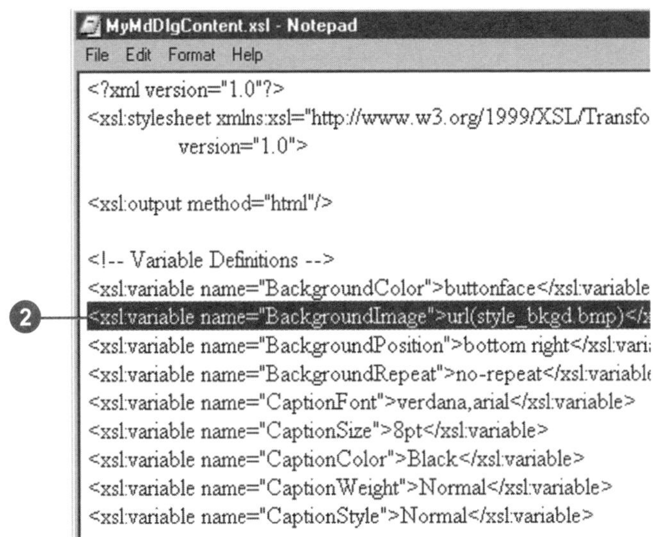

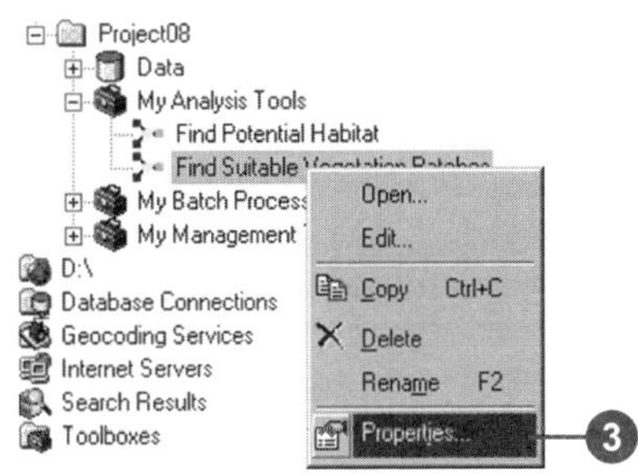

When entering the custom stylesheets into the Stylesheet text box, there is an order for the placement of the stylesheets. If you want to override both MdDlgContent.xsl and MdDlgHelp.xsl, type the path to each alternative stylesheet, separating them with a semicolon in the tool's properties dialog box. The path to the Content stylesheet must be placed first.

If you only want to apply a customized Help stylesheet, simply place a semicolon in place of the Content stylesheet and the default MdDlgContent.xsl will be used.

When just the Content stylesheet is modified, a semicolon is not necessary to make sure the default MdDlgHelp.xsl is applied. ►

See Also

See the appendix for more information on customizing your dialog boxes by modifying the default stylesheets.

4. Click the General tab.
5. Click the Browse button next to the Stylesheet text box.
6. Click the Look in dropdown arrow and navigate to the location of the Contents stylesheet.
7. Click the .xsl file and click Open.

   If you want to add a custom Help stylesheet also, type a semicolon, then the path to the Help stylesheet.
8. Check Store relative path names, instead of absolute paths.
9. Click OK on the Model Properties dialog box. ►

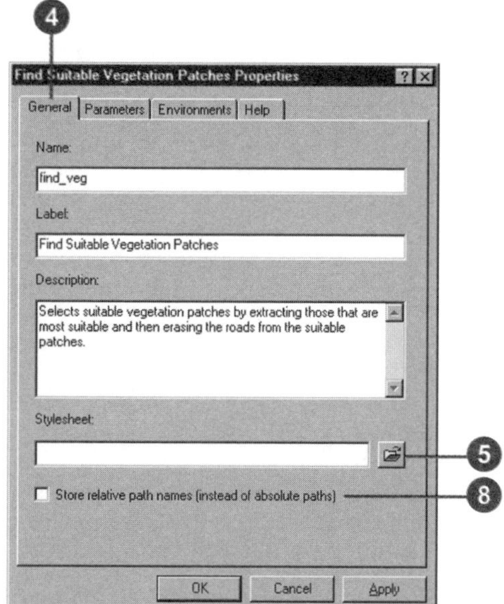

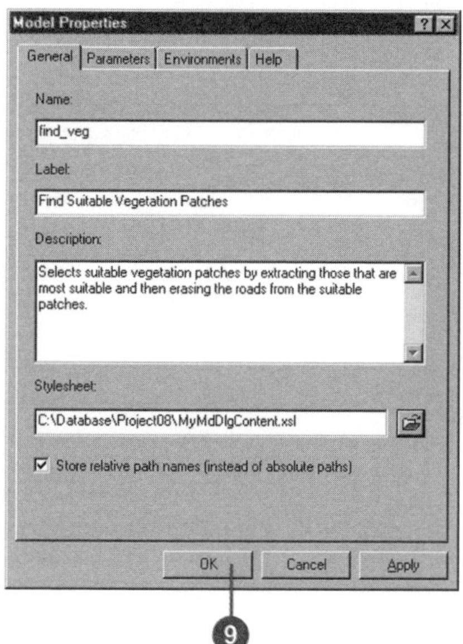

See Also

See 'Sharing your geoprocessing work' in Chapter 3 for more information on sending your work.

See Also

See 'Storing relative pathnames' in this chapter for information on setting relative paths for your tools.

10. Right-click the tool and click Open to see the changes you have made to the dialog box.

    By placing your modified version of the stylesheet at the same level as the toolbox and by setting relative paths for your tool, if you send the folder—containing the tool, the data the tool requires, and the stylesheet—to someone else, your custom stylesheet will be applied when the user of your tool opens the tool's dialog box. One exception is any graphic that will display on the tool's dialog box. These have to be placed in the correct location on the receiver's machine, either in the Dlg folder, if only the name is specified in the XSL file, or in the exact location specified in the XSL file.

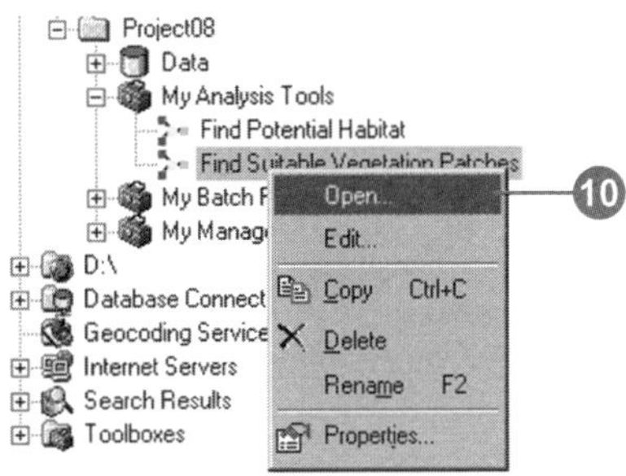

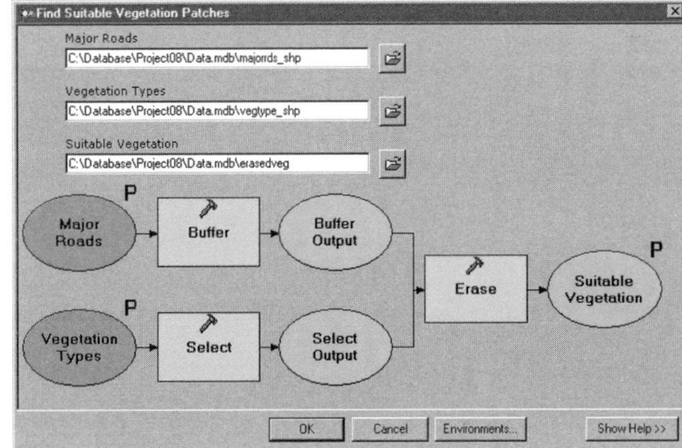

When you add data to a model, add a script to a toolbox, or specify a different stylesheet to use for a tool, references to these sources of information are stored with the tool. The next time you edit or open the tool, the source information is located based on the references. If the tool can't find a source of information it needs, you'll have to locate the source yourself or ignore the reference, in which case you may have difficulties running the tool.

If you plan on distributing your tools to others, they'll need access to the sources of information they reference. To make it easier to distribute all information sources with your tool, you can store relative pathnames to sources of information. All pathnames within the tool will be stored relative to the toolbox containing the tool. For example, if you set relative pathnames for your tool, then distributed it along with the data in the same directory, the references stored in the tool would be correct regardless of their placement on disk. ▸

See Also

See 'Sharing your geoprocessing work' in Chapter 3 for more information on sharing tools.

Storing relative pathnames

1. Right-click the tool and click Properties.
2. Click the General tab.
3. Check Store relative path names, instead of absolute paths.
4. Click OK on the properties dialog box.

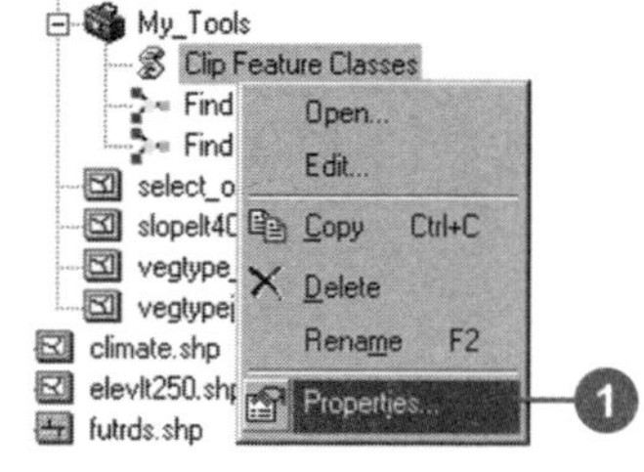

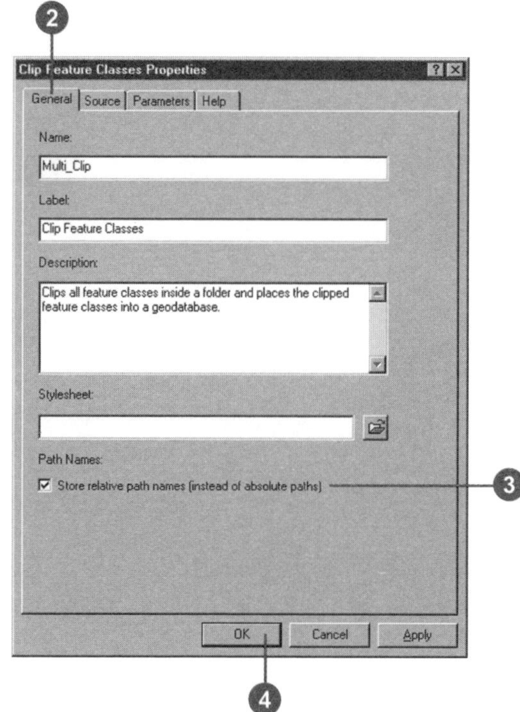

When a script is initially added to a toolbox, the name and location of the script that will run is specified.

There are many reasons you might want to alter the name or location of a script file. If you are sent a toolbox containing a script and the script was not saved with relative paths, the path to the script will need to be updated to point to the location of the script on your machine. Alternatively, renaming the script that the tool references on disk will mean it will have to be updated in the tool's Properties dialog box. ▸

See Also

See Writing Geoprocessing Scripts With ArcGIS *for information on writing scripts.*

See Also

See 'Adding a script to a toolbox' earlier in this chapter to learn how to add a script to a toolbox.

Changing the source for a script

1. Right-click the script and click Properties.
2. Click the Source tab.
3. Click the Browse button next to the Script File text box.
4. Click the Look in dropdown arrow and navigate to the location of the script you want to add.
5. Click the script and click Open.
6. Click OK on the properties dialog box.

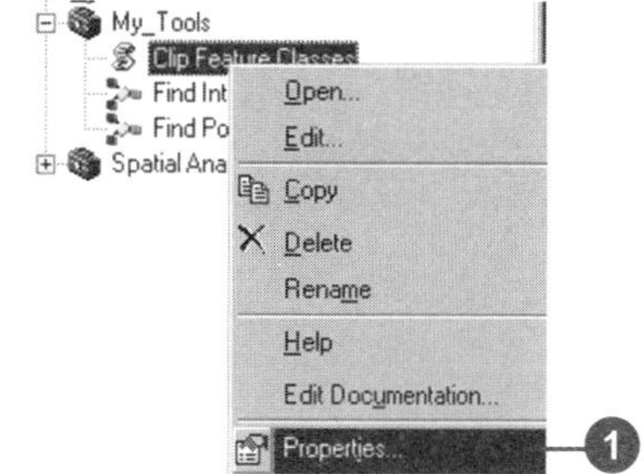

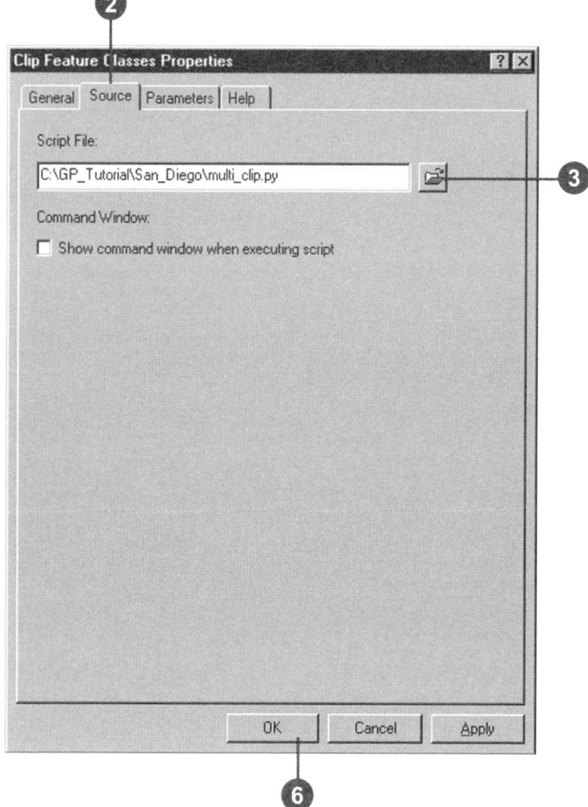

As the author of a script or a model, in the Parameters tab you can add parameters, remove them, or change their order.

As the user of a tool, you can remove, add, and change the order of parameters. Removing a parameter for a model or script, however, will remove the parameter from the model or the script's dialog box. If the model or script depends on user input for the value of the removed parameter, it will not run. ►

See Also

See 'Setting parameters' earlier in this chapter for information on setting parameters for scripts or models.

See Also

You can also set and remove parameters for a model inside the ModelBuilder window. It is sometimes easier to set model parameters in this way, as you can see all the processes that make up the model, making it easier to decide which variables should be set as model parameters. See Chapter 9, 'Using the ModelBuilder window', for more information.

Removing parameters for a model

1. Right-click the model and click Properties.
2. Click the Parameters tab.
3. Click the model parameter you want to remove and either press Delete on the keyboard or click the Remove button.
4. Click OK.

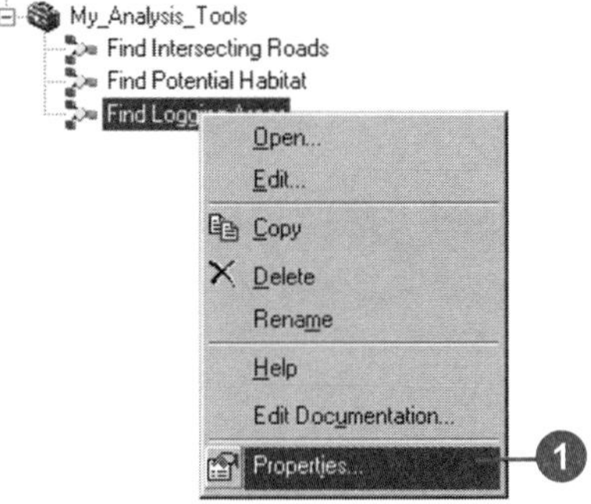

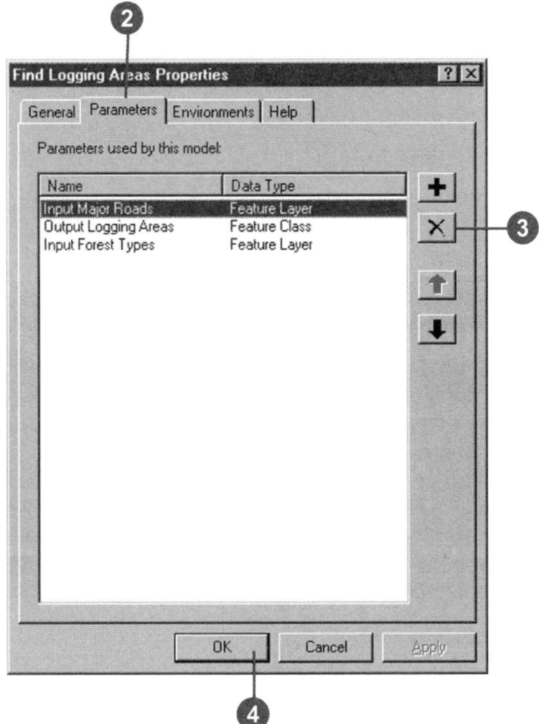

Removing parameters for a script

1. Right-click the script and click Properties.
2. Click the Parameters tab.
3. Click to the left of the display name for the parameter you want to remove, then press Delete on your keyboard.

   The parameter is removed.
4. Click OK.

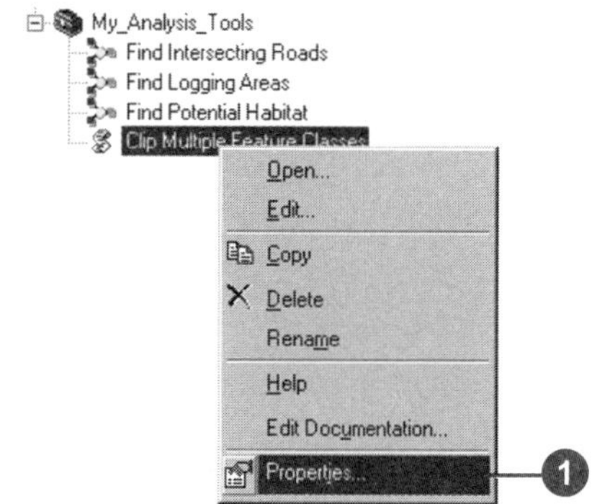

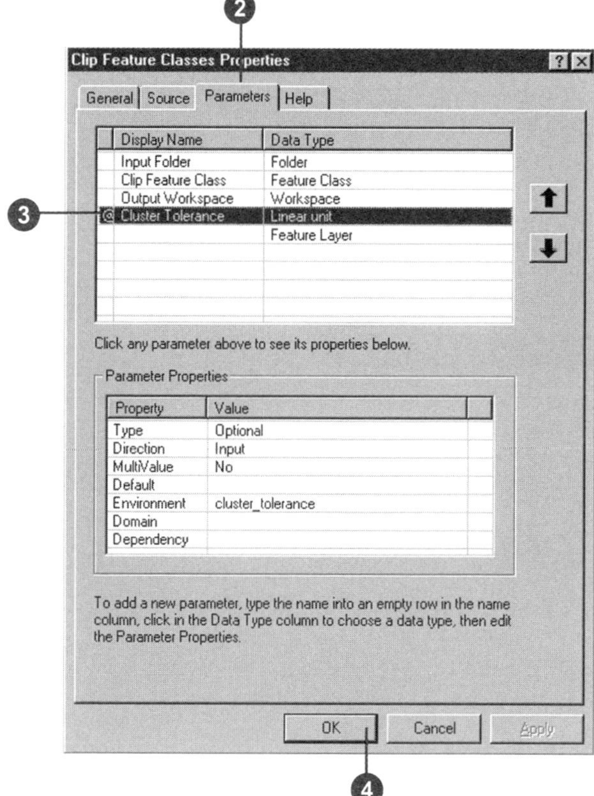

The order of parameters depends on the desired position for parameters on the dialog box of the model or script. Usually, input parameters are at the top, then output parameters, and optional parameters are at the bottom of the dialog box.

Changing the order of parameters for a model so they display in your chosen position on the model's dialog box will not have an adverse effect on the execution of the model. If you change the order of parameters on the Parameters tab of a script, you must change the order of system arguments (sys.argv [ ]) inside your script to mimic the order set in the Parameters tab; otherwise, your script will not execute correctly.

See Also

See 'Setting parameters' earlier in this chapter for information on setting parameters for scripts and models.

See Also

See 'Adding a script to a toolbox' earlier in this chapter to learn how to add a script to a toolbox.

Changing the order of parameters

1. Right-click the tool (either a model or a script) and click Properties.
2. Click the Parameters tab.
3. Click the row for the parameter you want to reorder.
4. Click the move up or move down button depending on where you want to position the parameter in the dialog box for your tool.
5. Click OK on the properties dialog box.
6. Right-click the tool and click Open to see the changes that have been made to the order of parameters on the tool's dialog box.

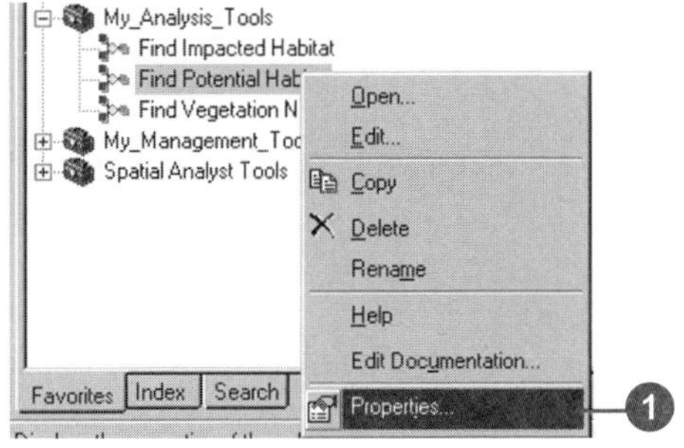

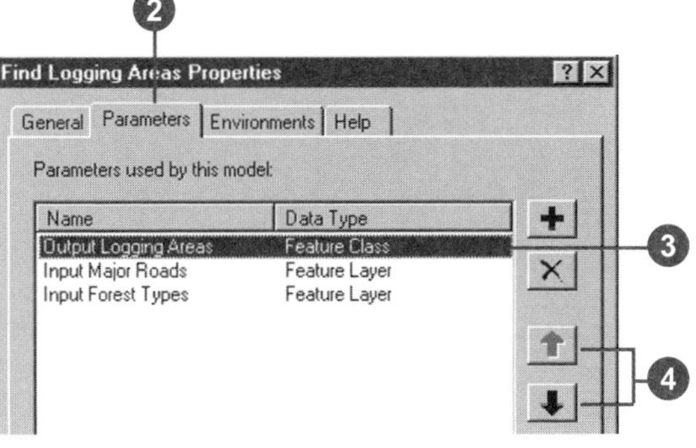

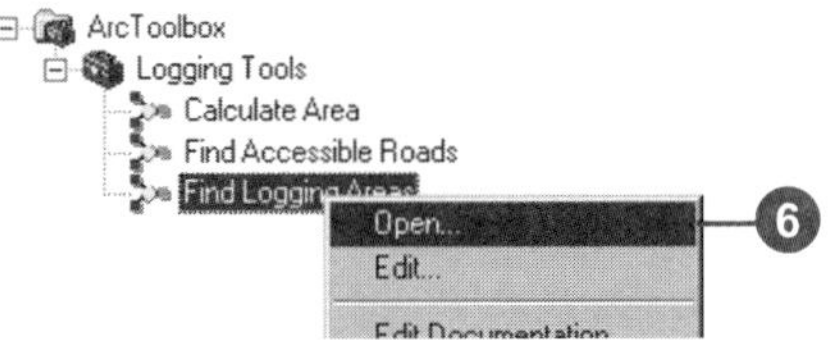

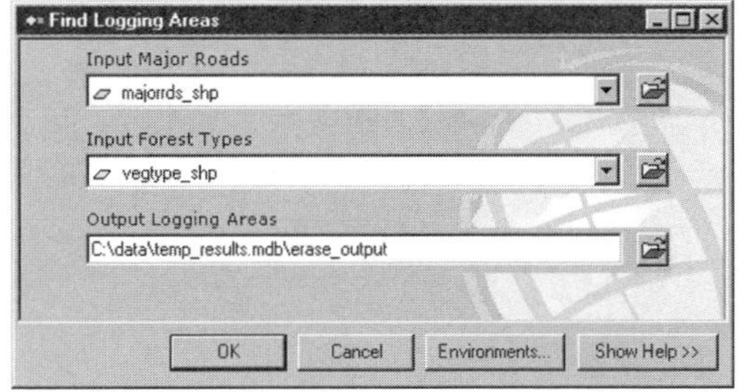

Adding documentation to tools

In a similar way to storing metadata for a dataset, you generally have documentation you would like to store alongside your tools.

The entries in the General Information section of the Documentation Editor will appear as the tool's metadata. The abstract provides a paragraph of text describing the tool's contents. Keywords can be used when searching for a tool in ArcCatalog. Author information is fundamental if you want users of your tool to be able to contact you, and you should document any constraints as to the use of the tool.

Entries in the Help section of the Documentation Editor appear in the Help page accessed via the ArcToolbox window or the tool's dialog box. You can include summary information, describing who wrote the tool and what the tool does; an illustration; usage tips; documentation for each tool parameter; and example code for running the tool from the command line or inside a script. ►

Adding documentation to a tool via the ArcToolbox window

1. Right-click the tool and click Edit Documentation.

   The Documentation Editor opens so you can add documentation to your tool.

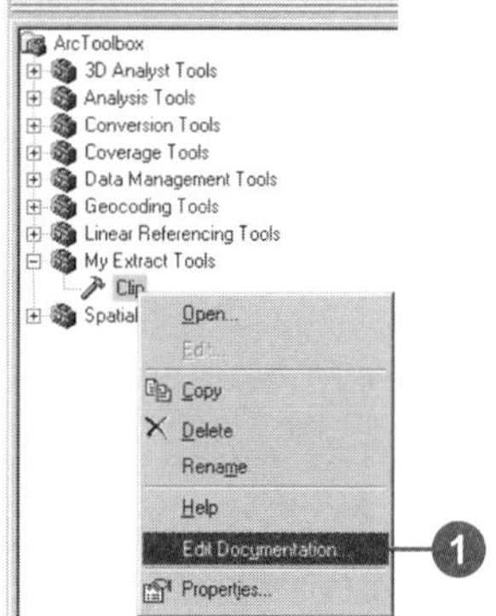

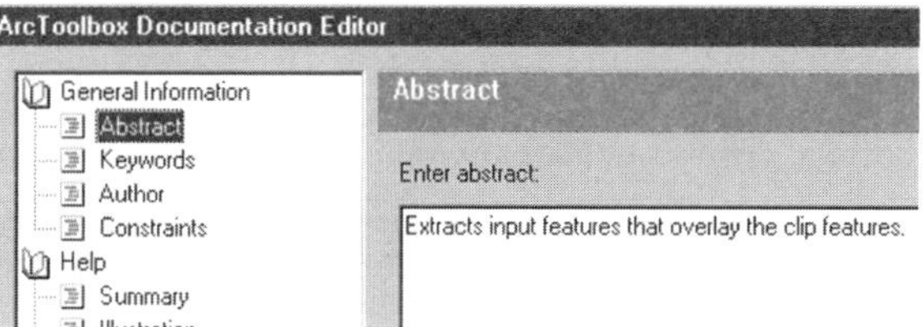

Adding documentation to a tool via the ArcCatalog tree

1. Click the tool in the ArcCatalog tree, then click the Metadata tab.
2. Click the Edit Metadata button on the Metadata toolbar.

   The Documentation Editor opens so you can add documentation to your tool.

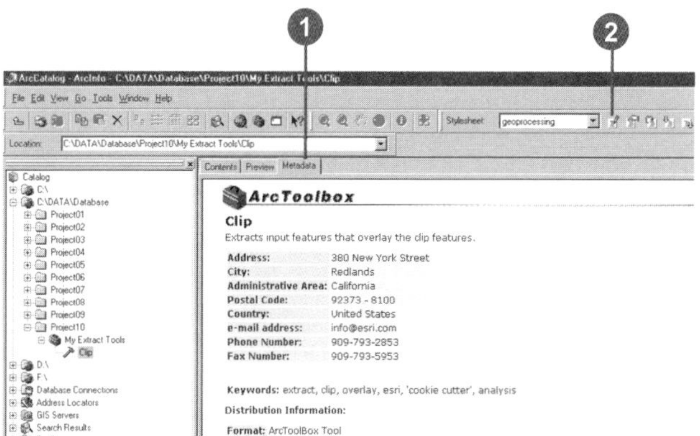

You can add an abstract to describe what the tool does. The abstract will appear when the tool's help is accessed via the metadata tab and also as the description in the Help panel of the tool's dialog box. It will take precedence over any text added as the tool's description in the General tab of the tool's Properties dialog box.

Keywords can make it easy to find tools using either the ArcCatalog Search tool or the Search tab of the ArcToolbox window. If your tools are spread out on disk in different toolboxes, you can simply search for all tools with the same keyword.

Keywords you might add include:

- The tool name
- The supported data types
- The name of the toolbox or toolset the tool belongs to
- Special words associated with the tool ▸

Tip

Writing documentation for tools

Use the documentation written for the system tools as a guide to help you write documentation for your own tools.

Adding an abstract for the tool's Help page

1. Open the Documentation Editor.
2. Click Abstract in the General Information section.
3. Enter a paragraph of text for the abstract.
4. Click Finish.

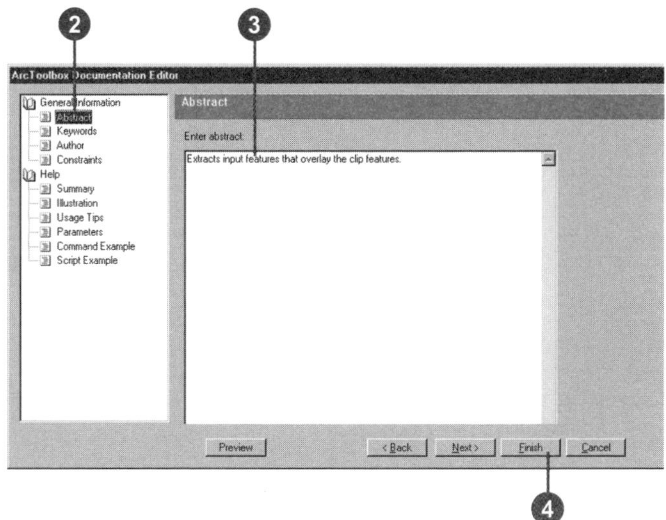

Adding keywords

1. Open the Documentation Editor.
2. Click Keywords.
3. Click the first row and enter the name for the first keyword.
4. Continue adding keywords as appropriate, then click Finish.

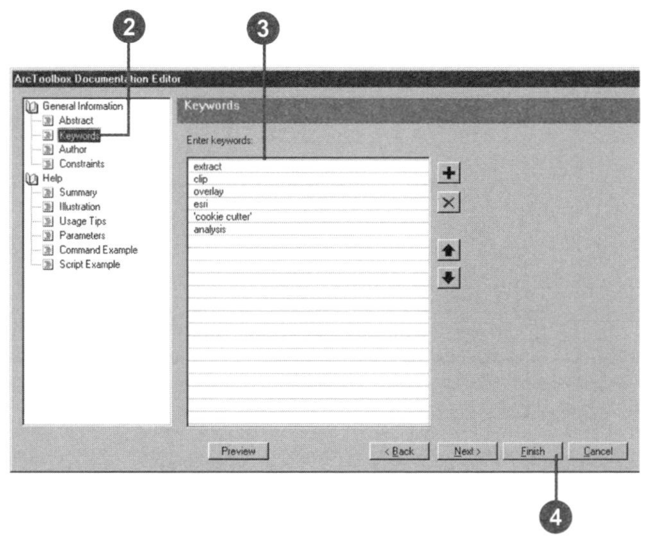

If you are distributing your tool, it is important to document your contact details. Users might, for instance, want to credit your work or contribute information that might increase the usefulness of your tool.

You can also list any constraints of the tool. Constraints might be related to the proper use or distribution of the tool and could also include a liability statement. ▸

See Also

See 'Viewing documentation for tools' in this chapter for information on viewing your created Help page.

See Also

For information on documenting toolboxes and toolsets, see 'Adding documentation to toolboxes' in Chapter 4.

Adding information about the author

1. Open the Documentation Editor.
2. Click Author in the General Information section.
3. Enter your name, organization, and position.
4. Enter your address details.
5. Click Finish.

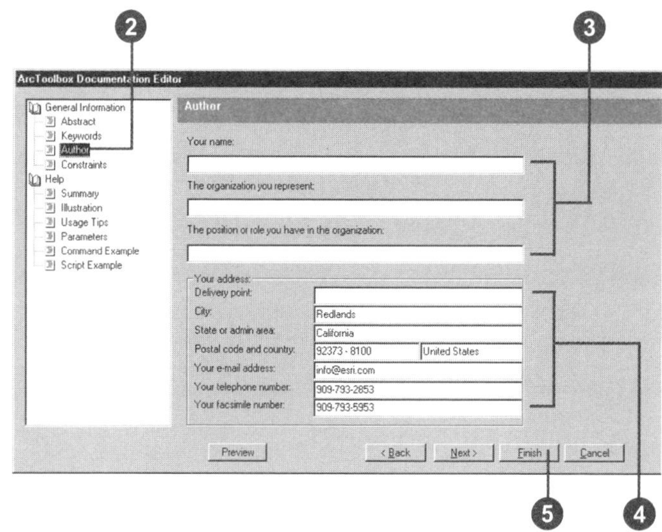

Adding information about constraints

1. Open the Documentation Editor.
2. Click Constraints in the General Information section.
3. Type information regarding the constraints of using the tool.
4. Click Finish.

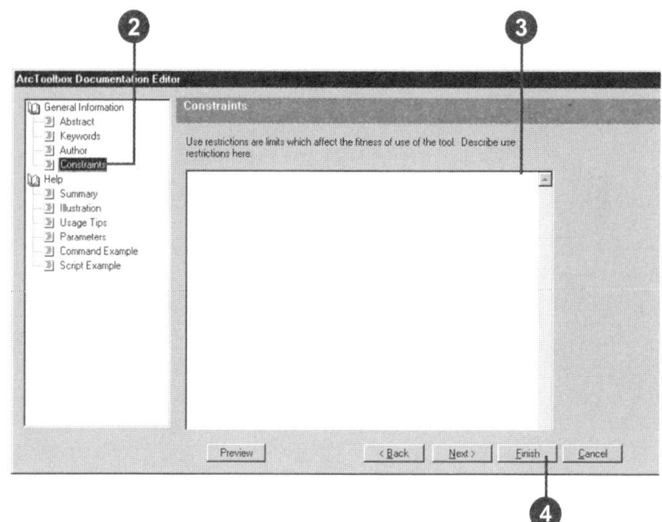

In the Help section of the Documentation Editor, you can add a summary describing what the tool does. Summary information can take the form of paragraphs, bullet points, hyperlinks, illustrations, subsections, and indented text. This information is displayed in the Help page accessed via the tool's context menu in the ArcToolbox window.

Add paragraphs of text to describe a tool's behavior, and bullet points to add lists of information. Hyperlinks allow the user of your Help page to click the link to open another document. It could be another Help page, a graphic, or a Web page. Hyperlinks allow related information to be accessible from a Help page. Graphics help to illustrate the information given in the text. Subsections are expandable sections of text that can include any items, such as illustrations, bullet points, paragraphs, and hyperlinks. They help to order the information in your Help page. Indents are useful for repositioning text or graphics in the Help page. They can be nested inside other indents in order to nudge text or graphics away from the left side of the Help page. ►

Adding a summary

1. Open the Documentation Editor.
2. Click Summary.
3. Click the desired button, depending on what information you want to add.

   Click Paragraph, then click the Paragraph item in the Contents list and type text for the paragraph on the right side of the Documentation Editor. Click Bullet to insert a bullet point. Click Hyperlink to insert a link to another document—for example, another Help page, a Web page, or a graphic. Click Illustration to insert a graphic. Click Subsection to add an expandable subsection, and click Indent, then another button, such as Paragraph, to insert indented text.
4. Click Finish.

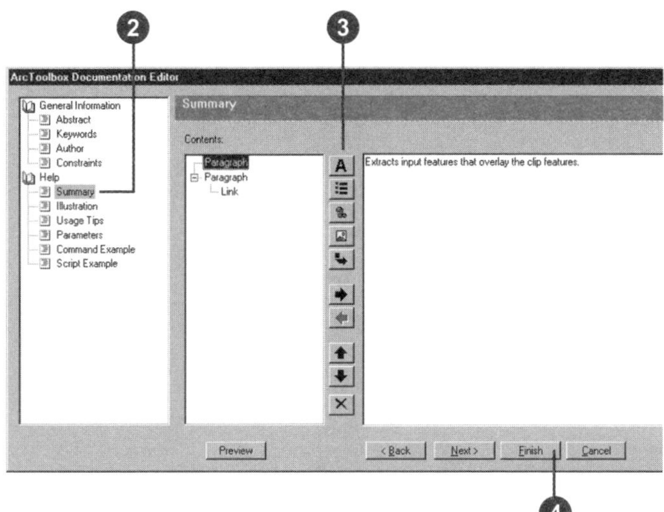

The parameters for which values are supplied on the dialog box of the tool can be documented to provide information about each one. This documentation will appear in the Help page in the Command line syntax section and the Scripting syntax section if you add documentation under the Command Reference section. It will appear when the user of your tool clicks the Help panel in the tool's dialog box, then clicks the documented parameters in the dialog box if you add documentation under the Dialog Reference section. Parameter documentation can take the form of paragraphs, bullet points, hyperlinks, illustrations, subsections, and indented text. This information is displayed in the Help page accessed via the tool's context menu in the ArcToolbox window. ►

Tip

Reordering content

The order of the Contents list in the Documentation Editor is reflected in the Help page. To reorder the contents, click an item, for instance, Paragraph, in the Contents list and click the down or up arrow to reposition the item in the desired location.

Adding documentation for parameters

1. Open the Documentation Editor.
2. Click Parameters.
3. Expand the parameter for which you want to add documentation and click Command Reference or Dialog Reference, depending on whether you want your help to display in the Help page or the tool's dialog box.
4. Click the desired button, depending on what information you want to add.

   Click Paragraph, then click the Paragraph item in the Contents list and type text for the paragraph on the right side of the Documentation Editor. Click Bullet to insert a bullet point. Click Hyperlink to insert a link to another document. Click Illustration to insert a graphic. Click Code to add example code. Click Subsection to add an expandable subsection, and click Indent, then another button, such as Paragraph, to insert indented text.
5. Click Finish.

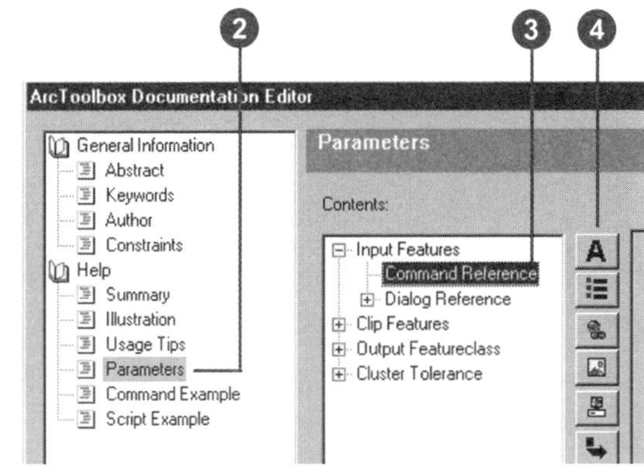

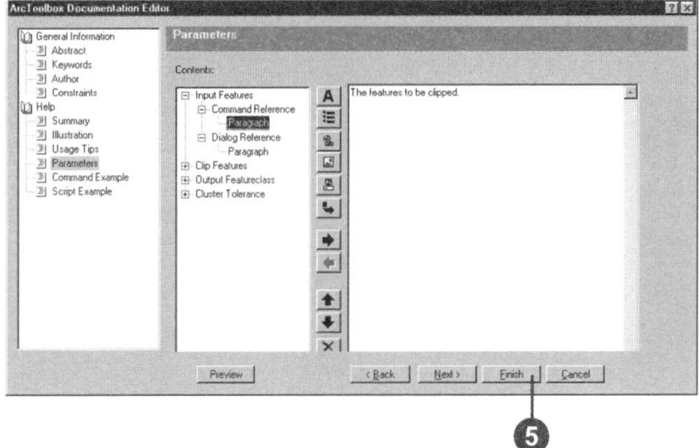

In the Help section of the Documentation Editor, you can add usage tips that can provide useful information to help the user run your tool. Usage tip information can take the form of paragraphs, bullet points, hyperlinks, illustrations, subsections, and indented text. This information is displayed in the Help page accessed via the tool's context menu in the ArcToolbox window.

Add paragraphs of text to describe a tool's behavior, and bullet points to add lists of information. Hyperlinks allow the user of your Help page to click the link to open another document. It could be another Help page, a graphic, or a Web page. Hyperlinks allow related information to be accessible from a Help page. Graphics help illustrate the information given in the text. Subsections are expandable sections of text that can include any items, such as illustrations, bullet points, paragraphs, and hyperlinks. They help to order the information in your Help page. Indents are useful for repositioning text or graphics in the Help page. They can be nested inside other indents to nudge text or graphics away from the left side of the Help page. ▸

Adding usage tips

1. Open the Documentation Editor.
2. Click Usage Tips.
3. Click the desired button, depending on what information you want to add.

   Click Paragraph, then click the Paragraph item in the Contents list and type text for the paragraph on the right side of the Documentation Editor. Click Bullet to insert a bullet point. Click Hyperlink to insert a link to another document—for example, another Help page, a Web page, or a graphic. Click Illustration to insert a graphic. Click Code to add example code. Click Subsection to add an expandable subsection, and click Indent then another button, such as Paragraph, to insert indented text.
4. Click Finish.

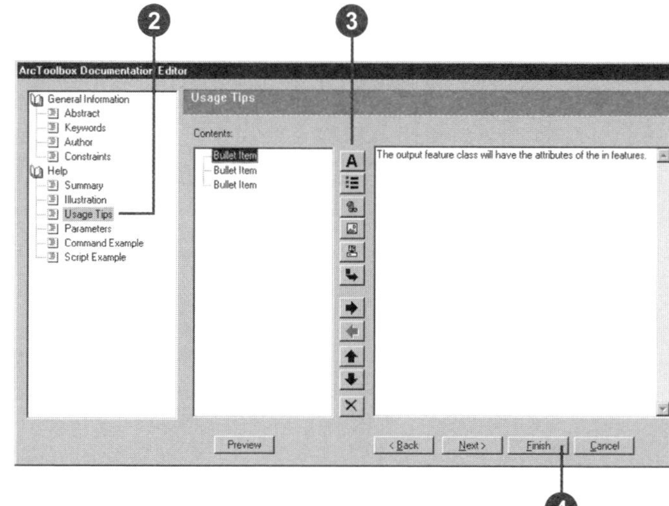

Illustrations can be powerful aids that can complement your tool summary and help explain what the tool does. You can add a name that will display when the mouse pointer is held over the illustration. The name describes the contents of the illustration. You can also add a caption, a single paragraph of text, that will display underneath the illustration. ▸

Tip

Previewing Help files

As you are adding documentation, click the Preview button in the Documentation Editor to see how your help will display in HTM format.

Adding an illustration

1. Open the Documentation Editor.
2. Click Illustration.
3. Type the path to the illustration or click the Browse button, click the Look in dropdown arrow in the Open dialog box, and navigate to the location of the illustration on disk. Click the illustration and click Open.
4. Type a name that describes the contents of the illustration.

   The name will appear when the cursor is held over the illustration.
5. Type a caption for the illustration.

   The caption will display under the illustration.
6. Click Finish.

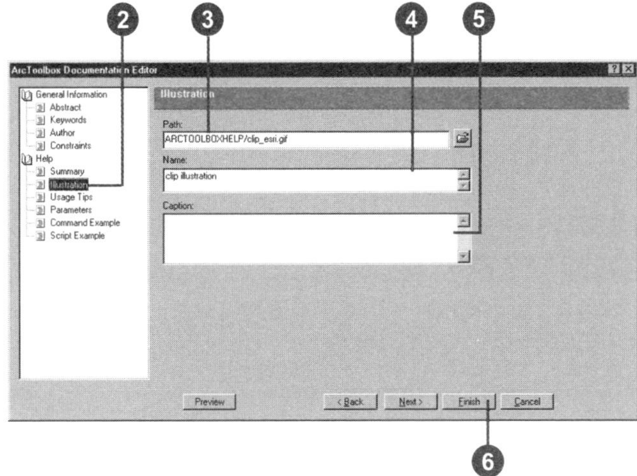

In addition to running via a dialog box, your tools can be run from a command line or a script. You can add an example to show how to run the tool at the command line or from within a script. ►

See Also

See 'Setting parameters' earlier in this chapter for information on setting variables as parameters so the parameters display on your tool's dialog box. Values for set parameters can be entered via the tool's dialog box, at the command line, or within a script.

Adding a command line example

1. Open the Documentation Editor.
2. Click Command Example.
3. Type the code to show an example of how to run the tool at the command line.
4. Click Finish.

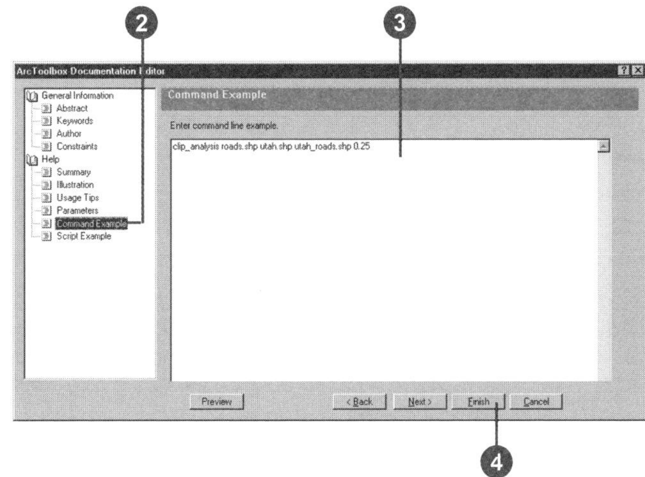

Adding a script example

1. Open the Documentation Editor.
2. Click Script Example.
3. Type the code to show an example of how to run the tool from within a script.
4. Click Finish.

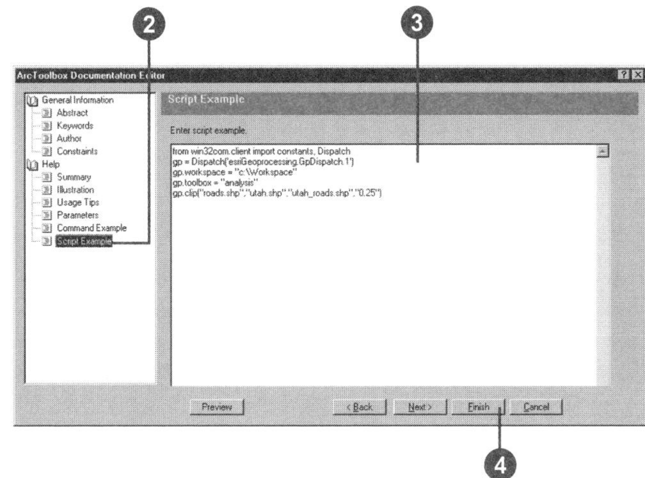

The documentation you add to a tool using the Documentation Editor can be exported to an HTML file.

Doing so enables you to use your created HTML files when compiling a Help file (.chm). ►

See Also

See 'Referencing a compiled Help file' in the task that follows for information on compiling Help files.

Tip

Alternative export options

Click the Metadata tab for a tool, then click Export metadata. Numerous export formats are available. Save to XML to alter the information between the XML tags, then import the XML file with the Import Metadata button to update the documentation for a tool.

Exporting tool documentation to an HTML file

1. Right-click the tool and click Properties.
2. Click the Help tab.
3. Click Export.
4. Click the Save in dropdown arrow and navigate to the location into which you want to save the HTML file.
5. Type a File name and click Save.
6. Click OK on the tool's properties dialog box.

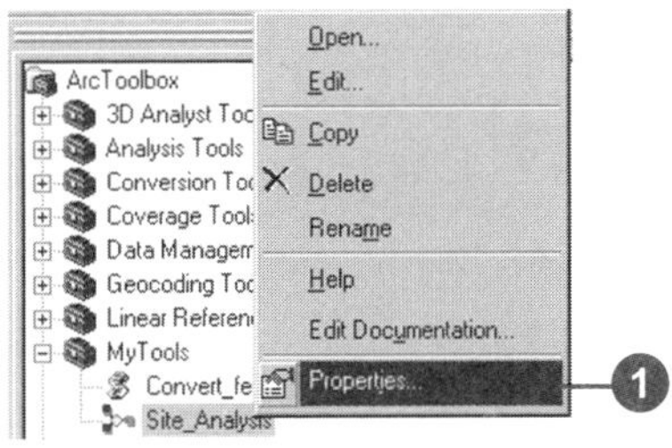

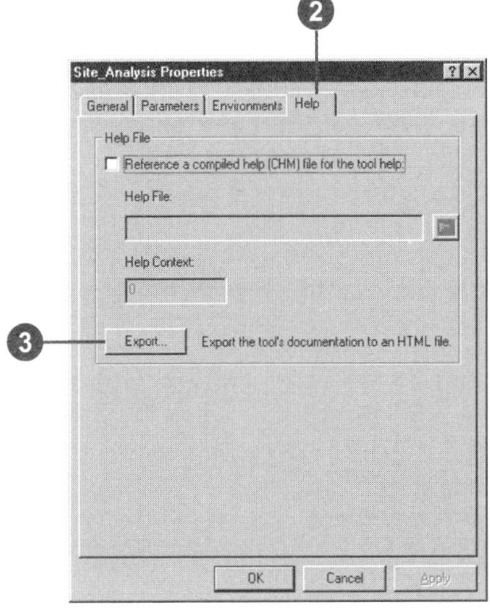

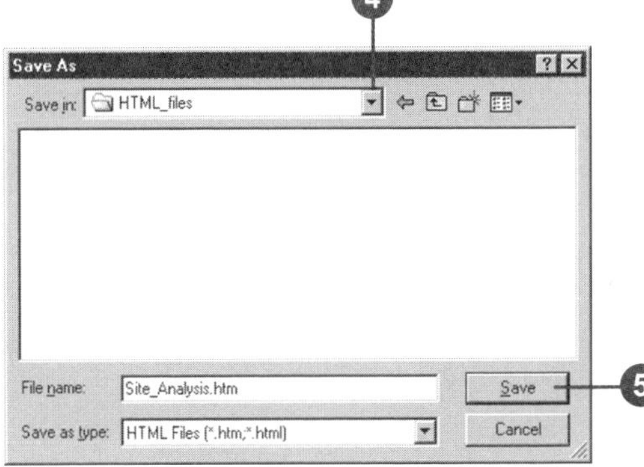

If you have a compiled Help file (.chm), your tools can reference it. A complied Help file is used for viewing a snapshot of a tool's documentation.

By providing the Help Context (the HTML topic ID), when the user of your tool clicks Help on the tool's context menu, the help topic associated with the Help Context ID in the CHM file will be displayed. Help written in the tool's Documentation Editor is always stored with the toolbox. However, the CHM file takes precedence over this help.

If you do not supply a Help Context ID or you supply an incorrect Help Context ID, the CHM file will display with its default page.

There are many sources of information on creating and compiling HTML Help, or CHM files. A good starting point is the 'Microsoft HTML Help 1.4 SDK' topic available from the Microsoft library. Type "http://msdn.microsoft.com" in your Web browser and click the link to the Library. In the Library, you'll find information on creating and compiling CHM files listed under 'Visual Tools and Languages, HTML Help'.

Referencing a compiled Help file

1. Right-click the tool and click Properties.
2. Click the Help tab.
3. Check Reference a compiled help (CHM) file for the tool help.
4. Browse to the location of the CHM file or type the path.
5. Type a Help Context that corresponds to a topic in the CHM file if you want a certain topic to be displayed by default when the help for the toolbox is opened.
6. Click OK.

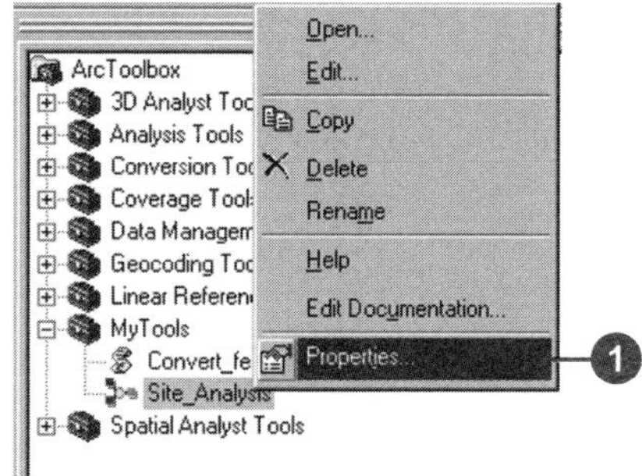

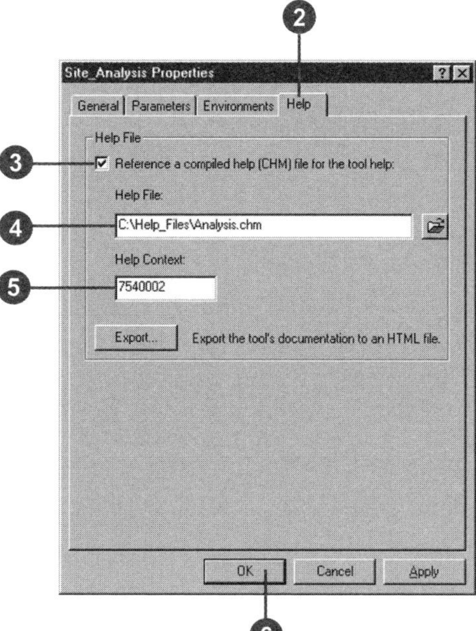

Viewing documentation for tools

A quick way to learn about a particular tool is to view the tool's documentation. You can view documentation for all system tools and for custom tools for which you have added documentation.

Each parameter within a tool's dialog box can have documentation associated with it, helping the user provide correct values for parameters. ►

See Also

See 'Adding documentation to tools' earlier in this chapter for information on documenting your own tools.

Viewing help for parameters on a tool's dialog box

1. Right-click the tool and click Open.

   A short description for the tool is displayed in the Help panel.

2. Click any parameter on the dialog box.

   Help for the parameter will display in the Help panel.

3. Click OK.

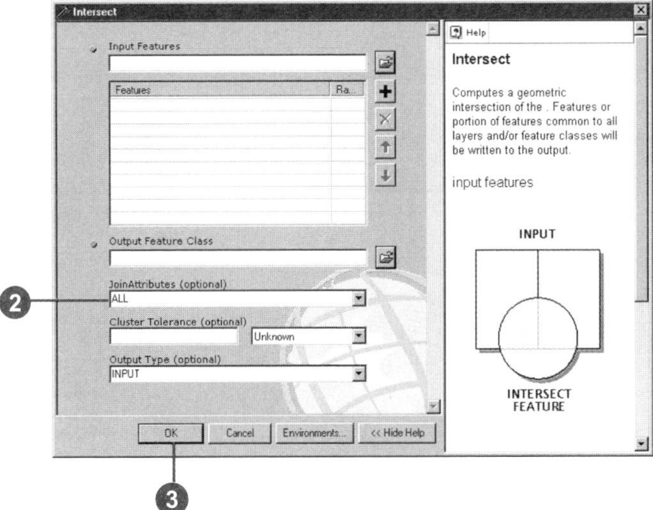

All documentation added to the Help section of the Documentation Editor will display when the Help for a tool is accessed through the ArcToolbox window or through the tool's dialog box.

Via the ArcToolbox window or the tool's dialog box, summary information, describing who wrote the tool and what the tool does; an illustration; usage tips; documentation for each tool parameter; and example code for running the tool from the command line or inside a script can be viewed.

There are three ways you can access documentation written in the Help section of the Documentation Editor—via the tool's dialog box, the Help option on the context menu of the tool, or the online Help system. ►

Viewing Help for a tool via its dialog box

1. Right-click the tool and click Open.
2. Click the Help button at the top of the Help panel to view Help for the entire tool.

   When you view Help for system tools in this way, the Desktop Help System is opened and Help for the particular system tool is displayed.

   When you view Help for your own tools, it will display in a Web browser.

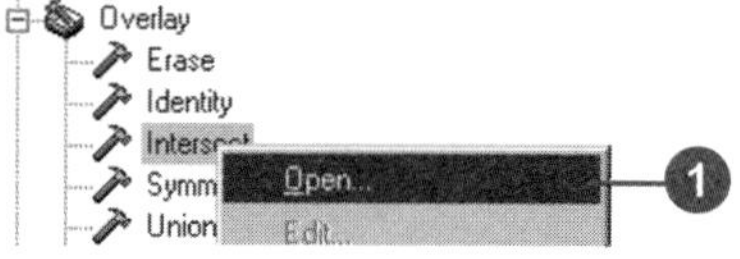

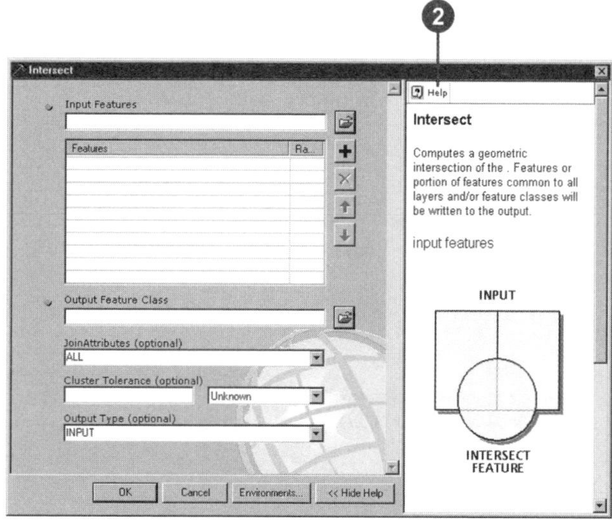

Tip

Using the online Help system

You can use the Contents tab to navigate to the Help file for a particular tool, or you can search the Index for specific tools. You can also use the Search tab to look up Help topics that have specific words or phrases.

Tip

Using the online Help system

The online Help system is useful for browsing through tool documentation, but you can also access other geoprocessing documentation through the online Help system including the contents of this book. Click the Extensions book in the Contents tab to access help for specific extensions you might have installed, such as ArcGIS Spatial Analyst or 3D Analyst.

Tip

Accessing tools

Click the Open Tool link on a tool's Help page to open the tool's dialog box.

See Also

See Using ArcCatalog *and* Using ArcMap *for more information on using the ArcGIS Desktop online Help system.*

Viewing help for a tool in the ArcToolbox window

1. Right-click the tool in the ArcToolbox window and click Help.

   The same documentation accessible from the Help button on the tool's dialog box is displayed.

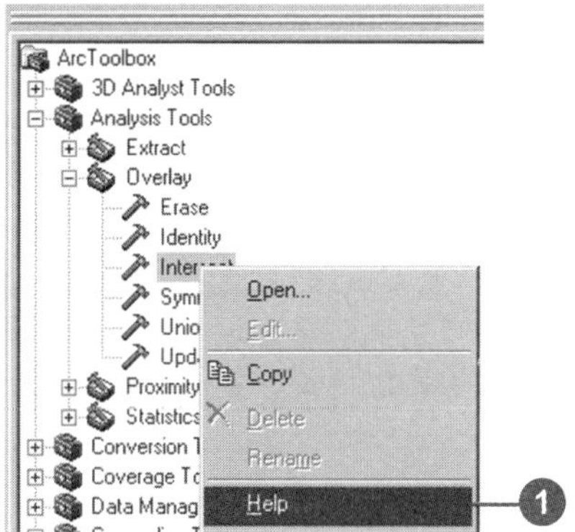

Viewing help for a tool via the online Help system

1. Click Help on the Main menu of the ArcGIS Desktop application you are running and click ArcGIS Desktop Help.
2. Click the Contents tab and double-click the Geoprocessing/ArcToolbox book to view its contents.
3. Double-click the Geoprocessing tool reference and navigate to the book you are interested in to see a list of the topics in that category.
4. Click the topic you want to read to display it.
5. Click expand all to open all sections within the topic at one time.

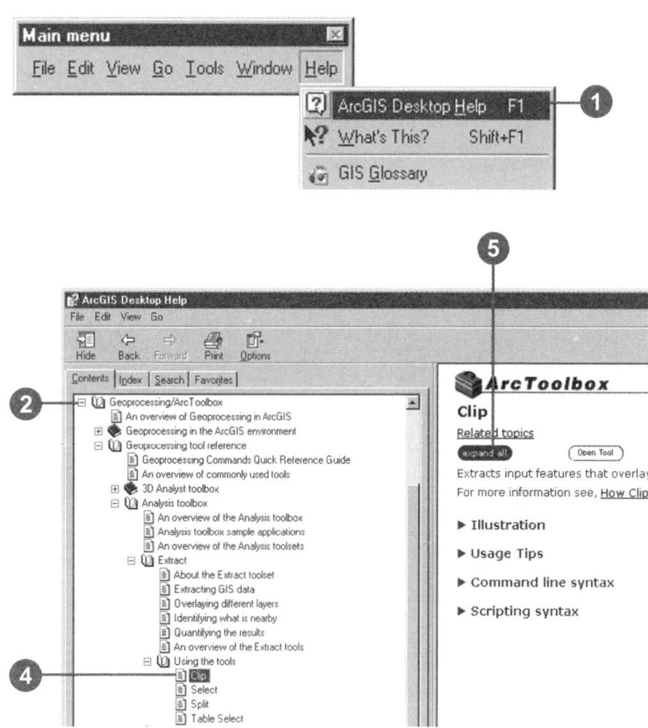

Metadata for a tool can be viewed in the tool's Metadata tab, via the ArcCatalog tree. Only the information added to the General Information section of the Documentation Editor is displayed. You can view an abstract describing what the tool does, keywords that can be used when searching for a tool in ArcCatalog, author information, and any constraints as to the use of the tool.

Viewing metadata for a tool

1. Click the tool in the ArcCatalog tree and click the Metadata tab.

   Documentation added in the General Information section of the Documentation Editor will display.

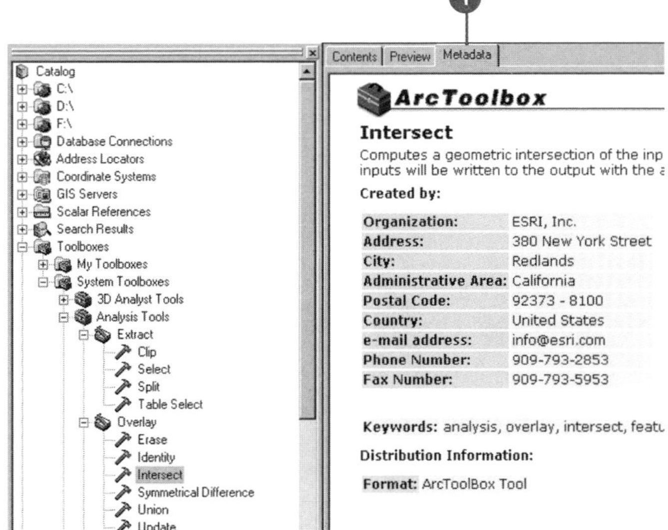

Finding tools

There are hundreds of geoprocessing tools that you can access, so finding a particular tool can be a daunting task. There are different ways you can search for tools to help you to find the tool you are looking for.

One way to search for tools is to search for a toolbox that contains a tool you want to use. You may have created toolboxes inside subfolders, buried deep within your system. You can use the Search capabilities within ArcCatalog to search for toolboxes. ►

Tip

An alternative way to find tools in ArcCatalog

Click the Search tool and click the Advanced tab. Click the Metadata element dropdown arrow and click Full Text. Click the Value input box and type a word in the tool's metadata, such as a keyword, then click Find Now. All tools that have the keyword in their metadata will be located.

Finding toolboxes in the ArcCatalog tree

1. Click Edit on the Main menu of ArcCatalog and click Search.
2. Click the Name & location tab.
3. To search by name, type all or part of the toolbox name you're looking for into the Name text box. Use an asterisk (*) to represent one or more letters. You can also type just an asterisk to search for all toolboxes in a certain location.
4. To search by type, scroll through the Data type list and click Toolbox.
5. Click the Search dropdown arrow and click where you want to search.
6. Click the Browse button; navigate to and click the folder, database connection, or Internet server in which to start searching; then click Open.
7. Type a name for your search in the Save as text box.
8. Click Find Now.

   The Search is saved in the Search Results folder and is selected in the ArcCatalog tree.
9. Click Close.

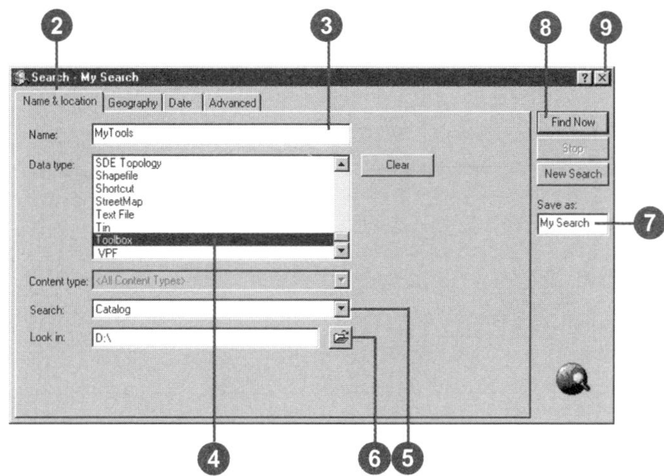

A second way you can search for tools is to use the ArcToolbox index. If you know the name of the tool you are looking for, this is a quick way to locate the tool.

A third way you can search for tools is to use the ArcToolbox Search tab. This is especially useful if you aren't sure of the tool name you are looking for, but you know the data type you want to use as input. In the lower example, typing "Coverage" finds all tools that have Coverage set as a keyword in the tool's documentation.

This method of finding tools is also useful for locating tools that have a different name from that which you might be used to. Keywords associate old names for system tools with their new names, so if you type the name of a tool that has been renamed, you will be able to locate the tool you want to use. For example, typing "GreaterThan" would locate the ArcGIS Spatial Analyst GreaterThanFrequency tool.

Tip

Searching for system tools using the online Help system

Use the Search tab to find Help topics for specific tools.

Using the ArcToolbox index to find tools

1. Click the Index tab at the bottom of the ArcToolbox window.
2. If it is known, type the name of the tool you want to use, or scroll through the list to find the tool you are looking for.
3. Select the tool, then either double-click the tool to open its dialog box or click Locate to switch to the list of Favorite toolboxes and view the location of the tool in its toolbox.

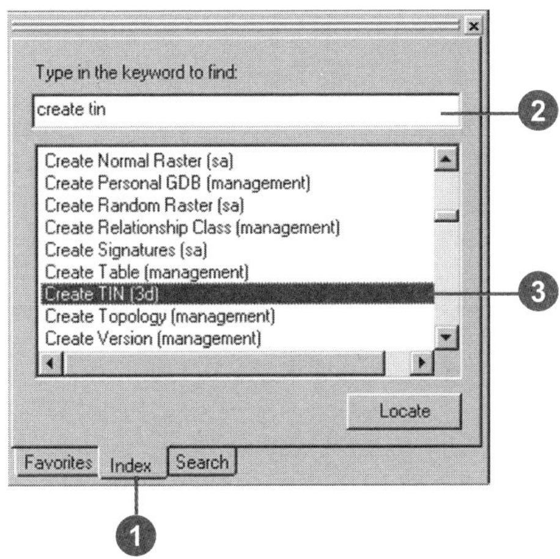

Searching ArcToolbox to find tools

1. Click the Search tab in the ArcToolbox window.
2. Type a word to search for tools.
3. Click Search.

   All tools containing that word are listed.
4. Select a tool, then either double-click the tool to open its dialog box or click Locate to switch to the list of Favorite toolboxes and view the location of the tool in its toolbox.

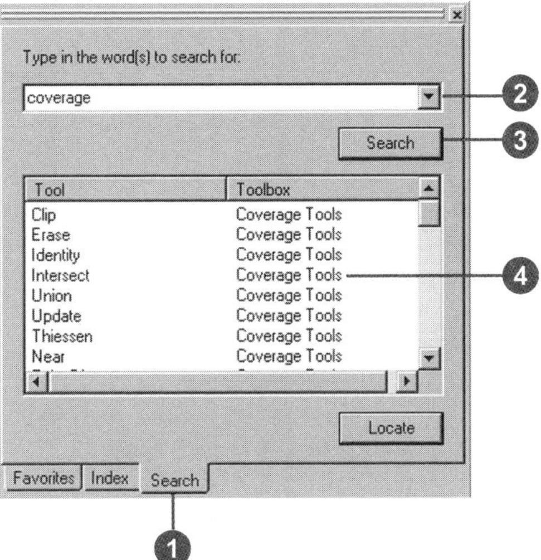

Understanding tool licensing

The system tools that are available for use depend on the ArcGIS software product (ArcView, ArcEditor, or ArcInfo) you are licensed to use. Extension products, such as ArcGIS Spatial Analyst, give you access to even more tools.

By default, in the ArcCatalog tree you will see all tools, regardless of whether or not you are licensed to use a particular tool. If you are not licensed to use a tool, a lock icon will appear over the tool.

In the ArcToolbox window, you will not see the tools that you are not licensed to use by default. If you want to view these tools, right-click the ArcToolbox window and click Show/Hide Locked Tools. Tools you are not licensed to use will appear with a lock icon over them.

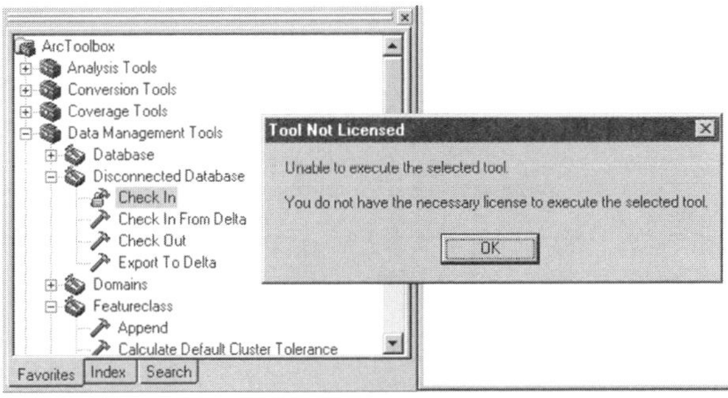

A lock icon will appear over unlicensed tools. You must have the appropriate license to run the tool.

For documentation on tools that are licensed with each ArcGIS software product, type "Quick Reference Guide" in the Search tab of the online Help system and double-click the link to the 'Geoprocessing Commands Quick Reference Guide'.

When a model is created using tools accessible only with an ArcInfo license and the model is edited on a machine that has ArcView or ArcEditor installed, you will see the lock icon over tools in the ModelBuilder window that are not available with the license you are using. The appropriate license (ArcInfo in this case) is needed for the tools within the model to run.

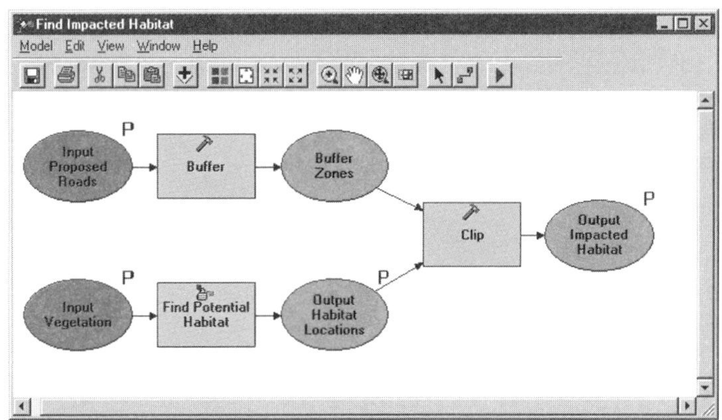

Within the ModelBuilder window, lock icons appear over tools that you are not licensed to use.

As with models, if a script is created using tools accessible only with an ArcInfo license and the script is run on a machine that has ArcView or ArcEditor installed, an error will occur when running the script. The appropriate ArcGIS product (ArcInfo in this case) needs to be installed for the script to run. See the section on licensing and extensions in *Writing Geoprocessing Scripts With ArcGIS* for information on license handling when writing scripts.

Similar behavior will be experienced for models or scripts that contain tools from an extension product when you do not have a license to run the extension product's tools. You need to check out a license in the Extensions dialog box of the application you are working in by clicking the Tools menu, then clicking Extensions to be able to work with a particular extension's tools.

Note that the number of tools available for an extension is the same regardless of the ArcGIS software product you have licensed. Note also that even if you install ArcInfo, you can only use the tools within the Coverage Tools toolbox if you have also installed ArcInfo Workstation. The same lock icon will appear over coverage tools if ArcInfo Workstation is not installed.

Specifying environment settings

6

IN THIS CHAPTER

Environment settings are values that may be used by tools. They are similar to a parameter and can apply to multiple tools. Examples of environment settings include the current workspace from which to take input data and place results or the extent to apply to results.

When geoprocessing tools are run, default environment settings set for the application are applied to all applicable tools. These settings can be changed in the Environment Settings dialog box.

There will be times when the environment settings specified for the application are not appropriate for a particular tool. For instance, you want to generate z-values for the output from one model, but you do not want to specify this for the application because doing so will result in z-values being created for all results from running tools. For such times, you can specify environment settings specifically for a particular model. When the model is run, these settings will always override environment settings specified for the application.

Within a model, there may be processes that require different environment settings from those set for a model or the application. You can specify environment settings for a process in the model if different environment settings are necessary. Environment settings specified for a process override those set for the model or the application.

This chapter explains how to specify environment settings for the application, a model, and individual processes within a model. You'll also learn about the various environment settings that are available.

About environment settings

Environment settings are hierarchical in nature. There are three levels at which they can be set. They can be set for the application you are working in so they apply to all tools, for a model so they apply to all processes within the model, or for a particular process within a model. Settings specified for a process override those set for the model and the application. Settings specified for the model override those set for the application.

Application environment settings

When you run a tool, there are certain environment settings that may apply, such as the current workspace from which to take input data and to place results. To determine which environment settings are used by a particular tool, consult Help for the tool you are using. Values for environment settings can be specified in the Environment Settings dialog box within the application you are using—either via the Geoprocessing tab of the Options dialog box, accessed via the Tools menu on the Main menu of the application you are running, or by right-clicking the ArcToolbox window—and they will be used by all appropriate tools.

When you specify a value for an environment setting, such as the current workspace, the value specified is automatically used by appropriate tools. In the case of the current workspace, if the value -the path to the workspace—is set in the Environment Settings dialog box, you can type the name of the input data to a tool and the name of the output data produced by running a tool, and the path to the current workspace will be supplied for you.

Application environment settings persist when altered in ArcCatalog, so you don't have to set environment settings for each ArcCatalog session. In ArcMap, you can store environment settings by saving the map document. You can also save your geoprocessing settings, which include settings specified in the application's Environment Settings dialog box, so they can be loaded at any time. This is useful if you are working on multiple projects that require different environment settings. See 'Geoprocessing settings' in Chapter 3 for more information.

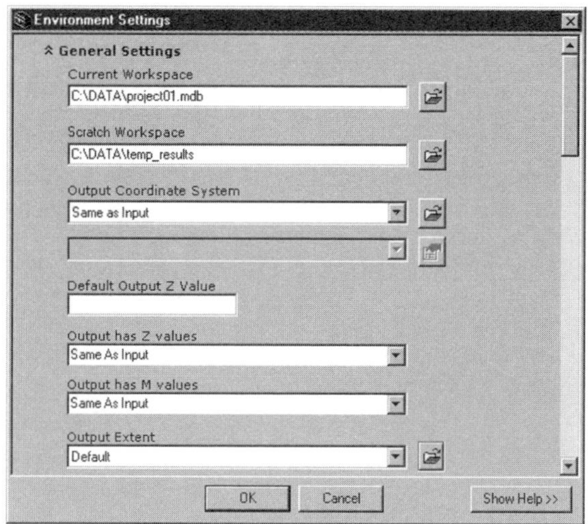

The application's Environment Settings dialog box with the General Settings section expanded and some values supplied. It is used to set environment settings for the application. Settings specified here apply to all appropriate tools.

You can alter environment settings within a tool's dialog box for a particular instance of running a tool. Environment settings specified for the application will be overwritten. Note that if environment settings are specified for a custom tool (a model or a script) or a process within a model, these settings will always override application-level settings—regardless of whether they are set in the application's Environment Settings dialog box or via the tool's dialog box at the time of execution. See Chapter 5, 'Working with toolsets and tools', for more information on setting environments at the time of running a tool from its dialog box.

Model/Script environment settings

Environment settings specified for the application are passed down to, and are used by, models and scripts, unless environment settings are specified for a specific model or script. Settings specified for a model or a script override settings specified for the application.

Environment settings for the model are specified in the Environments tab of the Model Properties dialog box. This dialog box is accessible via the Model menu within the ModelBuilder window or by right-clicking the model in its toolbox and clicking Properties.

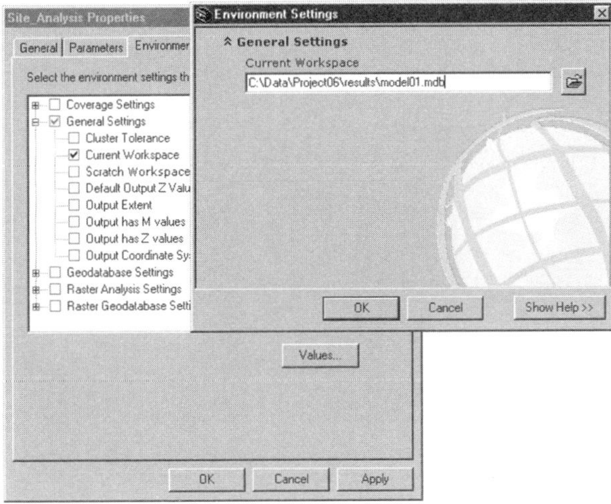

Within a script, environment settings can be specified and will be applied to all tools run within the script, as the following example shows.

```
# Set the workspace environment setting
GP.Workspace = "D:\\St_Johns\\data.mdb"
# Set the cluster tolerance environment setting
GP.ClusterTolerance = 2.5
# Calculate the default spatial grid index, half
#it and then set the spatial grid 1 environment
#setting
GP.SpatialGrid1 = pGP.CalculateDefaultGridIndex
("roads") / 2
# Clips the roads by the urban area feature class
GP.Toolbox = "Analysis"
GP.Clip("roads","urban_area","urban_roads")
```

Process environment settings

Processes in a model or script can have their own environment settings that will override those settings specified for the entire model or script as well as for the application.

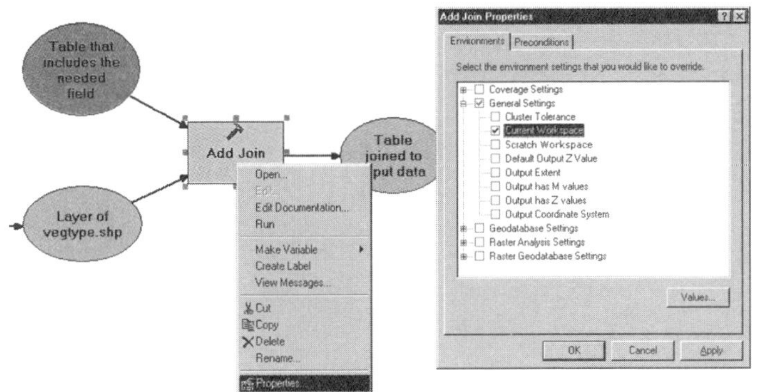

Accessing the Environment Settings dialog box for a process in a model. Settings specified either for the model or for the application will be overridden. In the example above, the inputs to and result from this process only will be placed in the workspace specified.

Within a script, you can set environment settings for one process, then reset or alter settings for another process if desired.

For more information

See the tasks that follow for more information on how to specify environment settings for the application, a model, and a process within a model and for information on each environment setting that can be specified. See Chapter 3 of *Writing Geoprocessing Scripts With ArcGIS* for more information on specifying environment settings within a script.

Specifying environment settings

Default environment settings are applied when you run tools. Rather than taking the default, you may want to specify your own environment settings.

Changing the default settings that will be used is a prerequisite to performing geoprocessing tasks. You may only be interested in analyzing a small piece of a geographic area, such as changing the extent for results, or you may want to write all results to a specific location—for example, changing the current workspace or the scratch workspace.

Environment settings can be set hierarchically, meaning that they can be set for the application so they will apply to all appropriate tools, be set within a particular model or script, or be set for a process within a model. ►

Tip

An alternative way to access the Environment Settings dialog box

Right-click the ArcToolbox window and click Environments.

Specifying settings for the application

1. Click Tools on the Main menu of the application you are using, and click Options.
2. Click the Geoprocessing tab.
3. Click the Environments button.

   The Environment Settings dialog box is opened.
4. Expand the contents of the section containing the environment settings you wish to alter.
5. Set values for appropriate environment settings.
6. Click OK.

   Environment settings specified will be applied to all tools that accept them.

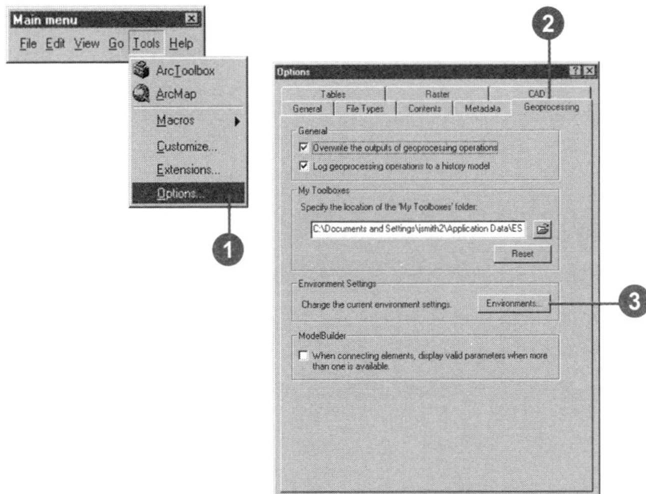

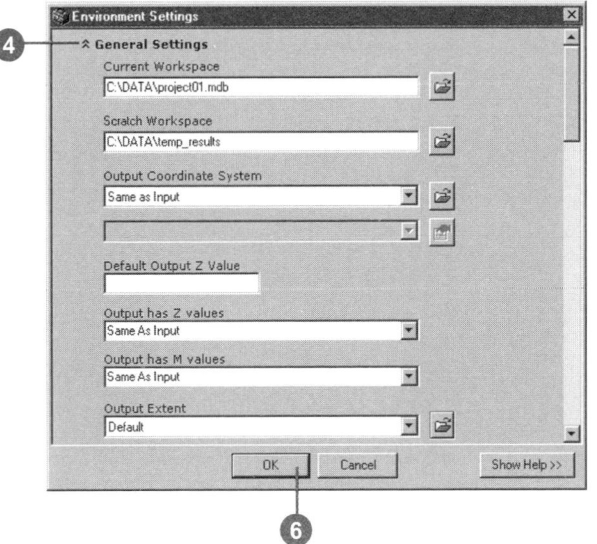

All tools will utilize the environment values set for the application unless they are manually set elsewhere. If you set environment values for a model, those values will supersede the ones set for the application.

If you set environment settings for a process inside a model, those settings will override settings for the model and settings for the application.

When specifying environment settings for a model or a process within a model, you can check a group of settings to override all values in that group, or you can expand the contents of a particular group of settings and check specific ones for which you want to override the default value.

Tip

An alternative way to access a model's Environment Settings dialog box

Right-click the model in its toolbox and click Properties, then click the Environments tab.

Specifying environment settings for a model

1. Right-click the model in its toolbox and click Edit.
2. Click the Model menu and click Model Properties.
3. Click the Environments tab.
4. Expand the settings you would like to override.
5. If you want to override all settings for a particular group of environment settings, such as all General Settings, simply check the box to the left of the General Settings section.

   If you want to override a specific setting within a particular group of environment settings, expand the group of settings that contains the setting you wish to override, then check the setting.
6. Click Values.

   The checked settings are displayed. ▸

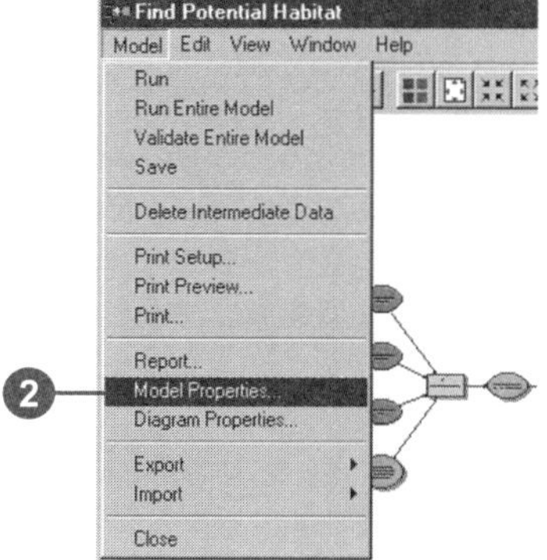

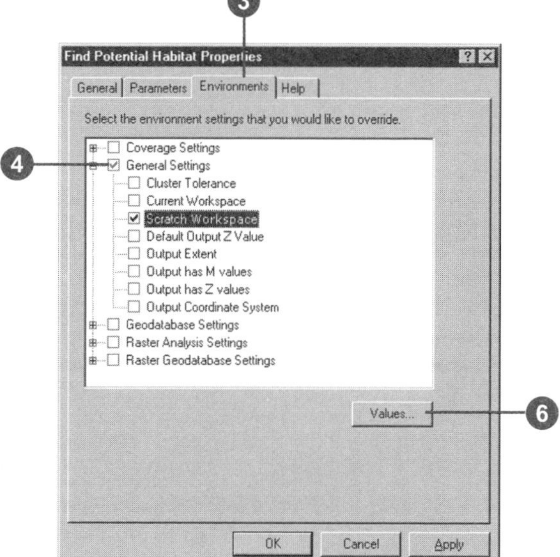

See Also

If you want to set values for environment settings within a standalone script, refer to Writing Geoprocessing Scripts With ArcGIS *for more information.*

7. Expand the categories of settings displayed and set an alternative value for the chosen settings.

   The values you specify will apply to all applicable tools within the model and will override any settings specified for the application.

8. Click OK on the Environment Settings dialog box.

9. Click OK on the tool's properties dialog box.

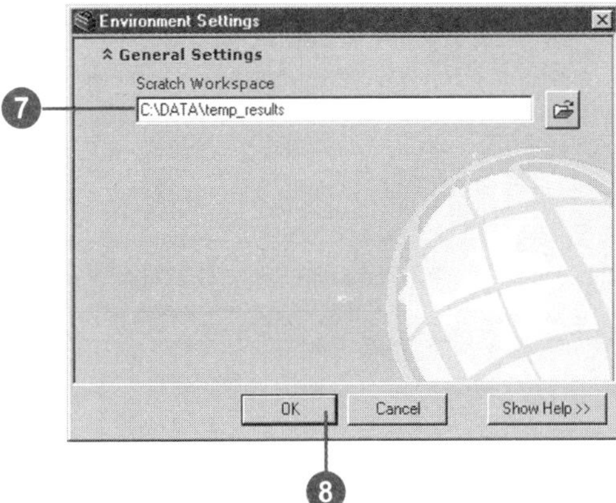

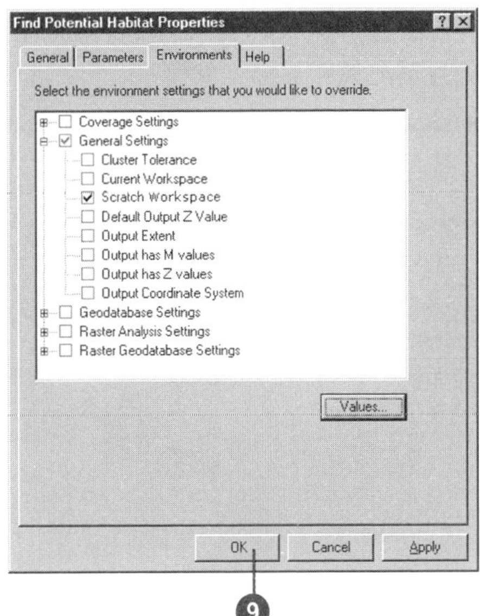

Specifying environment settings for a process

1. In the ModelBuilder window, right-click the tool of the process you want to change default environment settings for and click Properties.
2. Click the Environments tab.
3. Either check a group of settings if you wish to override all settings in the group or expand a group of settings and click a particular setting.
4. Click Values.

   The settings you checked are displayed in an Environment Settings dialog box.
5. Change the default value set for each environment setting to be more appropriate for the process.
6. Click OK, then click OK on the properties dialog box.

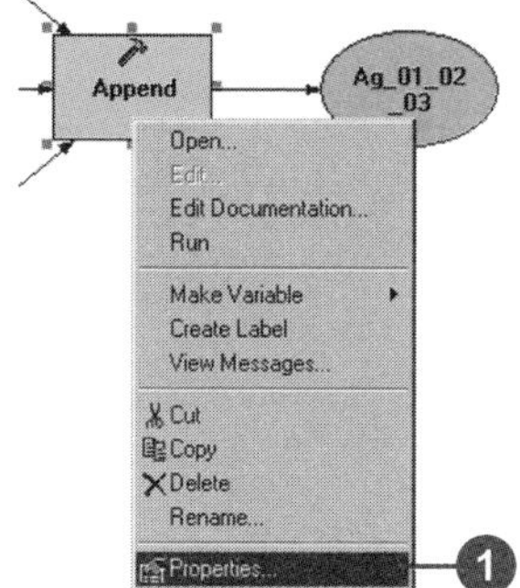

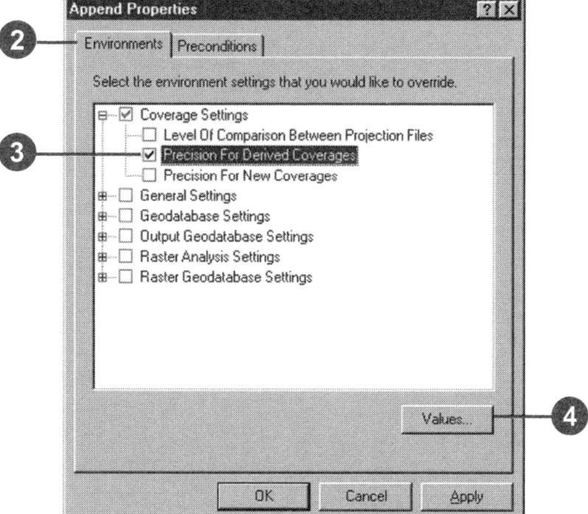

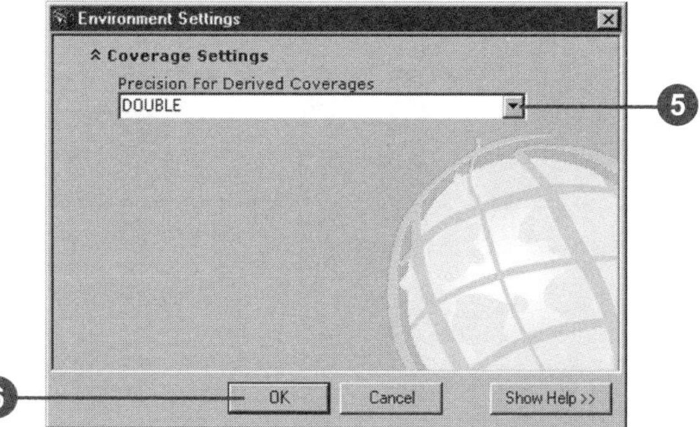

Specifying general settings

The General Settings section of the Environment Settings dialog box contains settings that are applicable to most output data types. Included in this section are the following settings that you can change: the current workspace, the scratch workspace, the output coordinate system, the default output z-value, whether outputs will have z- or m-values, the extent and snap raster, and the cluster tolerance. Each of these settings is discussed in the pages that follow.

Tip

An alternative way to access the Environment Settings dialog box

Right-click the ArcToolbox window and click Environments.

See Also

See 'About environment settings' earlier in this chapter for information on setting environments for the application, for a model, or for a process within a model.

Viewing general settings for the application

1. Click Tools on the Main menu of the application you are using, then click Options.
2. Click the Geoprocessing tab.
3. Click the Environments button.

   The Environment Settings dialog box is opened.
4. Click General Settings to view the settings.
5. Click OK.

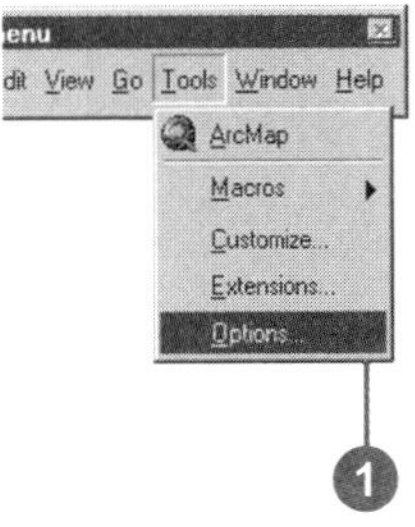

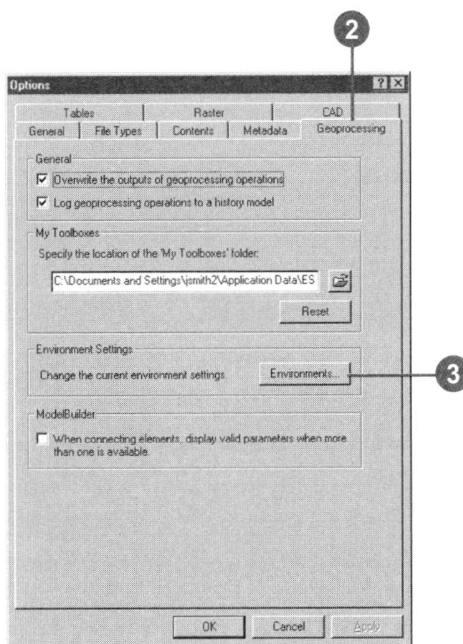

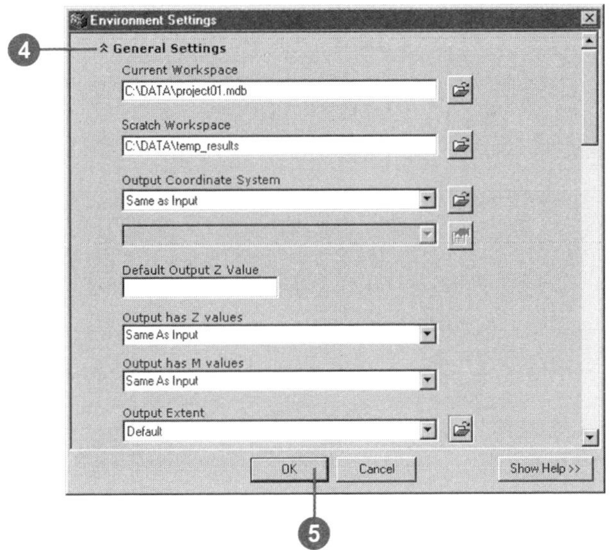

Working with workspaces

A workspace is the work area used during a geoprocessing session. Within the computer file system, the workspace is a directory containing geographic data. There are three directory structures that are supported: folders, geodatabases, and geodatabase feature datasets. There are two workspaces that can be set in the Environment Settings dialog box: the current workspace and the scratch workspace.

The current workspace

Setting the current workspace allows you to set the location from which inputs are taken and outputs are placed when running tools. It allows you to type the name of inputs and outputs for tools rather than also having to type the path or browse to the location of the data each time you run a tool.

The scratch workspace

If you have a scratch workspace set, outputs from tools will be placed in the location specified with a default name. If you remove the path given for the output parameter and simply type a name, the result will be saved into your current workspace if it is set or to the location of the input data if it is not set.

In your day-to-day work flow, you can accept the default path and name for results that you do not care to keep so they are placed in the scratch workspace, and type a meaningful name for outputs from tools that you want to keep so they are placed in the location of the current workspace.

If the current workspace and the scratch workspace are not set, the outputs from tools will, by default, be placed in the location of the input data. The exception is the Feature Class to Coverage tool, when the input is a geodatabase feature class. In this case, with no workspaces set, the output will be placed in the same location as the input geodatabase.

Using workspaces in ArcMap

When running individual tools in an application with a display via the ArcToolbox window, such as ArcMap, you can simplify the process of running tools by following the steps below.

1. Set workspaces

Set a current workspace and a scratch workspace in the Environment Settings dialog box. See the task 'Viewing general settings for the application' earlier in this chapter for information on accessing the Environment Settings dialog box.

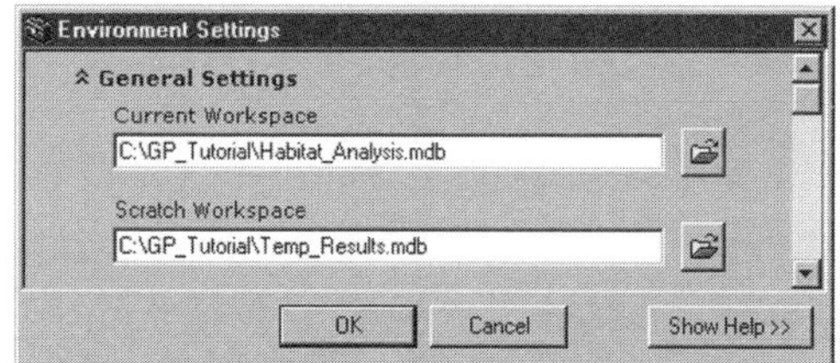

2. Check to create temporary results

Optionally, on the Geoprocessing tab of the Options dialog box, check to create temporary results when running tools from their dialog boxes or from the command line. You should do this if you know that you will be generating results from tools that you will likely not want to keep after the application is closed, such as when experimenting with different outcomes to produce a desirable result.

Note that the Add results of geoprocessing operations to the display check box must first be checked in order to check Results are temporary by default. If results are not set to be added to the display, they will always be permanently created.

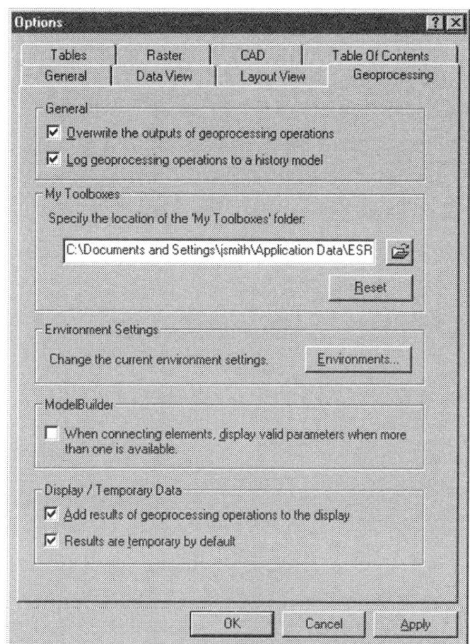

The Geoprocessing tab of the Options dialog box, accessed by clicking Options on the Tools menu

3. Run tools

After setting the workspaces (current and scratch) and, optionally, checking to create temporary results, you are ready to run tools.

Open a tool's dialog box and supply the name of a dataset contained within the current workspace, then click another parameter in the dialog box. The path to the current workspace set will be supplied for the input parameter, and the scratch workspace path will be supplied for the value of the output parameter, with a default name.

Fill in the rest of the parameters and click OK to run the tool.

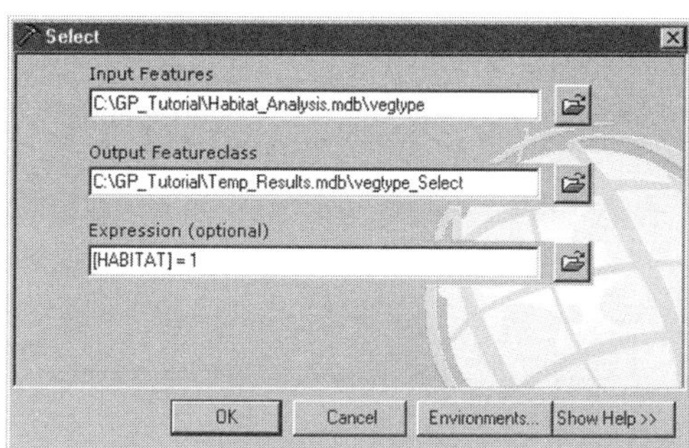

All results are temporarily created and placed in the location set for the scratch workspace.

When you want to create a result you want to keep, simply clear the default path and name given for the output parameter and type a name. It will be saved in the location set for the current workspace. As all your important results are saved to the location of the current workspace, you can simply delete the contents of the folder or geodatabase that you have specified as the scratch workspace.

4. Make results permanent

If you want to permanently save a certain temporary result, right-click the layer in the table of contents and click Make Permanent. The result will be permanently saved in the location of the scratch workspace.

Alternatively, uncheck Results are temporary by default in the Geoprocessing tab of the Options dialog box, and always create permanent results in your scratch workspace.

Using workspaces in a model

Within a model, you may want your intermediate results to be saved in one directory and your final results to be saved in another. In the Environment Settings dialog box for the application or the model, set the current workspace to one location and the scratch workspace to another location. In the model, leave the default path and name supplied for all intermediate outputs from tools, then simply type the name for the final result, without specifying a path. It will be saved to the location of the current workspace. See 'About intermediate data' in Chapter 9 for more information on managing intermediate data.

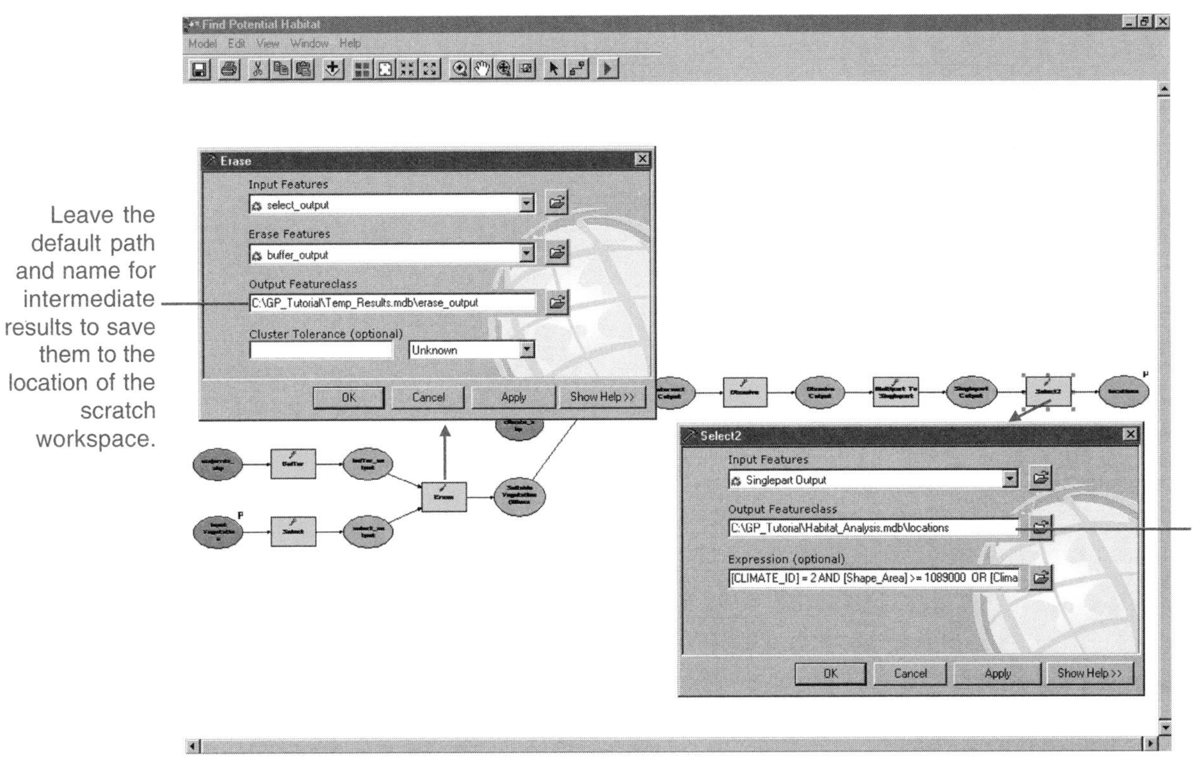

Leave the default path and name for intermediate results to save them to the location of the scratch workspace.

Right-click the final derived data element and uncheck intermediate. Type a name for the final output. It will be placed in the location of the current workspace. When you delete intermediate data within the model or run the model via its dialog box, this final result will not be deleted.

See Also

See 'Working with workspaces' earlier in this chapter for more information on setting the current and the scratch workspaces.

Setting the current workspace

1. Follow steps 1–4 of 'Viewing general settings for the application'.
2. Type the path to the workspace you want to use or click the Browse button to navigate to a location on disk.

   The workspace can be a geodatabase, a folder, or a geodatabase feature dataset.
3. Click OK.

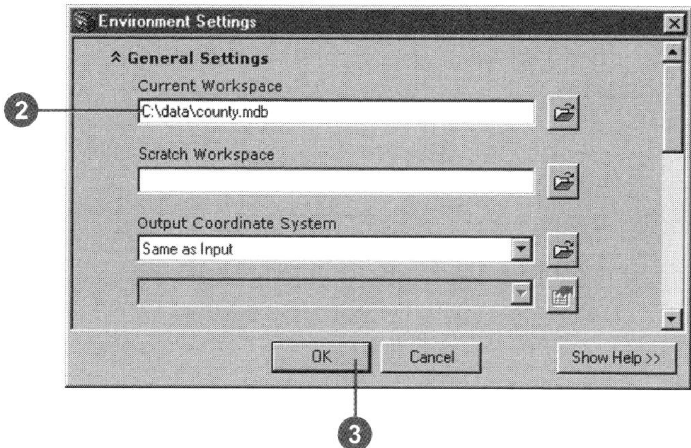

See Also

See 'Working with workspaces' earlier in this chapter for more information on setting the current and the scratch workspaces.

Setting the scratch workspace

1. Follow steps 1–4 of 'Viewing general settings for the application'.
2. Type the path to the Scratch Workspace you want to use, or click the Browse button to navigate to a location on disk.

   The workspace can be a geodatabase, a folder, or a geodatabase feature dataset.
3. Click OK.

   All output results from running tools will automatically be placed in the scratch workspace with a default name.

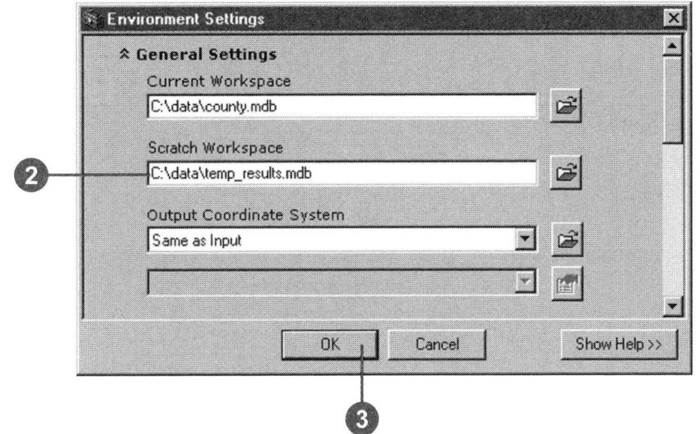

You can specify the coordinate system that will be applied to spatial data created by running tools. The *coordinate system* (geographic or projected) defines the location of the spatial data on the earth.

Precedence rules dictate which coordinate system is applied to outputs from running tools:

- If the output resides inside a feature dataset, the coordinate system of the feature dataset will always be used.
- If the coordinate system is set in the Environment Settings dialog box and the output does not reside inside a feature dataset, the coordinate system set in the Environment Settings dialog box is used.
- If there is no value set for the Output Coordinate System environment setting and the output does not reside inside a feature dataset, the coordinate system of the first input to the tool is used.

When setting the coordinate system to apply in the Environment Settings dialog box, the default specifies that the coordinate system of outputs be the same as the first input to the tool (Same as Input). You can change the default setting so that outputs are projected ▸

Setting the output coordinate system

1. Follow steps 1–4 of 'Viewing general settings for the application'.
2. Click the Output Coordinate System dropdown arrow and click the desired option—Same as Input, Same as Display, As Specified Below, or Same as Layer "layer name".

   If you choose As Specified Below, type the coordinate system information into the Output Coordinate System text box if it is known, or click the Browse button to set up the coordinate system information.

   Same as Display takes the coordinate system from the active data frame in ArcMap. In ArcCatalog you must be previewing data to see this option.

   Same as Layer "layer name" is only available in ArcGIS Desktop applications with a display, such as ArcMap or ArcScene.

   Alternatively, click the Browse button next to the dropdown list to apply the coordinate system information from an existing dataset.
3. Click OK on the Environment Settings dialog box.

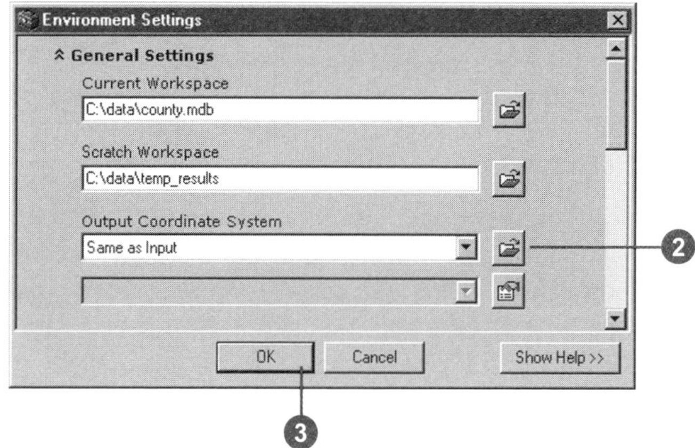

on the fly from the coordinate system of the input to the coordinate system specified (using As Specified Below) or so that the outputs are projected on the fly from the coordinate system of the input to that of the display, which is the active data frame in ArcMap. Alternatively, set the coordinate system for outputs to be the same as a layer in the table of contents of the display (using Same as Layer), or click the Browse button next to the dropdown list to take the coordinate system information from an existing dataset. ►

See Also

Before running geoprocessing tools that create feature datasets or feature classes within a geodatabase, you can set up spatial domain information—the allowable coordinate range for x,y coordinates, m (measure)-values, z-values, and precision information. See 'Specifying geodatabase settings' in this chapter for more information.

If your input feature classes contain *z-values* in the feature geometry or you have set Output has Z values to ENABLED in the General Settings section of the Environment Settings dialog box, you can set up a default output z-value that will be given to each vertex in the output feature class after a tool is run if no z-value can be attained from the input.

Each vertex in the feature class will contain an x, y, and z coordinate. The value for the z coordinate will be based on the input to the tool. If the input does not contain z-values, the value set for the Default Output Z Value setting in the Environment Settings dialog box will be applied as the z-value to all vertices in the output feature class. If no default output z-value is set, the value will be the minimum value set for the z-domain in the Environment Settings dialog box. If there is no z-domain set in the Environment Settings dialog box, the z-value will be taken from the minimum value set for the z-domain of the input. If the input does not contain a z-domain, zero is used as the z-value. ►

Setting the default output z-value

1. Follow steps 1–4 of 'Viewing general settings for the application'.
2. Type a z-value to apply to all features in feature classes created by running applicable tools. This value will be applied if a z-value cannot be used from the input to the tool.
3. Click OK.

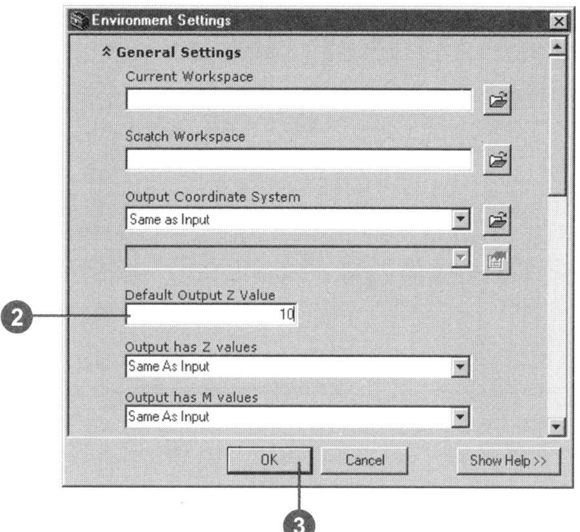

Z-values, such as building heights, may be represented on the z-axis in a three-dimensional x,y,z coordinate system.

If you want your feature class outputs to be able to store z-values for each vertex, you must specify this before running a tool. This capability cannot subsequently be added.

The default (Same As Input) takes the state of the input data to a tool—if the input can store z-values, the output will also be able to store z-values.

By setting the option to Enabled, output feature classes will be able to store z-values, regardless of whether the input data to a tool can.

By setting the option to Disabled, outputs from tools won't be able to store z-values, even if the input to a tool can.

If you set this option to Enabled or Same As Input, the z-values applied to each vertex in the output will be based on the input if they are present. If z-values are not present on the input and there is no value set for the Default Output Z Value in the Environment Settings dialog box, the minimum value in the z-domain is used as the z-value. If there is no z-domain in the input, zero is given as the z-value. ►

Specifying whether the output has z-values

1. Follow steps 1–4 of 'Viewing general settings for the application'.
2. Click the Output has Z values dropdown arrow and click Enabled, Disabled, or Same As Input.
3. Click OK.

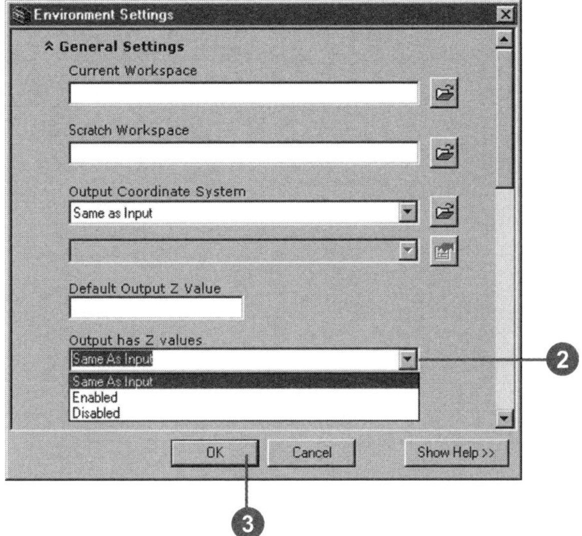

M-values on polylines are used in linear referencing to imply a route measure. Route measures can be used to represent a location along a route, such as mileages along a highway.

If you want your output feature classes to be able to store m-values for each vertex, you must specify this before running a tool. The ability to store m-values cannot be subsequently added.

The default (Same As Input) takes the state of the input data to a tool—if the input can store m-values, the output will also be able to store m-values.

By setting the option to Enabled, output feature classes will be able to store m-values, regardless of whether the input data to a tool can.

By setting the option to Disabled, outputs from tools won't be able to store m-values, even if the input data can.

If you set this option to Enabled or Same As Input, the m-values applied to each vertex in the output will be based on the input if they are present, unless the tool will set or reset measure values, in which case input measures will be ignored and new ones will be calculated.

If m-values are not present in the input, they will be set to Not a Number (NaN). ►

Specifying whether the output has m-values

1. Follow steps 1–4 of 'Viewing general settings for the application'.
2. Click the Output has M values dropdown arrow and click Enabled, Disabled, or Same As Input.
3. Click OK.

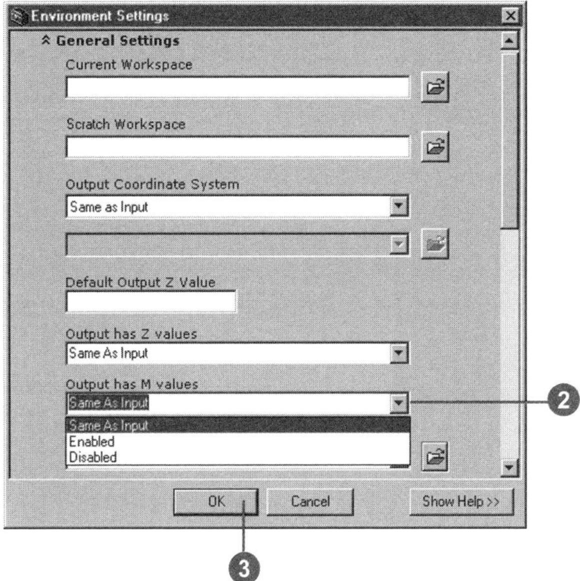

By specifying an extent, you define the area of interest for results from running tools.

In the case of feature data, all input features to a tool that pass through the area of interest will be included in the calculation.

For raster data, results from running tools will be contained within the extent set.

The extent is a rectangle, specified by identifying the coordinates of the window in map space.

The default extent is set to Default. With this extent set, the tool determines the extent of the output based on the extent of the input.

Intersection of Inputs sets geoprocessing to only be performed where all layers overlay, which is the minimum of the inputs.

Union of Inputs sets the extent of the results to be the same as the combined extent of inputs to a tool.

Same as Layer sets the extent to be the same as an existing layer in the table of contents.

As Specified Below sets a custom extent. Specify the four values, in map units, for the extent you wish to apply. You must specify the Top, Right, ▸

Changing the default extent

1. Follow steps 1–4 of 'Viewing general settings for the application'.
2. Click the Output Extent dropdown arrow and select the required option.

   If you click As Specified Below, type the x,y coordinates that will be used as the extent for all results from geoprocessing operations (Left = minimum x coordinate, Bottom = minimum y coordinate, Right = maximum x coordinate, and Top = maximum y coordinate).

   Alternatively, click the Browse button and browse to a dataset from which to take the extent.

   Same as Layer is only available in ArcGIS Desktop applications with a display, such as ArcMap or ArcScene.
3. Click OK.

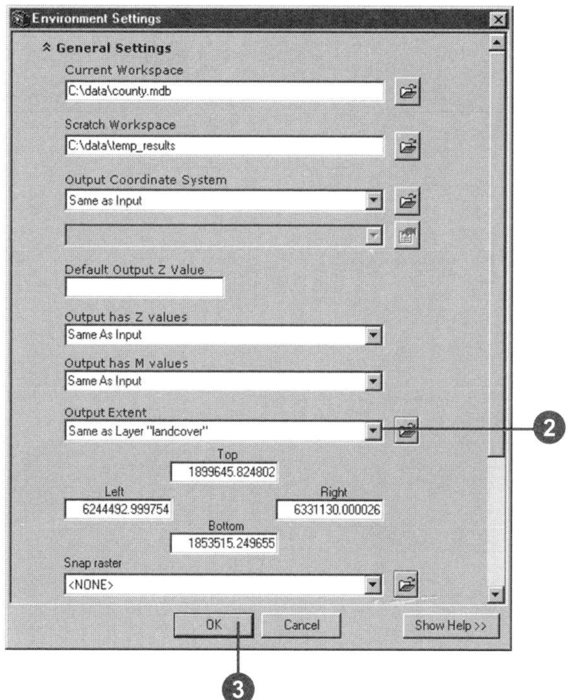

Bottom, and Left values. An easy way to do this is to set the extent to that of an existing dataset, then modify the values to the required extent.

Same as Display sets the extent to be the same as the area visible in the display of the application. If you are working in ArcCatalog, you have to be previewing data to see this option in the dropdown list.

Alternatively, you can click the Browse button and browse to a dataset on disk to take the extent from. ▸

Setting the extent to Same as Display

1. Click the Zoom In tool and zoom in on the display to the area in which you wish to perform geoprocessing.

   Note: If you are working in ArcCatalog, click a dataset, then click the Preview tab first.

2. Follow steps 1–4 of 'Viewing general settings for the application'.
3. Click the Output Extent dropdown arrow and click Same as Display.
4. Click OK.

   The extent of all results will be set to that of the display.

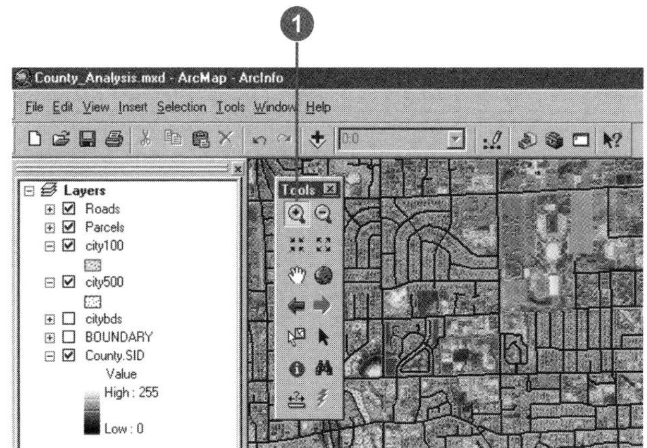

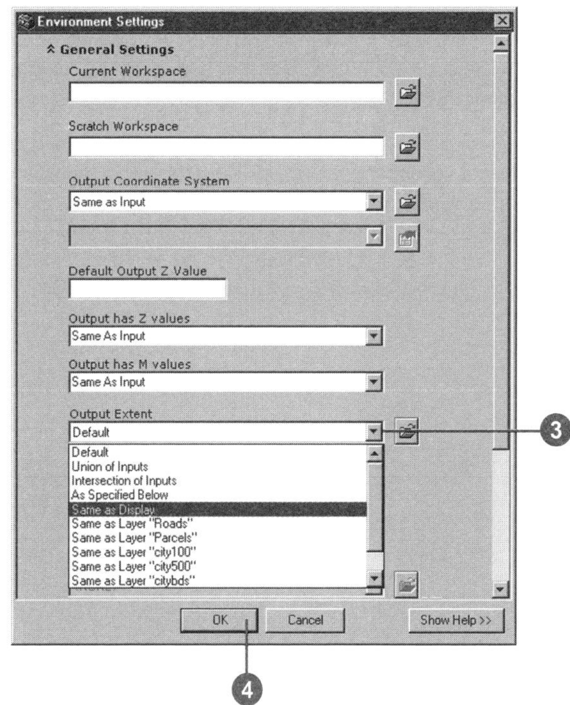

Setting a snap raster ensures that the cell alignment of the analysis extent will match accurately with an existing raster. This is done by snapping the lower-left corner of the specified analysis extent to the lower-left corner of the nearest cell in the snap raster and snapping the upper-right corner of the specified analysis extent to the upper-right corner of the nearest cell in the snap raster. ▸

Tip

Working with raster data

When using raster data with different cell alignments together in the same tool, nearest neighbor interpolation will be used to match the different cell alignments during analysis. This could cause unwanted artifacts with continuous data sources and is not recommended. To ensure this will not happen, it is always best to try to create raster data with the same cell alignments.

Setting a raster to snap all results to

1. Follow steps 1–4 of 'Viewing general settings for the application'.
2. Click the Output Extent dropdown arrow and click either Same as Display or As Specified Below, or click the Browse button to browse to a dataset to set the extent for analysis results. Alternatively, click the dropdown list and click a layer to take the extent from.

   The snap raster setting is only enabled when one of the above options is set for the extent. This is because these are the only options in which the analysis extent specified may not have the same cell alignment as your raster data.
3. Type the path and name of the raster to snap to.

   Alternatively, click the Browse button, click the Look in dropdown arrow, and navigate to the location of the raster to use. Click the raster, then click Add.
4. Click OK.

   All raster results will be snapped to the cell alignment of the raster specified.

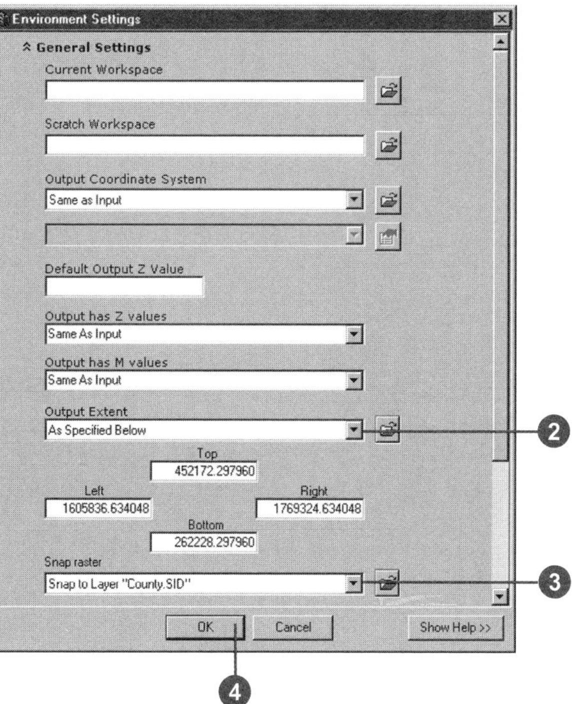

The cluster tolerance is the distance range in which all vertices and boundaries in a feature class are considered identical, or coincident.

To minimize error, the cluster tolerance you choose should be as small as possible, depending on the precision level of your data.

By default, the minimum possible tolerance value is calculated in the units of the spatial reference of the input. This default value is applied if you don't specify a cluster tolerance before running a tool.

Tip

Calculating the default cluster tolerance

You can calculate the default cluster tolerance for a feature class using the Calculate Default Cluster Tolerance tool in the Featureclass toolset of the Data Management Tools toolbox.

See Also

Type "Topology Basics" in the Search tab of the online Help system for more information on the cluster tolerance.

Setting the cluster tolerance

1. Follow steps 1–4 of 'Viewing general settings for the application'.
2. Type a value for the Cluster Tolerance and set the units.

   When running subsequent tools, the cluster tolerance applied to the results will first be converted to the units of the input data if the units of the cluster tolerance and the input data are different.
3. Click OK.

   The cluster tolerance set here will be used by all tools that use a cluster tolerance.

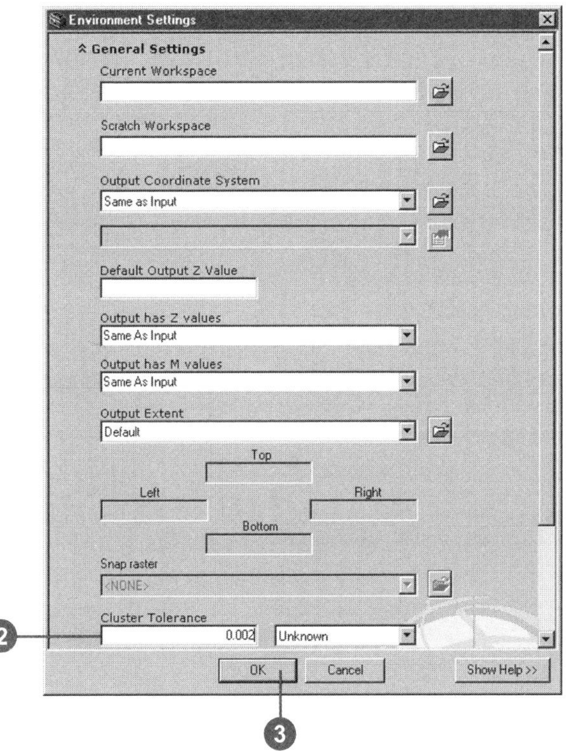

Specifying coverage settings

The environment settings in the Coverage Settings section of the Environment Settings dialog box apply to coverage tools only. The default value for the following settings can be changed: the precision for new or derived coverages and the level of comparison between projection files. Each of these settings is discussed in the pages that follow. ▸

Tip

An alternative way to access the Environment Settings dialog box

Right-click the ArcToolbox window and click Environments.

See Also

See 'About environment settings' earlier in this chapter for information on setting environments for the application, a model, or a process in a model.

Viewing coverage settings for the application

1. Click Tools on the Main menu of the application you are using and click Options.
2. Click the Geoprocessing tab.
3. Click the Environments button.

   The Environment Settings dialog box is opened.
4. Click Coverage **Settings** to view the settings for which values can be changed.
5. Click OK.

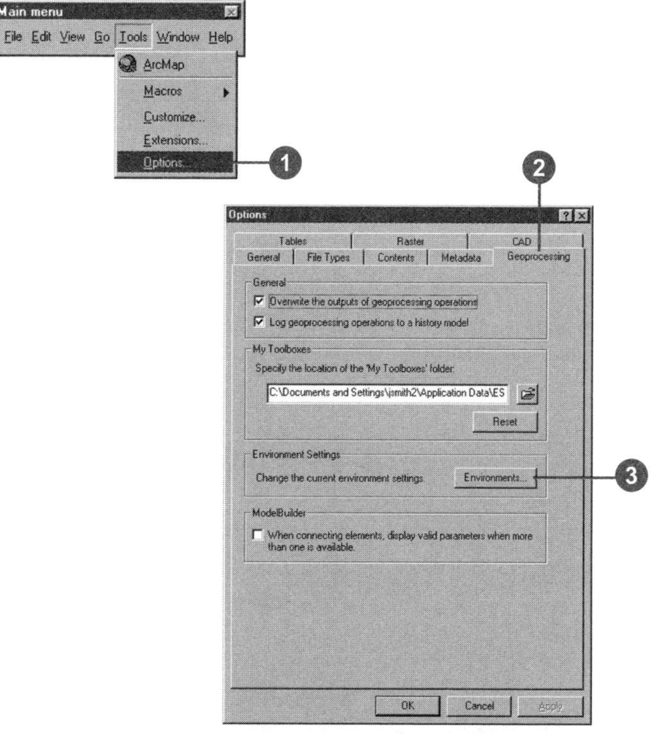

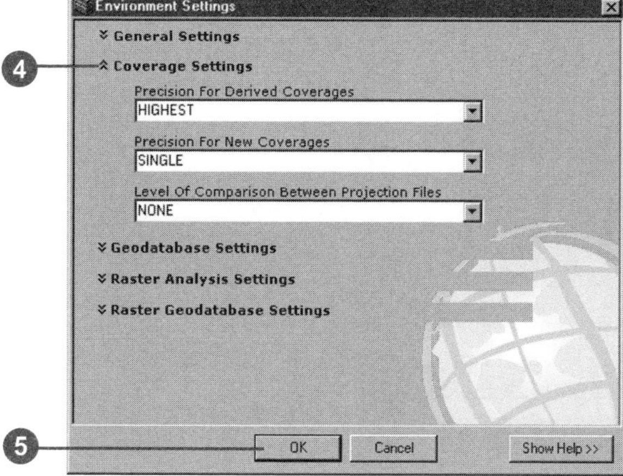

Before running coverage tools, you can set the degree of similarity between input projection files required for a match to occur. Use this option if you want to validate the input projections before running coverage tools.

The following settings are available:

NONE—no comparison of projection files is made. Any combination of projection information will result in a match. Tools will run regardless of the projection of the inputs. This is the default.

PARTIAL—at least one projection file must be defined; the others can be unknown and will result in a match. Defined projection files must be identical.

FULL—all projection information must be specified and identical in the projection file of each input dataset.

Similar projections and parameters that are defined in different ways will not match. For example, UNITS METERS and UNITS1, which are equal, will not match. Defining Universal Transverse Mercator (UTM) and State Plane projections by their central meridians or parallels will not match with the same projection defined using the ZONE option.

Setting the level of comparison between projection files

1. Follow steps 1–4 of 'Viewing coverage settings for the application'.
2. Click the Level Of Comparison Between Projection Files dropdown arrow and choose the level of comparison you require before tools are run—NONE, PARTIAL, or FULL.
3. Click OK.

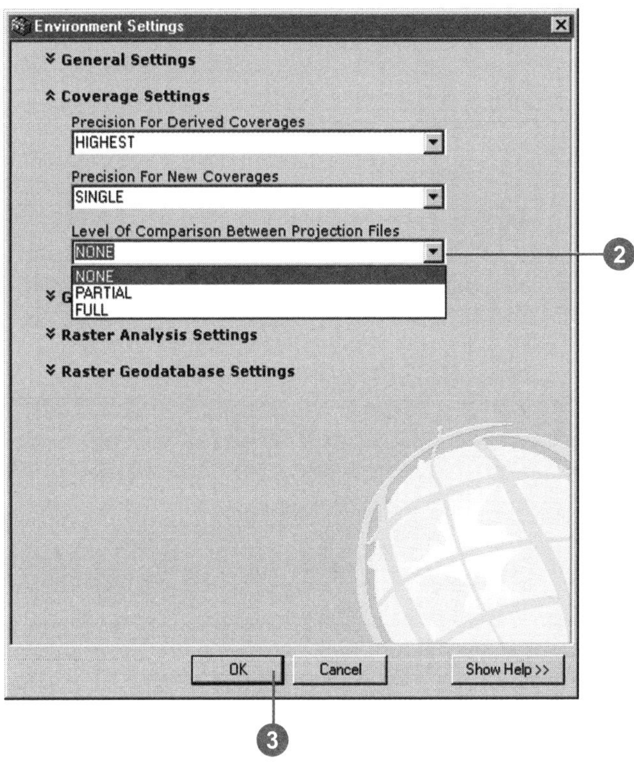

About coverage precision

What is coordinate precision?

Coordinate *precision* refers to the mathematical exactness of a coordinate and is based on the possible number of significant digits that can be stored for each coordinate. ESRI ArcInfo coverages are stored in either single- or double-precision coordinates. *Single precision* stores up to seven significant digits for each coordinate. This means that values of 4,999,999.6 units or 5,000,000.4 units are rounded off to 5,000,000 units when stored in single precision. *Double precision* stores up to 15 significant digits—typically 13 to 14—and retains a mathematical precision of much less than 1 meter at a global extent. Note that mathematical precision does not, by itself, define accuracy. It is, however, a major factor in coverage resolution.

When to use double precision

It is probably best to use double precision if the coordinate resolution of a coverage must be maintained past six significant digits. Here are some of the conditions that meet this requirement:

- Your coverage requires a high level of accuracy—for example, a parcels layer.
- You need to use a map projection whose coordinate values exceed the available coordinate precision.
- You are changing datums—for example, from NAD27 to NAD83.

How does precision work?

The computer can only discern a limited number of decimal places, depending on the precision being used. For single-precision coordinates, the computer assumes that values such as 1.2345678 and 1.23456789 are equal because numbers beyond the seventh digit are ignored. You might imagine a small halo surrounding each coordinate that is equal to the resolution of coordinate storage on the computer. Coordinates with overlapping halos are seen by the computer to represent the same location.

Coverages stored with either single- or double-precision coordinates can be used interchangeably; for example, you can display single- and double-precision coverages over each other, overlay them, merge them, and so on. This capability, however, does not replace the need to be clear and deliberate about how you encode, store, and manage each coverage. Base your decision of which type of precision to use on the desired level of coordinate accuracy to be maintained for each coverage.

The coordinate systems for many map projections use large coordinate values—for example, State Plane and UTM coordinates contain values in the 2 million to 6 million range. In these cases, you can use double-precision coordinates to maintain accuracy of less than one unit, extending beyond the decimal point.

As mentioned, double-precision coverages can store up to 15 significant digits. This is sufficient to map any point on the earth to better than a millimeter of accuracy. However, double-precision coverages require additional disk space for storing coordinates. Thus, it may be worthwhile to store coverages that require high levels of accuracy, such as parcels, in a double-precision coordinate system and coverages needing less accuracy, such as soils, in a single-precision coordinate system.

Precision rules

You can define the coordinate precision of new and derived coverages created. These two rules are creation and processing.

The creation rule specifies the precision with which to create all new coverages. Any time a new coverage is created, the coordinate precision of the new coverage is defined by the current creation rule. Single precision is the default. Thus, if you want to create double-precision coverages, you must set the precision for new coverages to DOUBLE.

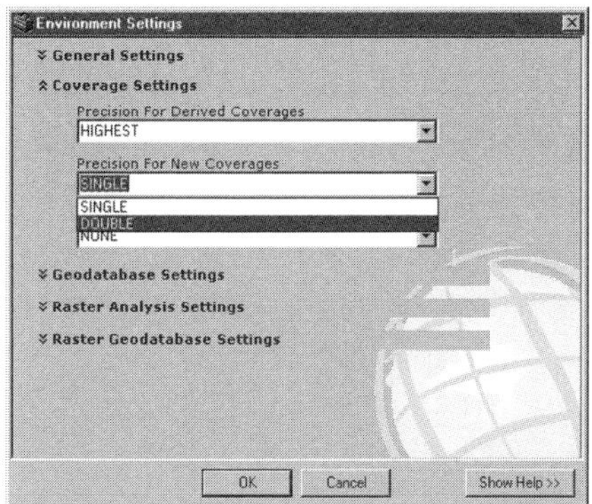

To always create new coverages in double precision, set the Precision For New Coverages to DOUBLE in the Environment Settings dialog box. All new coverages created by tools that create coverages, such as the Create Coverage tool, will be in double precision.

The processing rule specifies the precision with which to create all derived coverages. When a coverage is derived from one or more existing coverages, such as the result of the Buffer or Update tools, the coordinate precision of the derived coverage reflects the current processing rule. For example, the output coverage precision can be the highest of a set of input coverages. Thus, if all input coverages are in a single-precision coordinate system, the output coverage will be single precision; if at least one of the coverages is double precision, the output coverage is double precision. By setting LOWEST for the precision of derived coverages, the precision of derived coverages will be taken from the input coverage with the lowest precision. Thus, if all input coverages are in double precision except one that is in single precision, the derived coverage's precision will be single.

To create double-precision coverages regardless of the precision of the input coverages, set the processing rule to DOUBLE before you begin.

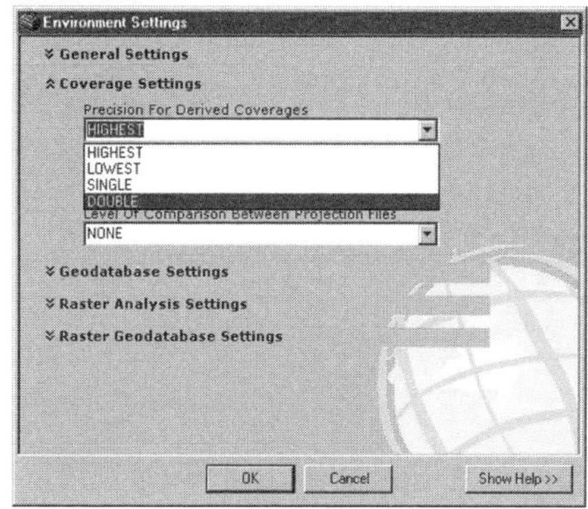

To always create your derived coverages in double precision, regardless of the precision of the input coverages, set the processing rule to DOUBLE in the Environment Settings dialog box before you run geoprocessing tools.

Setting precision for coverages

Before you begin using tools in the Coverage Tools toolbox, you should determine what the precision of your output coverages should be.

New coverages are those that are created, for example, using the Create Coverage tool. Derived coverages are those that are derived from tools that accept input data to derive the output coverage, for example, coverages created by running the Buffer tool.

The default precision is set to SINGLE for new coverages that are created and HIGHEST for derived coverages. If you want new coverages that are created to be in double precision, set the Precision For New Coverages to DOUBLE. If you always want derived coverages to be in double precision, regardless of the precision of the input data, set the precision for derived coverages to DOUBLE.

See Also

See Chapter 7, 'Using the Command Line window', for information on specifying environment settings at the command line.

Setting the precision for new coverages

1. Follow steps 1–4 of 'Viewing coverage settings for the application'.
2. Click the Precision For New Coverages dropdown arrow and choose the level of precision you require, either SINGLE or DOUBLE.
3. Click OK.

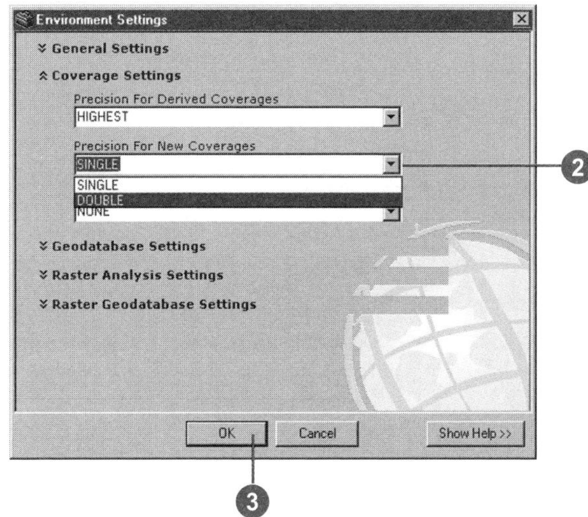

Setting the precision for derived coverages

1. Follow steps 1–4 of 'Viewing coverage settings for the application'.
2. Click the Precision For Derived Coverages dropdown arrow and choose the option for setting the precision— HIGHEST, LOWEST, SINGLE, or DOUBLE.
3. Click OK.

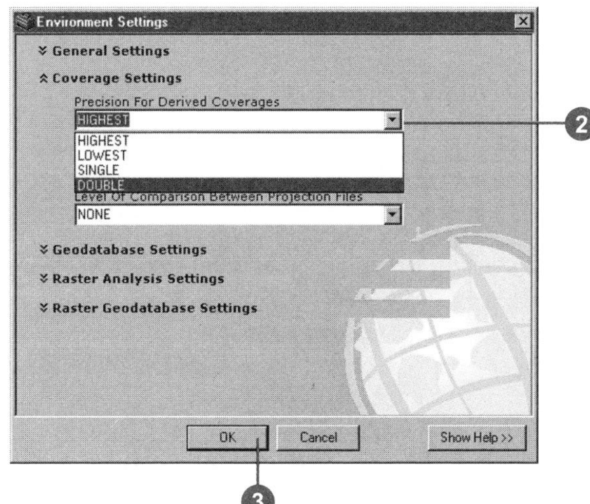

Specifying geodatabase settings

The Geodatabase Settings section of the Environment Settings dialog box provides environment settings for results placed in a geodatabase. Included are the following settings that can be changed: the output configuration keyword, output spatial grids, and spatial domain and precision information. Each of these settings is discussed in the pages that follow. ▸

Tip

An alternative way to access the Environment Settings dialog box

Right-click the ArcToolbox window and click Environments.

See Also

See 'About environment settings' earlier in this chapter for information on setting environments for the application, a model, or a process in a model.

Viewing geodatabase settings for the application

1. Click Tools on the Main menu of the application you are using and click Options.
2. Click the Geoprocessing tab.
3. Click the Environments button.

   The Environment Settings dialog box is opened.
4. Click Geodatabase Settings to view the settings for which values can be specified.
5. Click OK.

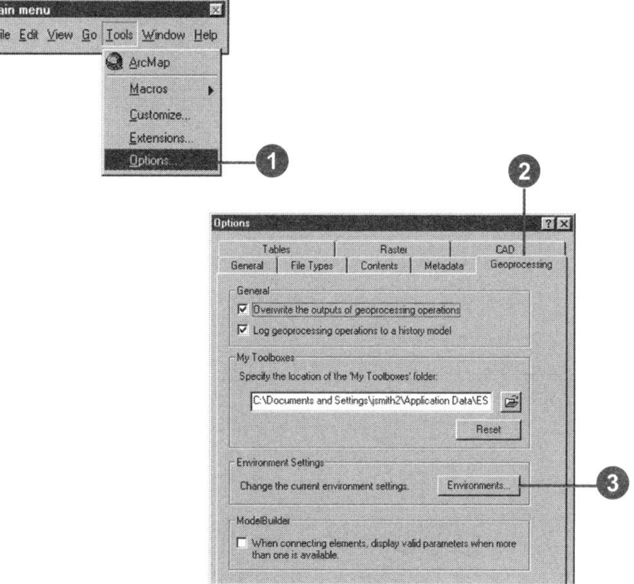

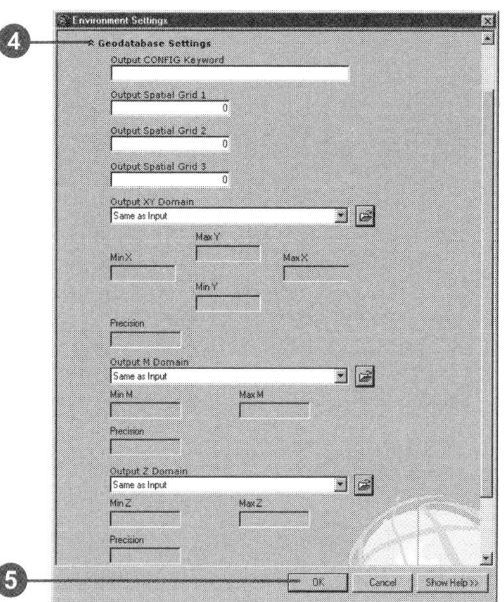

The configuration keyword specifies the storage parameters (configuration) for geodatabases in a Relational Database Management System (RDBMS)—ArcSDE only.

The database storage configuration enables your database administrator to fine-tune how the RDBMS in which your geodatabase is implemented stores data. The configuration parameters are grouped together into one or more configuration keywords, one of which is the default configuration keyword that specifies the default storage parameters.

When you create a dataset, such as a feature class, geometric network, or raster, you can tell the database which configuration keyword to use. Different configuration keywords can be used for different datasets. Normally, you can simply use the default configuration keyword, which geoprocessing tools will access automatically. In some cases, the database administrator may have created alternative configuration keywords for use when you create particular datasets or types of data in order to maximize their performance or fine-tune some aspect of how they are stored in the database. In this case, an alternative keyword can be specified. ▸

Setting the output configuration keyword

1. Follow steps 1–4 of 'Viewing geodatabase settings for the application'.
2. Type a configuration keyword.

   To use the default configuration keyword, leave the value for the setting blank.
3. Click OK.

   For more information, see the appropriate ArcSDE configuration and tuning guide.

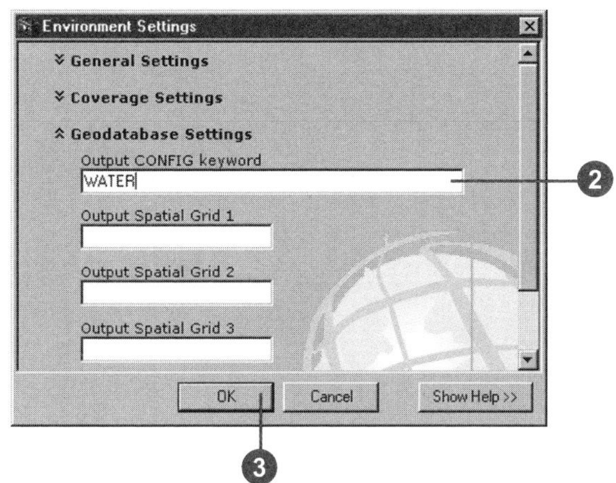

You can set up output *spatial grids* that are two-dimensional grid systems that span a feature class. When zoomed in on a feature class and performing a spatial search, only the features that fall in the necessary cells of the spatial grid are searched, enabling the quick location of features that might match the criteria of the spatial search.

Personal geodatabase feature classes require a single spatial grid. ArcSDE geodatabase feature classes can have up to three spatial grids. Each spatial grid must be at least three times the previous spatial grid. Feature size is an important factor in determining an optimum size for the spatial grid. Data that contains features of different sizes may require additional spatial grids to increase the speed of graphical queries.

If you are unfamiliar with creating spatial grids, use the default, which will calculate an appropriate size. A poorly defined grid size can increase the spatial search time. ►

Tip

Obtaining a value for Output Spatial Grid 1

Run the Calculate Default Spatial Grid Index tool located inside the Feature class toolset in the Data Management Tools toolbox.

Setting the output spatial grid

1. Follow steps 1–4 of 'Viewing geodatabase settings for the application'.
2. Type a value for the cell size of Output Spatial Grid 1 in map units.
3. If applicable, type values for the cell sizes of Output Spatial Grids 2 and 3 in map units, or leave the default (blank). The value for each spatial grid level should be at least three times larger than the previous level.

   These grids are only used if the input feature class used by a tool is an ArcSDE feature class.
4. Click OK.

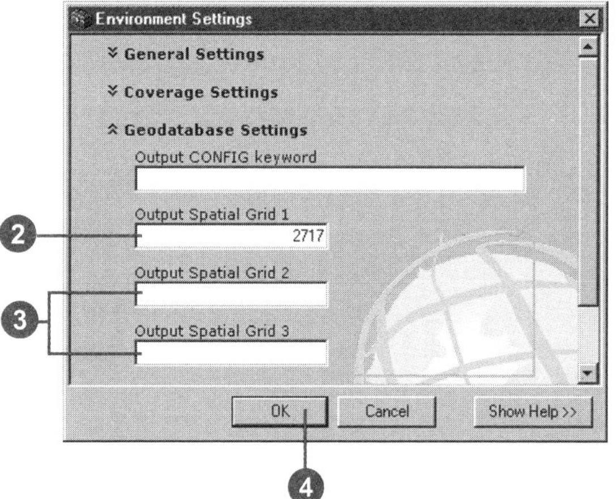

When results from running tools will be feature classes or feature datasets within a geodatabase, you can first set up the spatial domain, which includes the precision, which defines the level of detail that is maintained when data values are stored in a geodatabase, and the allowable coordinate range for x,y coordinates, m (measure)-values, such as mileages along a highway, and z-values, such as building heights.

Precedence rules dictate where the spatial domain information will come from when running tools:

- If the output will be placed inside a feature dataset, the x,y domain set for the feature dataset will always be applied—if an x,y domain is set in the Environment Settings dialog box it will be ignored in this case. M- and z-domain information is not determined by the feature dataset. This information can be set in the Environment Settings dialog box. If it is not set there, the m- and z-domains will be calculated by the tool, based on its inputs.
- If spatial domain information is set in the Environment Settings dialog box and the output does not reside ►

Setting the x,y domain

1. Follow steps 1–4 of 'Viewing geodatabase settings for the application'.
2. Click the Output XY Domain dropdown arrow and click the desired option—Same as Input, As Specified Below, Same as Display, or Same as Layer "layer name".

   If you choose As Specified Below, type the x,y domain information into the text boxes if it is known.

   The Same as Display option in ArcCatalog takes the x,y domain information from the current dataset being previewed. In ArcMap, this option takes the x,y domain information from the active data frame.

   Same as Layer "layer name" is only available in ArcGIS Desktop applications with a display, such as ArcMap or ArcScene.

   Alternatively, click the Browse button to apply the x,y domain information from an existing dataset.
3. Click OK on the Environment Settings dialog box.

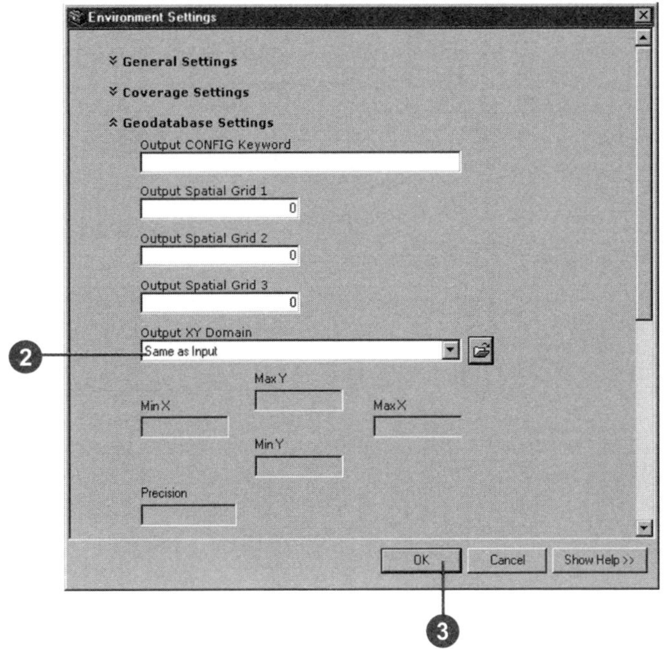

inside a feature dataset, the spatial domain information set in the Environment Settings dialog box is used.

- If there are no values set for the spatial domain settings in the Environment Settings dialog box and the output does not reside inside a feature dataset, the spatial domains will be calculated using the domains of the tool input.

When setting the spatial domain information to use in the Environment Settings dialog box, the default (for x,y coordinates or for m- and z-values) specifies that the spatial domain of outputs be the same as the first input to the tool (Same as Input). You can change the default setting so that outputs are projected on the fly from the spatial domain set for the input to the spatial domain set for the display (Same as Display), or to the spatial domain specified (As Specified Below). Alternatively, set the spatial domain for outputs to be the same as a layer in the table of contents of the display (Same as Layer).

See Also

For more information on the spatial domain, see Building a Geodatabase.

Setting the z-domain

1. Follow steps 1–4 of 'Viewing geodatabase settings for the application'.
2. Click the Output Z Domain dropdown arrow and click the desired option—Same as Input, As Specified Below, Same as Display, or Same as Layer "layer name".

   If you choose As Specified Below, type the z-domain information into the text box if it is known.

   The Same as Display option in ArcCatalog takes the z-domain information from the current dataset being previewed. In ArcMap, this option takes the z-domain information from the active data frame.

   Same as Layer "layer name" is only available in ArcGIS Desktop applications with a display, such as ArcMap or ArcScene. Only layers that have a z-domain will be available in the dropdown list.

   Alternatively, click the browse button to apply the z-domain information from an existing dataset.
3. Click OK on the Environment Settings dialog box.

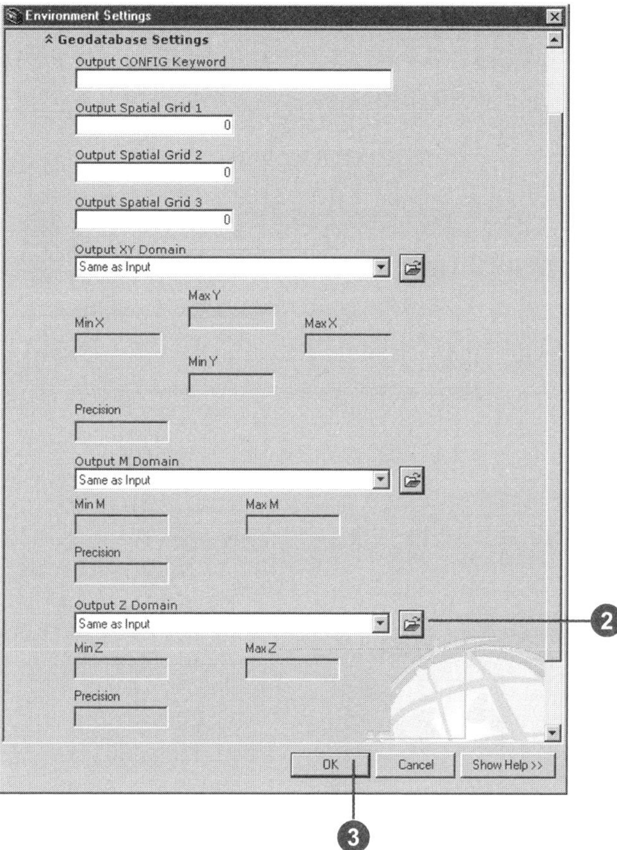

Tip

Storing m- and z-values

If the input to a tool does not contain z- or m-values and the default of Same as Input is set for Output has Z Values and Output has M Values in the General Settings section of the Environment Settings dialog box, regardless of whether you have a domain set for m-values and z-values, they cannot be stored in the output from running the tool. You must set Output has Z Values and Output has M Values to Enabled in the General Settings section of the Environment Settings dialog box before running tools using inputs that do not have the ability to store m-values and z-values.

See Also

The spatial domain information forms a part of the spatial reference you might want to set up before running tools. Spatial reference information also consists of the coordinate system to apply to results. For information on setting the coordinate system before running tools, see 'Setting the output coordinate system' earlier in this chapter.

Setting the m-domain

1. Follow steps 1–4 of 'Viewing geodatabase settings for the application'.
2. Click the Output M Domain dropdown arrow and click the desired option—Same as Input, As Specified Below, Same as Display, or Same as Layer "layer name".

   If you choose As Specified Below, type the spatial domain information into the text box if it is known.

   The Same as Display option in ArcCatalog takes the m-domain information from the current dataset being previewed. In ArcMap, this option takes the m-domain information from the active data frame.

   Same as Layer "layer name" is only available in ArcGIS Desktop applications with a display, such as ArcMap or ArcScene. Only layers that have an m-domain will be available in the dropdown list.

   Alternatively, click the Browse button to apply the m-domain information from an existing dataset.
3. Click OK on the Environment Settings dialog box.

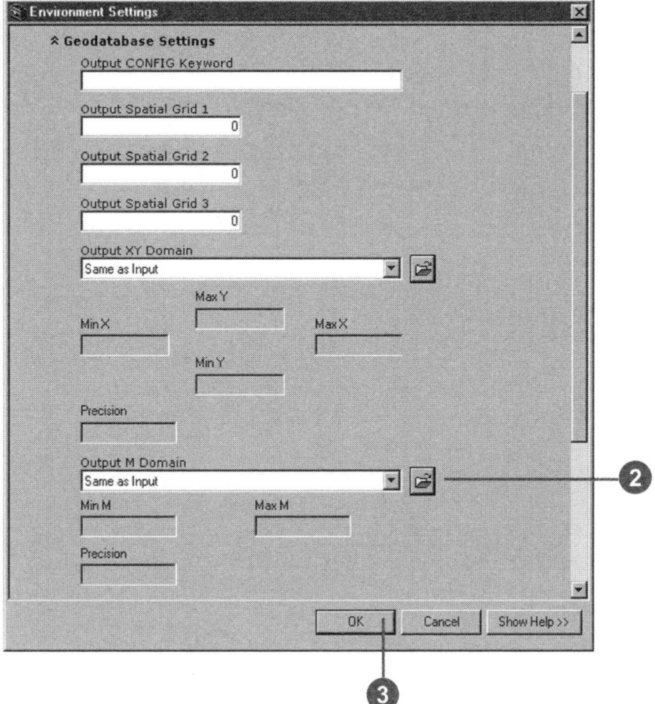

Specifying raster geodatabase settings

The raster geodatabase settings section of the Environment Settings dialog box provides settings for raster geodatabase results. Included are the following settings that can be changed: the default compression type for ArcSDE geodatabases, the default settings for pyramid creation and calculating statistics, and the tile size used when loading raster data to an existing raster dataset. Each of these settings is discussed in the pages that follow. ▸

Tip

An alternative way to access the Environment Settings dialog box

Right-click the ArcToolbox window and click Environments.

See Also

See 'About environment settings' earlier in this chapter for information on setting environments for the application, a model, or a process in a model.

Viewing raster geodatabase settings for the application

1. Click Tools on the Main menu of the application you are using and click Options.
2. Click the Geoprocessing tab.
3. Click the Environments button.

   The Environment Settings dialog box is opened.
4. Click Raster Geodatabase Settings to view the settings for which values can be specified.
5. Click OK.

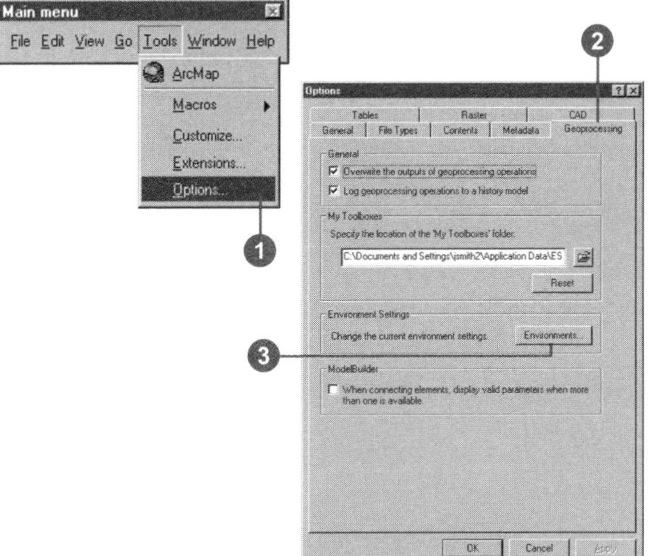

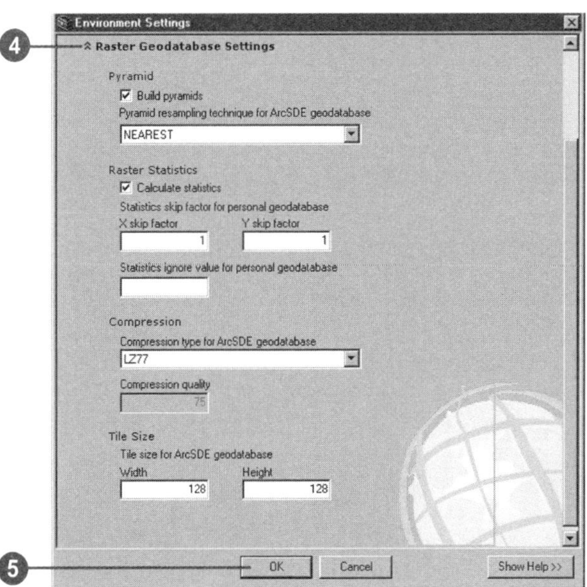

The compression type setting is used by any tools that load raster data into an ArcSDE geodatabase. The blocks of data in a raster can be compressed before being stored in an enterprise geodatabase. Three types of *compression* are supported, LZ77, JPEG, and JPEG2000.

LZ77 (the default) is a lossless compression that preserves all raster cell values. It uses the same compression algorithm as the PNG image format and one similar to ZIP compression. As you can rely on the pixels not changing their values after you compress them, use LZ77 for performing visual or algorithmic analysis.

JPEG is a lossy compression because raster cell values may not be preserved after compression and decompression. It uses the public domain JPEG (JFIF) compression algorithm and only works for unsigned 8 bit raster data (single band grayscale or three band raster data).

JPEG2000 uses wavelet technology to compress rasters so they visually appear lossless, meaning that although the cell values do get manipulated, the differences between the original and the same raster with compression are not easily distinguishable. ►

Setting the compression type

1. Follow steps 1–4 of 'Viewing raster geodatabase settings for the application'.
2. Click the Compression dropdown arrow and click the compression type you want to use.
3. If JPEG or JPEG2000 is selected for the Compression, enter a value for Compression quality.

   All raster dataset results created in an ArcSDE geodatabase will first be compressed in the manner specified.
4. Click OK.

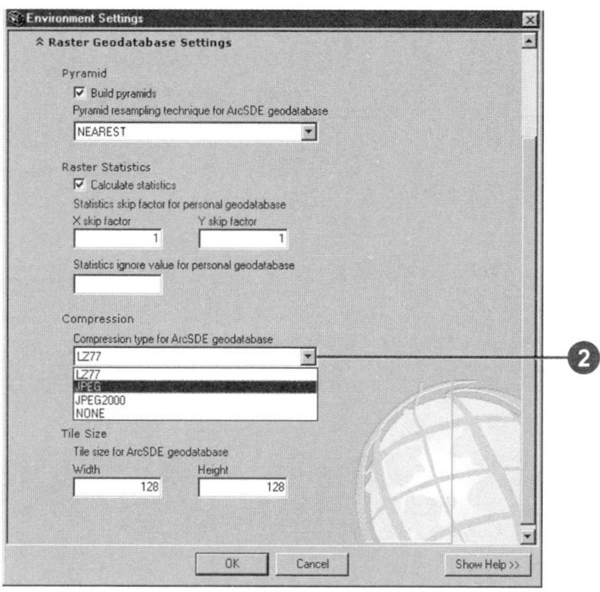

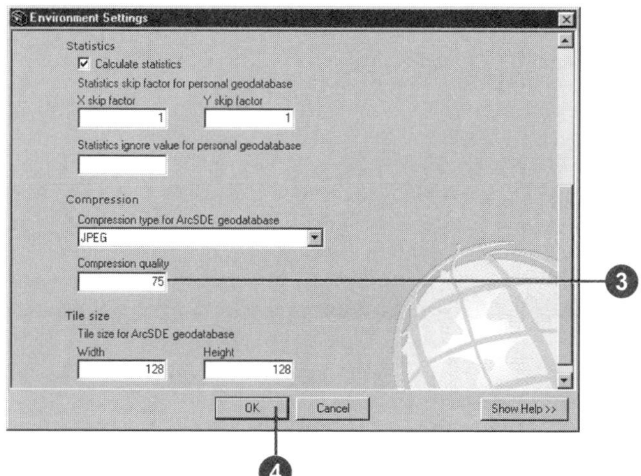

Use JPEG or JPEG2000 for rasters that are meant as pictures or backdrop imagery.

If JPEG or JPEG2000 is selected, you can also set the compression quality to control how much loss the image will be subjected to by the compression algorithm. The values of the pixels of an image compressed with a higher compression quality will be closer to those of the original image. Valid value ranges of compression quality are from five to 95. The default compression quality is 75. The amount of compression will depend upon the data and compression quality. The more homogeneous the data, the higher the compression ratio. The lower the compression quality, the higher the compression ratio. Lossy compression normally results in higher compression ratios when compared to lossless compression.

The primary benefits of compressing data are that compressed data requires less storage space and data display times will be quicker as there is less information to transmit. ▸

Pyramids are reduced resolution representations of your dataset. They can speed up display of raster datasets by retrieving only the data that is necessary at a specified resolution.

By default, pyramids are created for raster datasets in an ArcSDE geodatabase by resampling the original data. There are three resampling methods available: nearest neighbor, bilinear, and cubic.

The default is nearest neighbor. It works for any type of raster dataset. Use nearest neighbor for nominal data or raster datasets with colormaps, such as land use data, scanned maps, and pseudocolor images.

Use bilinear interpolation or cubic convolution for continuous data, such as satellite imagery or aerial photography.

If you uncheck Build pyramids, pyramids will not be created with the output raster. Not building pyramids saves storage space but will lead to slower display speeds, especially for larger raster datasets. ►

Creating pyramids

1. Follow steps 1–4 of 'Viewing raster geodatabase settings for the application'.
2. Build pyramids is checked by default. If you previously unchecked it, check it.
3. Click the Pyramid resampling technique dropdown arrow and click the resampling technique you wish to use to resample the data in order to build the pyramid layers.
4. Click OK.

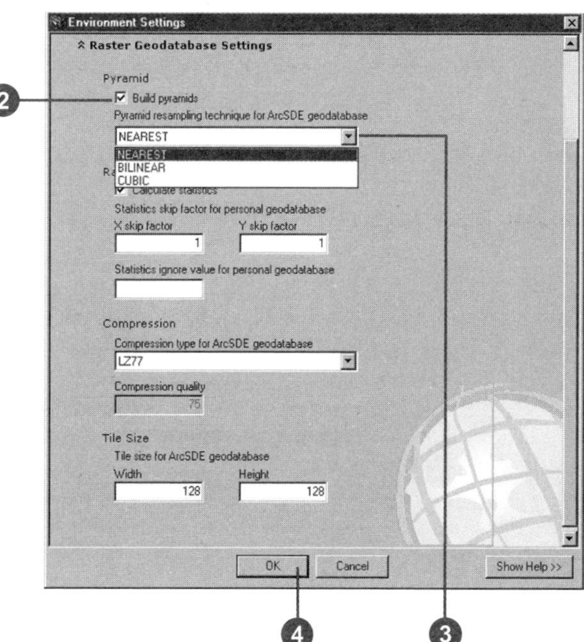

The Statistics option enables you to build statistics for output raster datasets created in a geodatabase. Statistics are required for your raster dataset to perform certain tasks in ArcMap or ArcCatalog, such as applying a contrast stretch or classifying your data. It is not essential to build statistics if they have not already been calculated, since they are calculated the first time they are needed. It is, however, recommended that you calculate statistics for your raster datasets before using them if you want to use certain features that require statistics. The default display of your raster will be improved in most cases if statistics have already been calculated, because a standard deviation stretch is applied if statistics are present.

Setting a Skip factor allows you to speed up the process of calculating statistics by skipping pixels. The Skip factor only applies for file-based or personal geodatabase raster datasets.

Values you set to ignore will not participate in the statistics calculation. Normally you may want to ignore the values of the background. This only applies for personal geodatabase raster datasets. ▶

Calculating statistics

1. Follow steps 1–4 of 'Viewing raster geodatabase settings for the application'.
2. Calculate statistics is checked by default. If you previously unchecked it, check it.
3. Type a value for the skip factor in the X and Y direction.

   If, for instance, you type a value of 2 for both, every other pixel will be skipped while calculating statistics. A skip factor of 1,1 ensures full, accurate statistics.
4. If there are values in the raster that you wish to ignore while calculating statistics for the raster, type their values, separated by a semicolon (;). These are usually values such as the background.
5. Click OK.

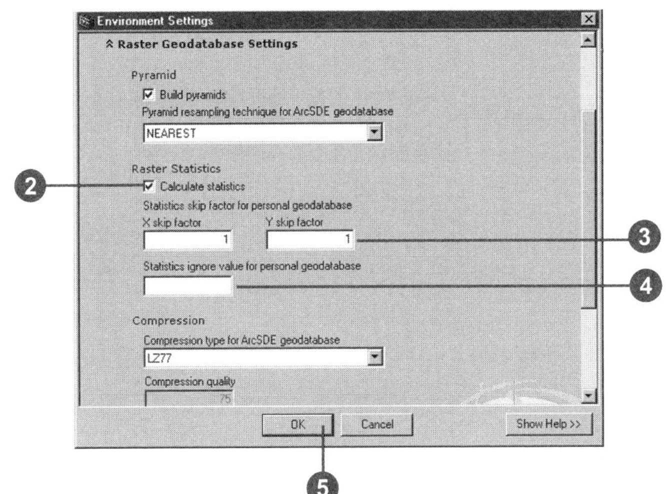

The tile size setting is used by any tools that create raster datasets in an ArcSDE geodatabase. ArcSDE geodatabases store raster datasets in a data type known as a binary large object, or BLOB.

The tile size option lets you control the number of pixels that are stored in each BLOB and, therefore, lets you control the size of each BLOB. It is specified as the number of pixels in X (tile width) and Y (tile height).

The default tile size is 128 by 128, which is good for most cases. However, if the tile size is too big, you will end up bringing up more data than is needed each time you access the data. For example, you want to display a window of 100 * 100 and it only covers one tile. If you set the tile size to be 512, you need to get the tile of 512 * 512 pixels. If your tile size is set to 128 * 128, you'll bring up less extra data if the display window is 100 * 100.

Setting the tile size

1. Follow steps 1–4 of 'Viewing raster geodatabase settings for the application'.
2. Type a value for the tile Width and the tile Height if you want to change the default.
3. Click OK.

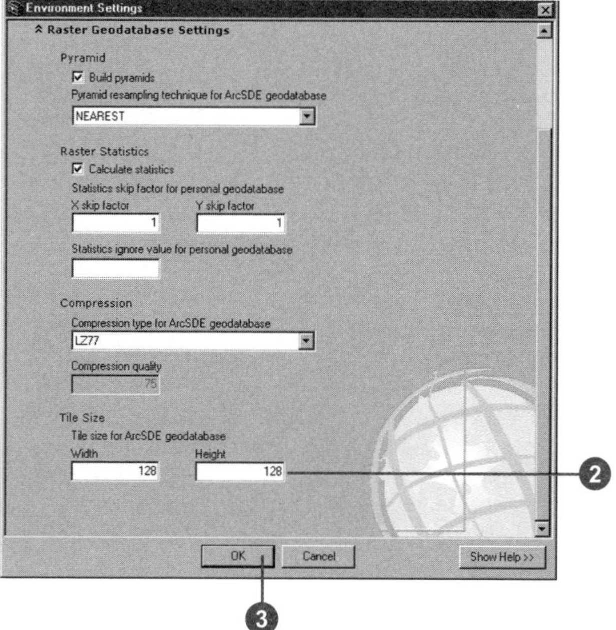

Specifying raster analysis settings

The Raster Analysis Settings section of the Environment Settings dialog box provides environment settings that will apply when working with tools that input or output a raster, either file-based or within a personal or ArcSDE geodatabase. Included in this section are the following settings that you can change: the cell size and the mask. These settings are discussed in the pages that follow. ►

Tip

An alternative way to access the Environment Settings dialog box

Right-click in the ArcToolbox window and click Environments.

See Also

See 'About environment settings' earlier in this chapter for information on setting environments for the application, a model, or a process in a model.

See Also

See 'Specifying general settings' earlier in this chapter for information on setting the extent and snap raster for results from running tools.

Viewing raster analysis settings for the application

1. Click Tools on the Main menu of the application you are using and click Options.
2. Click the Geoprocessing tab.
3. Click the Environments button.

   The Environment Settings dialog box is opened.
4. Click Raster Analysis Settings to view the parameters for which values can be set.
5. Click OK.

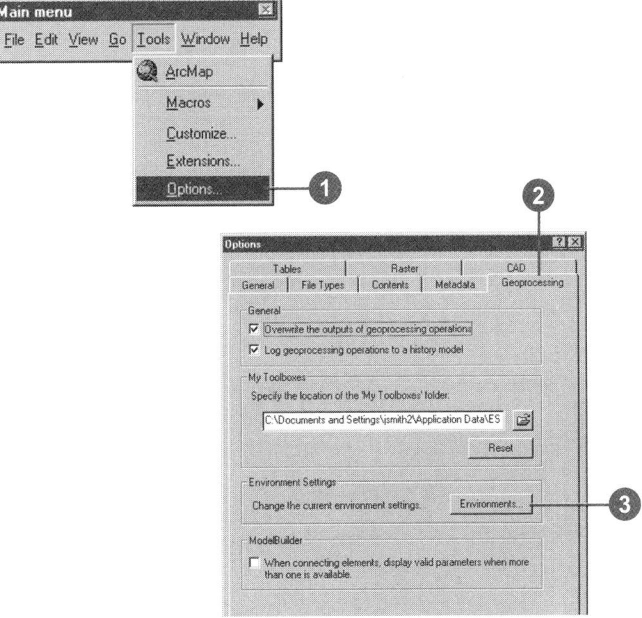

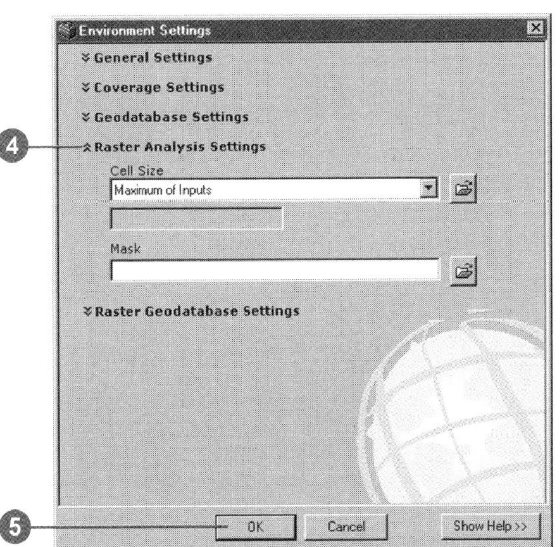

The default *cell size*, or resolution, for analysis results is set to the largest cell size of the input raster datasets for a tool, the maximum of the inputs.

The default cell size when a feature class is used as input to a tool is the width or the height (whichever is shortest) of the extent of the feature class divided by 250.

Exercise caution when specifying a cell size finer than the input raster datasets. No new data is created; cells are interpolated using nearest neighbor resampling. The result is as precise as the coarsest input.

Other options available: Minimum of Inputs sets the cell size of your analysis results to the input raster dataset with the smallest cell size; As Specified Below enables you to specify a cell size for analysis results; and Same as Layer enables you to select an input raster layer on which to base the cell size of your analysis results.

For tools that accept only features as input, you can specify the cell size for your output raster in the tool's dialog box. ►

Setting the cell size for outputs

1. Follow steps 1–4 of 'Viewing raster analysis settings for the application'.
2. Click the Cell Size dropdown arrow and click the required option.

   If you click As Specified Below, type a value for the cell size.

   Alternatively, click the Browse button and browse to a raster dataset from which to take the cell size.

   Note that the option Same as Layer is only available if you are accessing the Environment Settings dialog box from within an ArcMap session and you have added a layer to your ArcMap session.
3. Click OK.

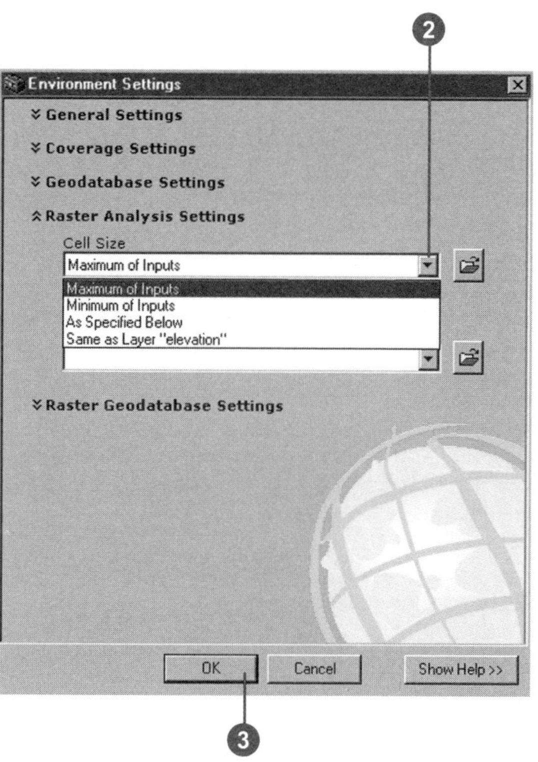

Sometimes you only want to perform analysis on cells in a particular area. The *mask* identifies those cells within the analysis extent that will be considered when running a tool. If the mask was overlaid on the input raster you want to use for analysis, only cells covered by the mask will be processed. All other cells will be assigned the NoData value in the result after running a tool.

Setting an analysis mask is a two-step process. The analysis mask must first be created if you do not already have one. An analysis mask can be feature (point, line, or polygon) or raster data. You could create an empty feature class in ArcCatalog, then define the area of interest (the mask) using the ArcMap Editor toolbar. Alternatively, you could create a mask raster using the Reclassify tool, giving a value to cells you wish to create results for and assigning NoData to the rest. The analysis mask must then be set in the Environment Settings dialog box.

Setting a mask

1. Follow steps 1–4 of 'Viewing raster analysis settings for the application'.
2. Type the path and the name of the Mask or click the Browse button, click the Look in dropdown arrow, and navigate to the location of the mask, then click Add.
3. Click OK.

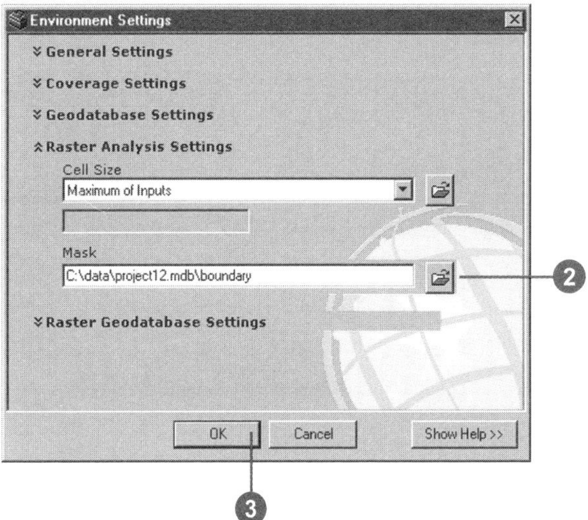

Using the Command Line window 7

IN THIS CHAPTER

- Starting the Command Line window
- Working with the command line
- About variables
- Creating variables
- Managing variables

The Command Line window contains a message section and a command line. When tools are executed—from a tool's dialog box, the command line, a ModelBuilder window, or a script—execution messages appear in the message section of the window. The command line offers an alternative to using a dialog box to run a tool. It allows you to type the name of the tool and its parameter values as a string, then execute the tool when you press Enter. It can be quicker to run a tool at the command line, especially if you are familiar with the tool and its parameter values and you have created variables for complex parameters you use frequently. For ArcInfo Workstation users, the command line usage will be familiar, as it is nearly identical to the usage displayed using the Usage command.

Through the Command Line window you can:

- Obtain a list of available tools and environment settings.
- View usage for tools to see the parameter values that must be specified.
- Run tools by entering their names and parameter values.
- Set values for environment settings.
- Create variables for parameter values that you can save and reuse.
- View output messages for environment settings specified or tools run.
- Open the dialog box of a previously run tool, edit parameter values, then rerun the tool.

Starting the Command Line window

The Command Line window consists of a command line, where you can execute tools and set values for environment settings, and a message section.

When you execute tools—from a dialog box, the command line, or within a ModelBuilder window or a script—messages are created. These messages include such information as when the tool began executing, what parameter values are being used, the progress of the tool's execution, and warnings of potential problems or errors.

Information messages indicate that an event happened, such as the progress of a tool's execution or the time a tool was executed. By default they display in black text.

Warning messages indicate issues, such as a problem with the execution of a tool. By default they display in green text.

Error messages indicate a critical event that will prevent a tool from executing. Errors are generated when one or more parameters have invalid values, such as a nonexistent path to ▸

Opening the Command Line window

1. Click the Show/Hide Command Line Window button on the Standard toolbar of the application you are working in.

   The Command Line window is opened and docked within the application.

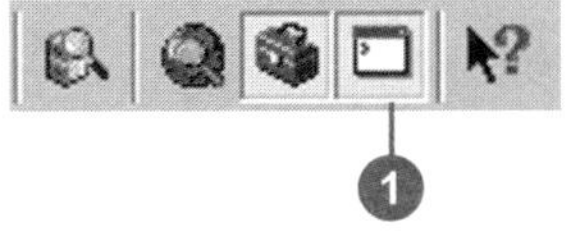

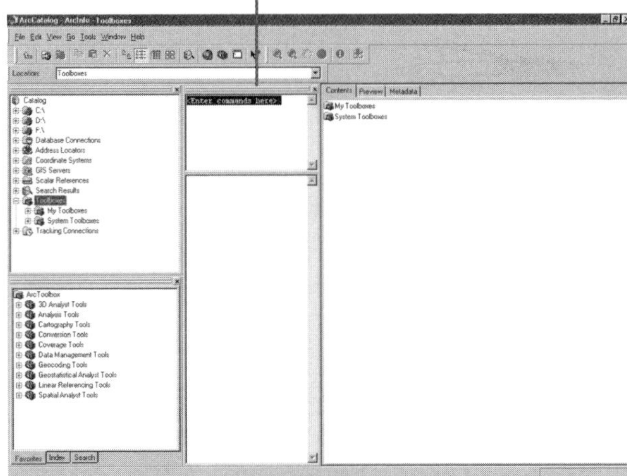

The Command Line window can be used inside any ArcGIS Desktop application and is a dockable window, so you can position it anywhere within the application you are working in or outside it.

data or an invalid keyword. By default they display in red text.

The default font and text color for messages (information messages, warnings, and errors), as well as for commands you type and processes run, can be altered by right-clicking the command line or the message window and clicking Format.

Tip

Positioning the window

Press and hold the Ctrl key while moving the Command Line window within the application to prevent it from docking itself until it is in the desired position.

Docking the Command Line window

1. Click and drag the bar at the top of the Command Line window to your preferred location within the application.
2. Drop the panel to dock the window.

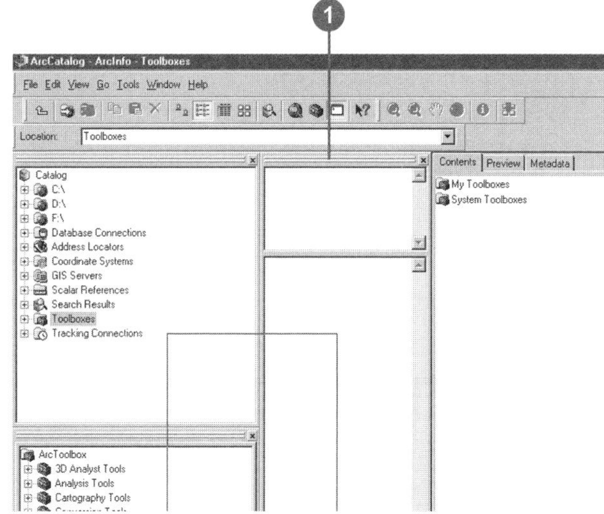

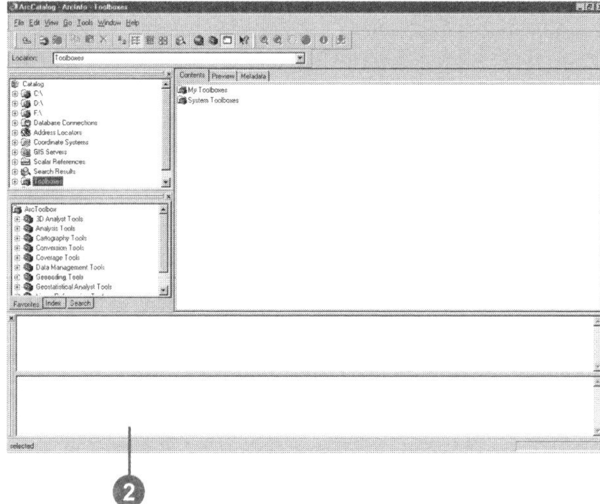

Working with the command line

Type any character at the command line to see a list of available *commands* (tools and environment settings) in a dropdown list. After typing the name of the command and typing a space, usage appears. Usage is a command's parameter *syntax*. It shows you the parameters for which values must be specified. ▸

Tip

An alternative way to scroll through the list of tools and environment settings

Use the up and down arrows on the keyboard to scroll through the list of tool names and environment settings.

See Also

Before running tools at the command line, you can choose whether, by default, you create a temporary result, add the result to the display, and overwrite existing outputs. See 'Results from running tools' in Chapter 3 for more information on these options.

Getting a list of available tools and environment settings

1. Type any character at the command line to see a list of tools and environment settings.

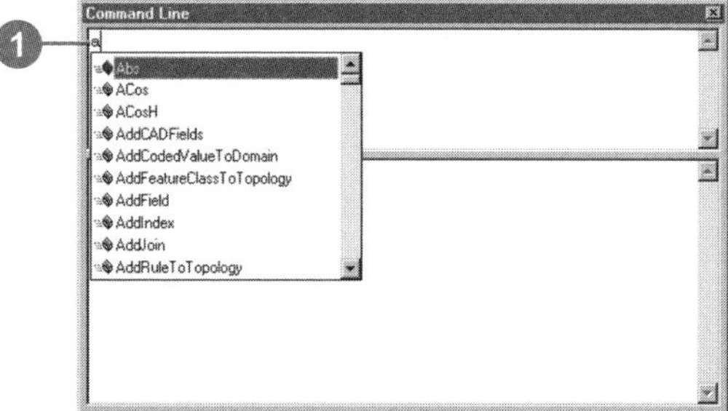

Typing tool or environment setting names

1. Type the first few letters of the tool or environment setting name, then use the mouse pointer to scroll through the list and select the tool or environment setting that you want to use.
2. Press the Space key to add the tool or environment setting name to the command line.

   By pressing the Space key, a space is automatically added after the tool or environment setting name and the usage is displayed.

   Usage helps you supply values for the tool's or the environment setting's parameters.

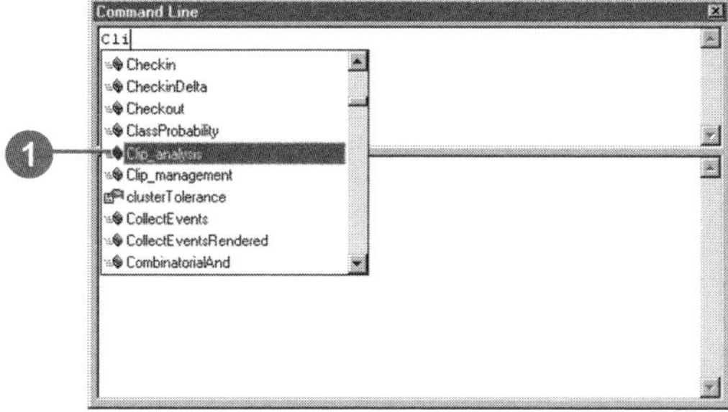

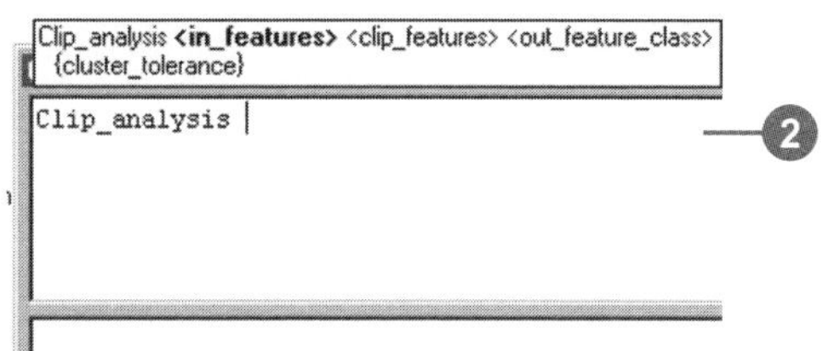

To execute a tool at the command line, the toolbox containing the tool must first be added to the ArcToolbox window.

To execute a tool you'll first type the tool name followed by a series of parameter values. After typing the tool name, type a space to view the usage for the tool.

The usage helps you supply appropriate values for parameters. Required parameters appear between <>, and optional parameters appear between {} in the usage. You must specify a value for required parameters, but you do not need to specify a value for optional parameters. You can simply type a # or "" in place of optional parameter values, and the default value will be used. ►

Tip

Optional parameters

To use the default value for all optional parameters, instead of typing a # or "" in place of a value, simply leave all optional parameter values blank.

See Also

See Chapter 4, 'Working with toolboxes', for information on adding toolboxes to the ArcToolbox window.

Executing a tool at the command line

1. Type the name of the tool and press the Space key to view the usage.

   Note that you can type the first few characters of the tool name and press the Space key to fill in the rest of the name.

2. Give a value for each parameter, separating each one with a space.

   The parameter that needs a value specified for it is highlighted in bold in the usage.

   If the tool takes multiple inputs, type parentheses around the inputs and a semicolon (;) between each input. If your inputs have spaces in the name, enclose each name with apostrophes.

3. When all parameters have appropriate values, press Enter to execute the tool.

4. Examine the output messages in the message section of the Command Line window.

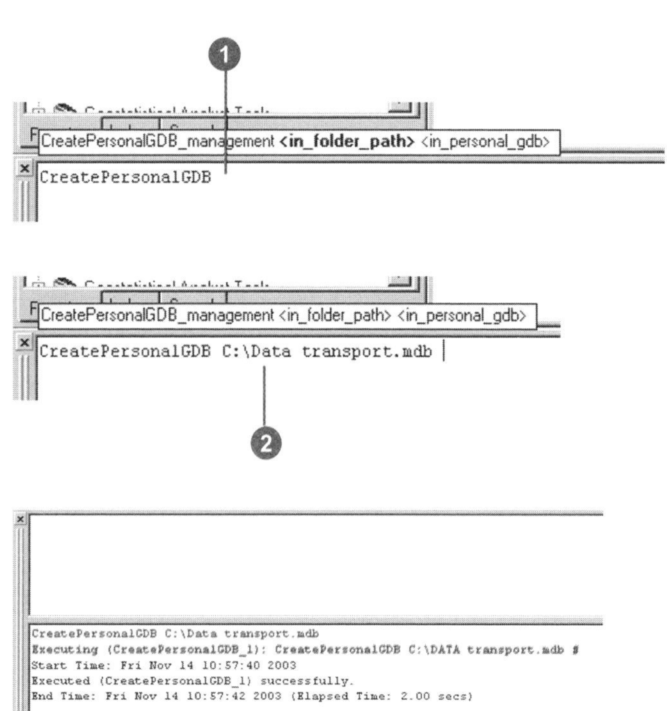

You can enter multiple commands at one time at the command line by pressing the Ctrl key then Enter after each line. Once you have entered all the commands you want to run, press Enter to execute them.

You can run separate commands at one time, or you can string commands together so the output from one command is the input for the next command. In the example, the workspace is set before running tools so that the path to inputs and outputs does not have to be given. Then the output from the MakeLayer tool (landuselyr) is used as input to the AddJoin tool, where a table is joined to the input layer. The layer is in memory and will be removed after the current session has ended, so the CopyFeatures tool copies the layer to a feature class. ►

Tip

Clearing messages

To clear messages from the message section of the Command Line window, right-click the window and click Clear All.

Tip

Finding and replacing text

Right-click the command line and click Replace to find and replace text you have entered.

Executing multiple commands

1. Enter a command you would like to run.
2. Press and hold the Ctrl key and press Enter to type another line.
3. Follow steps 1 and 2 until you have entered all the commands you would like to run.
4. Press Enter to execute the commands.
5. View the messages produced in the message section of the Command Line window to check for any problems with the execution of the commands.

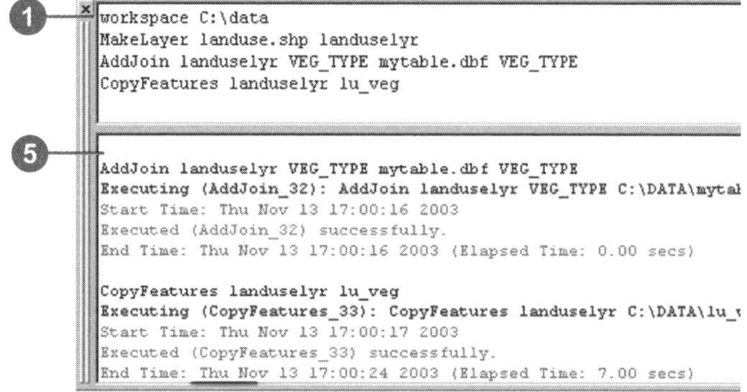

Working with nonunique tool names

If you have tools in the ArcToolbox window with the same name, you must concatenate the toolbox alias name as a suffix to the tool name, separated by an underscore, for example clip_analysis, then type a space to view the usage. See the section 'Managing toolboxes' in Chapter 4 for information on setting and getting the alias name.

Tip

Copying and pasting

Select the current string entered in the command line, right-click, and click Copy. In the location you wish to paste the string, click Paste. Alternatively, copy and paste a string into the command line.

See Also

See 'About variables' later in this chapter for information on creating variables for parameter values.

Executing a tool using layers as input data

1. Click the Add Data button in any ArcGIS Desktop application with a display and add the data that will be used as input parameter values.
2. Click Show/Hide ArcToolbox Window.
3. Right-click the ArcToolbox window and click Add Toolbox to add the toolbox containing the tool you want to run.
4. Click Show/Hide Command Line Window on the Standard toolbar.
5. At the command line, type the name of the tool, then a space to view the usage for the tool.
6. Press the down arrow on the keyboard to select the layer you want to use.
7. Press the Space key to see the next parameter in bold in the usage, for which a value must be specified.
8. Continue adding parameter values as necessary.
9. Press Enter to execute the tool.

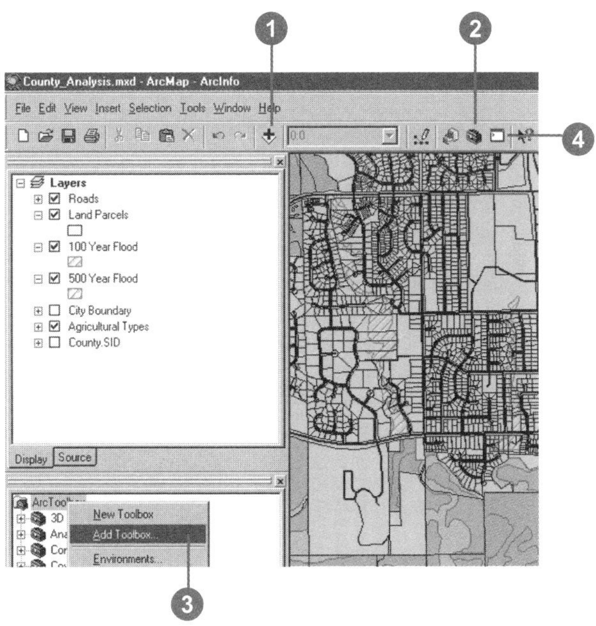

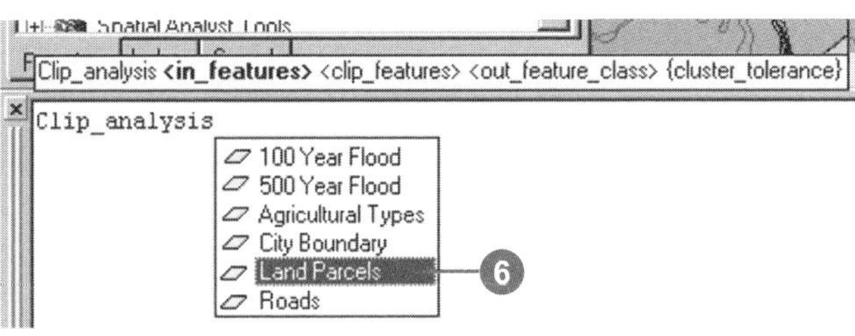

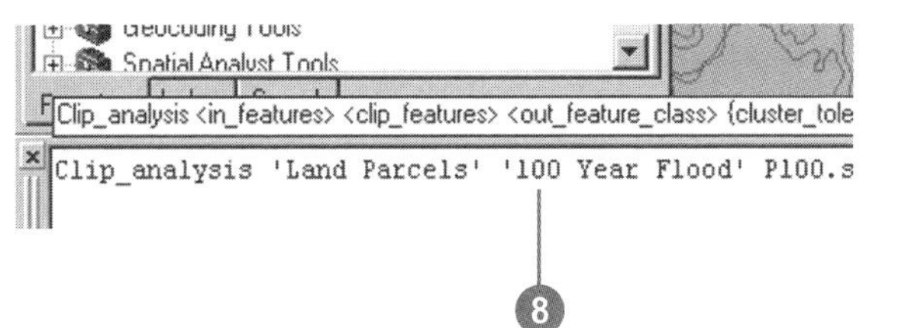

Before executing tools and as an alternative to specifying any relevant values for environment settings in the Environment Settings dialog box for the application, you can specify values for environment settings at the command line. Values set for environment settings at the command line will be passed to the application's Environment Settings dialog box and will be used by all tools.

The online help for Environment Settings gives examples of setting values for environment settings at the command line. ▸

See Also

See Chapter 6, 'Specifying environment settings', for information on specifying values for environment settings for the application, a model, and a process within a model.

See Also

When working in an ArcGIS Desktop application with a display, such as ArcMap, layers from the table of contents can be used as inputs to appropriate tools that are run at the command line. See 'Executing a tool using layers as input data' earlier in this chapter for more information.

Setting values for environment settings

1. Type the name of the environment setting. For example, type "Workspace", then a value for the <Workspace or Feature Dataset> parameter (the pathname in this case).

2. Press Enter.

   In the example, the workspace is set to the pathname specified. This workspace will be set as the current workspace in the application's Environment Settings dialog box and used by all relevant tools.

3. Type the name of the tool and press the Space key to view the usage for the tool.

4. The first parameter after the tool name is usually input data. In the example, as the value for the workspace parameter is set to the location of the input data, only the name or names of the input data need to be supplied.

5. When all parameters have appropriate values, press Enter to execute the tool.

   In the example, the result (P_100.shp) is placed in the location on disk set for the workspace.

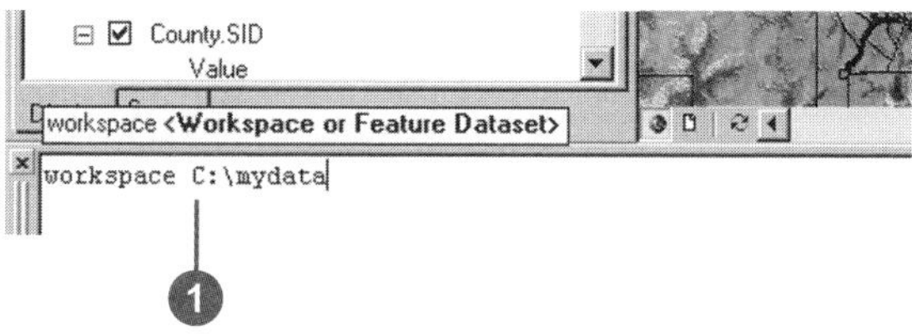


```
workspace C:\mydata
Set C:\mydata into workspace
```

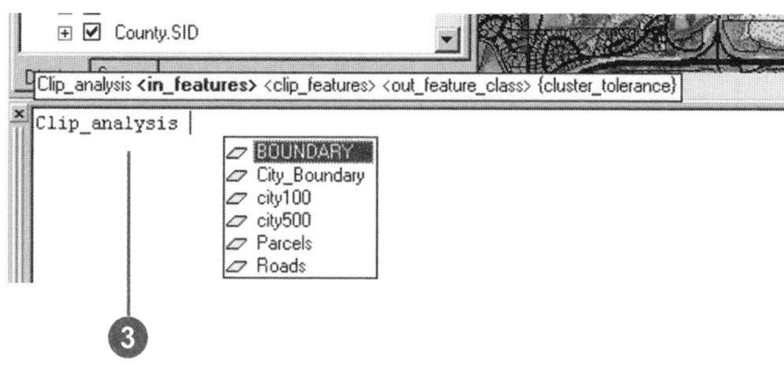

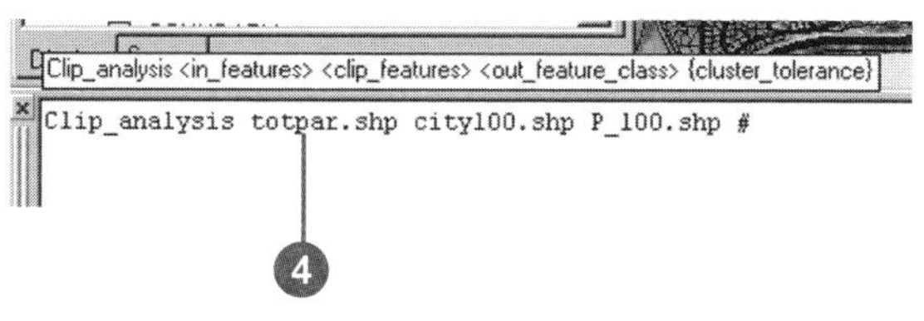


```
Clip_analysis totpar.shp city100.shp P_100.shp #
Executing: Clip C:\mydata\totpar.shp C:\mydata\city100.shp C:\mydata\P_100.shp
Start Time: Fri Nov 14 09:03:29 2003
Reading Features...
Cracking Features...
Assembling Features...
Executed (Clip) successfully.
End Time: Fri Nov 14 09:03:34 2003 (Elapsed Time: 5.00 secs)
```

Text you enter at the command line can be saved to a text file so you can load the text file at a later date. This is particularly useful if you use the same string of commands on a regular basis. Saving your commands to a text file means you can simply load it at the command line when you want to run the commands contained within the text file, alter parameter values that are set, and rerun the commands.

Any text entered at the command line can be saved, not only commands. So if you want to set up the commands that you will run and add notes to yourself or another person, simply type the text into the command line, then save the text to a file. Any text entered that is not part of a command will have to be removed when you load the text back into the command line window, but it may be helpful to have a text file containing pertinent information as well as the commands you will run. ▸

See Also

See 'Loading a text file' later in this chapter for information on loading the commands that you have saved.

Saving commands to a text file

1. After typing a command or a series of commands, right-click the command line and click Save As.
2. Click the Save in dropdown arrow and navigate to the location where you want to save the text file containing the commands.
3. Click the File name text box and type a name for the text file, with a .txt extension.
4. Click Save.

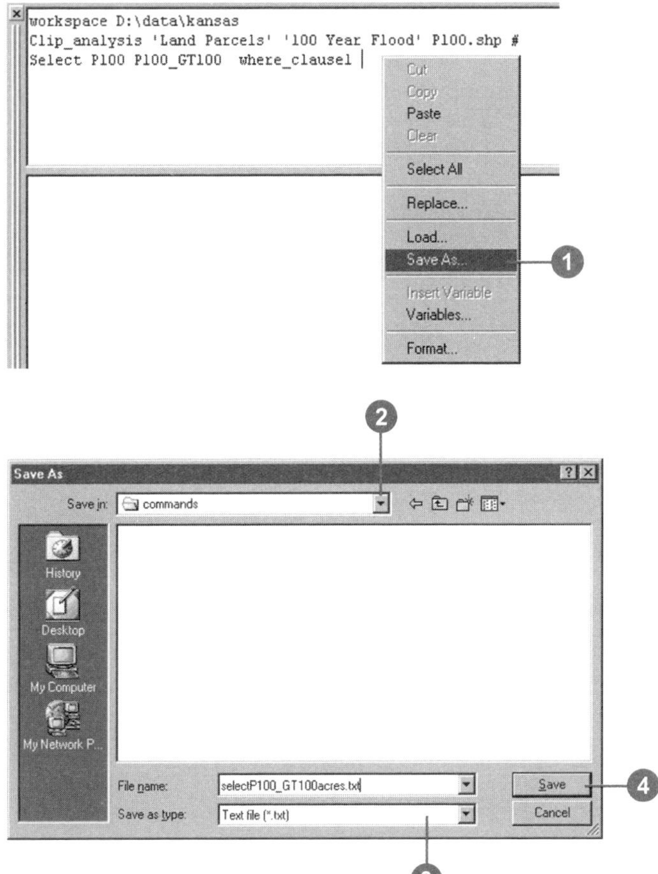

Any text file (.txt) can be loaded at the command line. However, you'll likely only load text files that you created by saving commands entered at the command line. Any other text entered into a text file that is not part of a command must be removed before executing the commands. ►

See Also

See 'Saving commands to a text file' earlier in this chapter for information on saving the commands you have entered at the command line to a text file.

Loading a text file

1. Right-click the command line and click Load.
2. Click the Look in dropdown arrow and navigate to the location of the text file containing commands you want to run.
3. Click the text file and click Open.

   The text is entered at the command line.

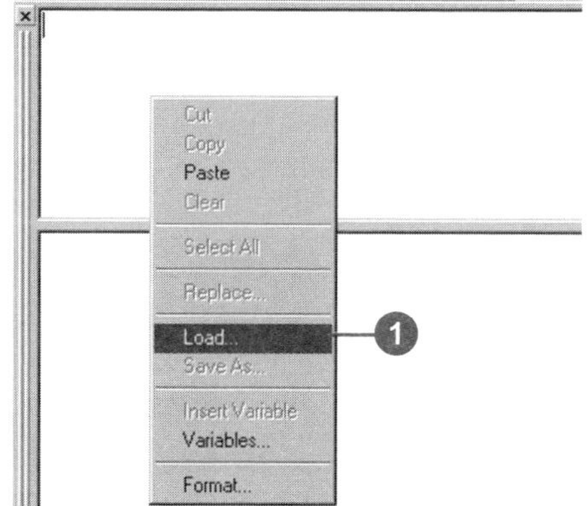

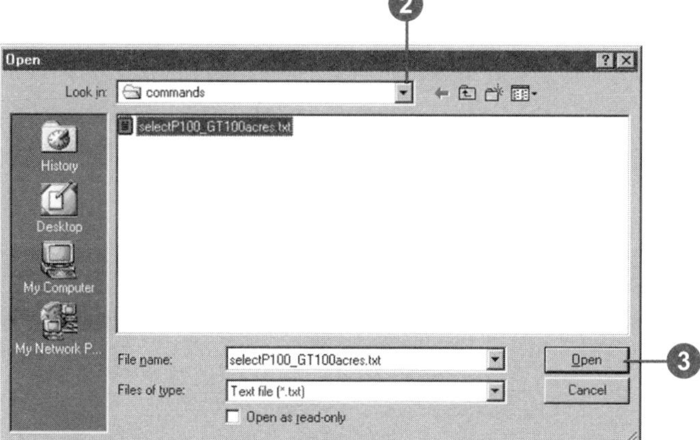


```
workspace D:\data\kansas
Clip_analysis 'Land Parcels' '100 Year Flood' P100.shp #
Select P100 P100_GT100  where_clause1
```

When a tool is executed, a copy of the tool is placed inside a history model by default, so you can keep track of the tools performed in each session. For more information on history models see the task that follows and 'Keeping track of geoprocessing operations' in Chapter 3.

When you open an executed tool, a copy of the run tool is taken from the history model and displayed via a dialog box, so you can edit the tool's parameter values, then reexecute the tool.

If the option to log geoprocessing operations to a history model is unchecked on the Geoprocessing tab of the Options dialog box (accessed via the Tools menu on the Main menu of the application you are running), you will not be able to reexecute your tools. ►

Tip

Shortcuts for recalling previously entered commands

Double-click a previously entered command in the message section of the Command Line window to enter the command at the command line.

At the command line, use the up and down arrows on the keyboard to scroll through previously entered commands.

Reexecuting a tool

1. If a tool has not already been executed, follow steps 1–4 of 'Executing a tool at the command line' earlier in this chapter. Alternatively, run a tool using its dialog box or from within a ModelBuilder window or a script.
2. In the message section of the command line, right-click the tool that was executed and click Open.

   A dialog box is opened, enabling you to modify values set for parameters.
3. Modify parameter values as necessary, then click OK to reexecute the tool.

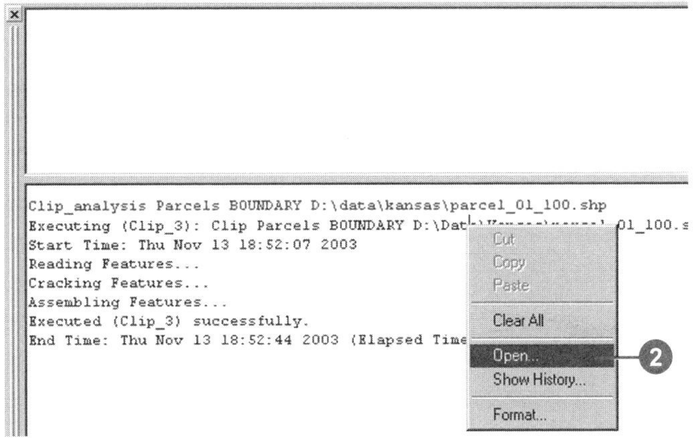

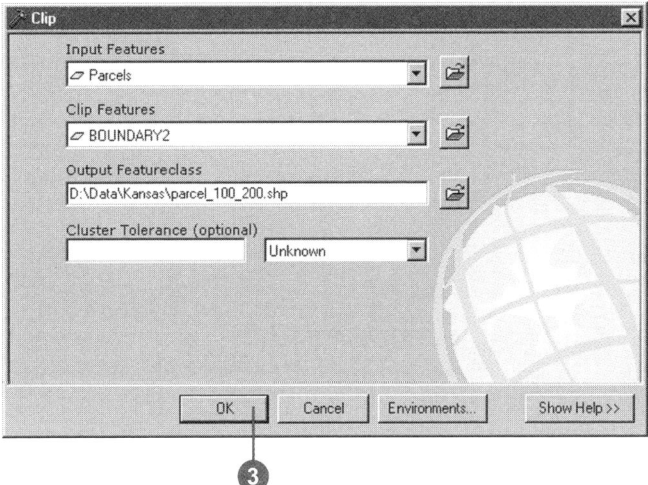

After executing tools in a session, then closing the session, a history model is generated in the History toolbox in the My Toolboxes folder. It contains tool elements that represent the tools run in the last session. By double-clicking a tool element to open its dialog box, you can view the parameter values that were set for the tool.

In the Command Line window, you can view the contents of the history model as it is being generated, allowing you to view the tools that have been run and parameter values that have been set to date in the current session.

Tip

Disabling history model creation

If you don't want a history model to be generated for each session, turn the option off. Click the Tools menu in the application you are running and click Options. Click the Geoprocessing tab and uncheck the Log geoprocessing operations to a history model check box.

See Also

See 'Keeping track of geoprocessing operations' in Chapter 3 for information on history models and alternative ways to keep track of performed geoprocessing tasks.

Viewing history in the current session

1. If a tool has not already been executed, follow steps 1–4 of 'Executing a tool at the command line' earlier in this chapter. Alternatively, run a tool using its dialog box or from within a ModelBuilder window or a script.
2. Right-click the execution string and click Show History.
3. A model is opened displaying tool elements for all tools run to date in the session.
4. Double-click a tool element to view the parameter values that were set.

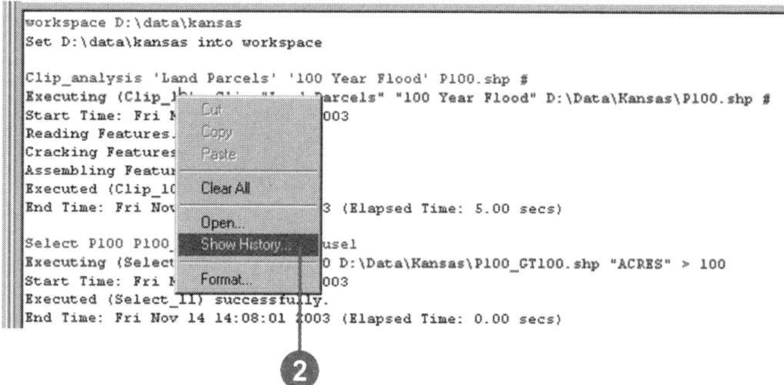

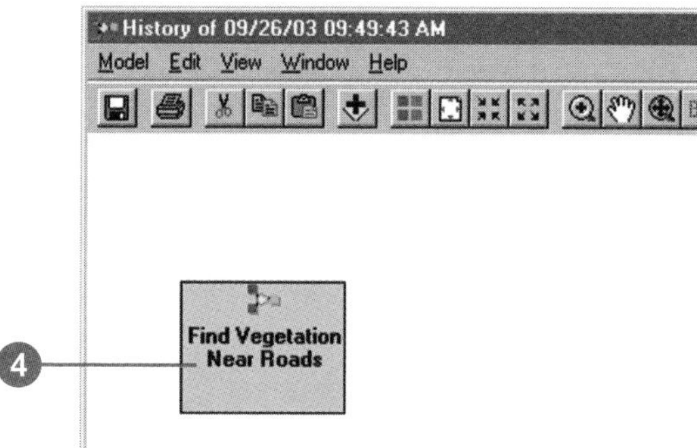

This history model displays the tools run in the session to date. In this particular session, only one tool was run. By double-clicking the tool element, you can open the tool's dialog box to view the parameter values that were set.

About variables

You can type a text string at the command line for all parameter values that must be specified for a tool to run. However, for parameters that require more complex values, it is easiest and most efficient to create a variable—by using a dialog box to define the parameter value—that can be referenced instead of typing the parameter's value each time you run a specific tool.

It's easy to create variables, and it is recommended to create a variable for a complex parameter value rather than typing a complicated text string. Once you have created a variable, you can reuse it again and again, reducing the amount of typing each time you reexecute a tool.

There are many times when it is useful to create variables for parameter values when running tools at the command line, such as:

- To avoid typing the pathname to input data.

  Instead of typing the pathname to input data each time you run a tool, you can create variables for each input data parameter value using a dialog box to specify the path to the input data.

- When projecting multiple feature classes.

  Creating a variable for the coordinate system parameter value can speed up the process of projecting multiple feature classes. Instead of typing the same text string for the value of the coordinate system parameter each time you use the Project tool, you can create a variable using a dialog box to set up the coordinate system information you want to apply for the parameter value.

- When reclassifying datasets.

  Instead of typing the old and new values for the remap parameter as a text string, you can create a variable for this parameter's value. When you create a variable it opens a dialog box where you can type in the new values, add values, or delete values in a similar way to using the ArcGIS Spatial Analyst Reclassify tool.

For information on how to create variables, see the tasks that follow.

Creating variables

There are two ways to create variables. You can create a variable as you type parameter values and come across a more complex parameter, or you can create variables at any time using the Variable Manager if you know the parameter values you'll be using.

Once you have created a variable, it will display in a dropdown list the next time you encounter the same parameter when typing a command. Variables can be distinguished by their icon in the dropdown list (▤).

See Also

See 'Saving variables' in this chapter for information on saving variables so they can be used between applications or for certain projects.

Creating a variable while typing parameter values

1. Type the name of the tool and press the Space key to view the usage for the tool.

   Note that you can type the first few characters of the tool name and press the Space key to fill in the rest of the name.

2. Give a value for each parameter, separating each with a space.

   The parameter that needs a value specified for it is highlighted in bold in the usage.

   In the example, the Reclassify tool and its parameter values are entered at the command line, using a raster layer called Landcover as the value for the <in raster> parameter and landuse as the value for the <reclass field> parameter.

3. When you come across a parameter, such as <remap> and you are not sure what the value should be, create a variable.

   Right-click the parameter and click Insert Variable or press F8 on the keyboard.

   The Insert Variable dialog box appears, enabling you to define the value for the ▸

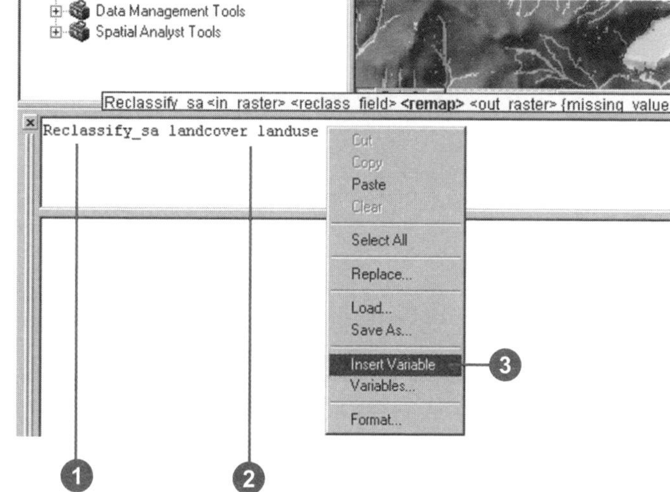

parameter using a dialog box.

In the example, the remap values can be entered.

4. When you have set the parameter definition, click OK.

   A variable with a default name is added as the parameter value (remap1 in the example).

5. Press the Space key and continue to fill in the rest of the parameter values.

6. Press Enter to execute the tool.

7. Examine the output messages in the message section of the Command Line window.

   When the tool is executed, the first message tells you that the tool is executing. You can see that the parameter value you defined using a dialog box is entered as a text string; in the example, it contains the text "Agriculture 4;Forest 6;", and so on. This would be the text string you would have had to type for this parameter value if you had not created a variable. You can see how much easier and quicker it is to create a variable for such complex parameters.

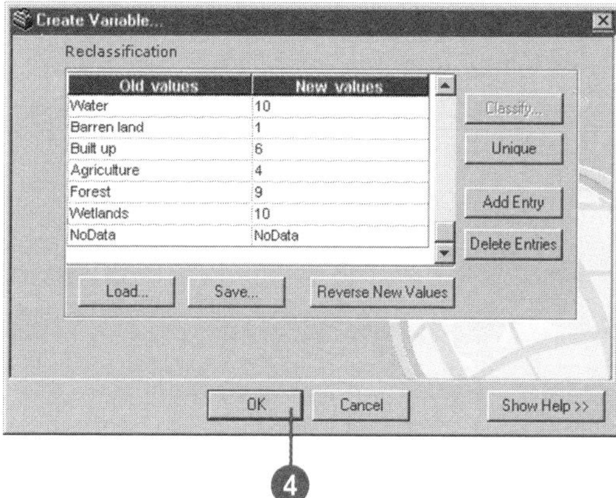

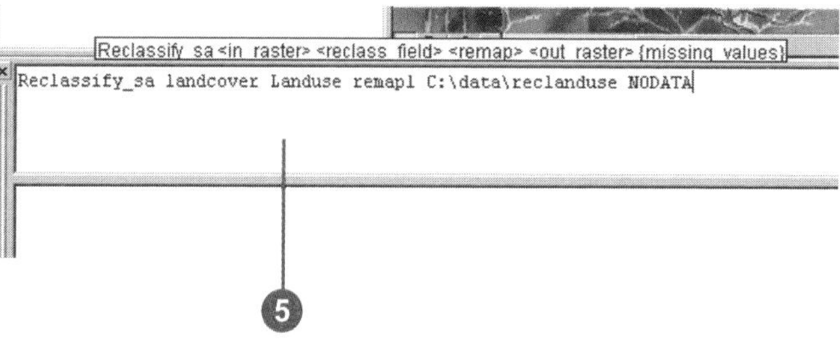


```
Reclassify_sa landcover Landuse remap1 C:\data\reclanduse NODATA
Executing (Reclassify_1): Reclassify landcover Landuse "Brush/transitio
1;Built up 6;Agriculture 4;Forest 9;Wetlands 10" C:\data\reclanduse tru
Start Time: Thu Nov 13 18:36:57 2003
  Validating...

Executing Reclassify...
  Processing...

Completed Reclassify.

Executed (Reclassify_1) successfully.
End Time: Thu Nov 13 18:37:03 2003 (Elapsed Time: 6.00 secs)
```

7

Creating variables using the Variable Manager

1. Right-click in the command line at any time and click Variables to open the Variable Manager.
2. Click the Add button.
3. Click the name Variable0 and rename the variable to something more meaningful.

   In the example, a variable (site1) is to be created for a shapefile stored on disk so it can be used as input data for multiple tools.
4. Click the data type String and scroll to the appropriate data type for the variable. In the example, the data type is a Shapefile.
5. Click the value Empty and click Edit to define the value for the variable.

   In the example, the path to the input data is defined.
6. Click OK, then click OK on the Variable Manager dialog box.

   The next time you come across a parameter that accepts this type of input data for its value, you'll be able to select the variable from the dropdown list. It will save you from having to type a text string each time you need to supply a value.

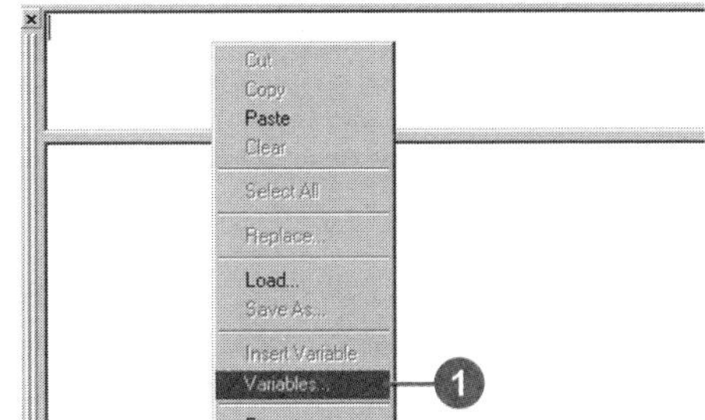

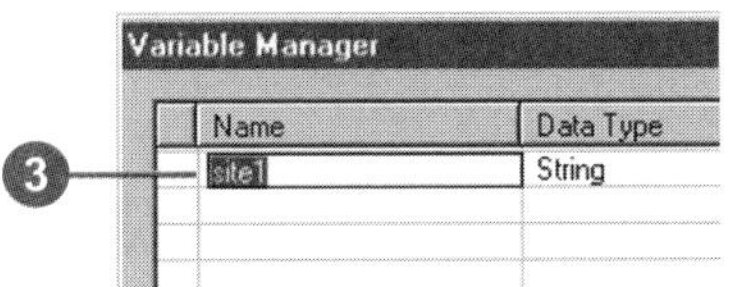

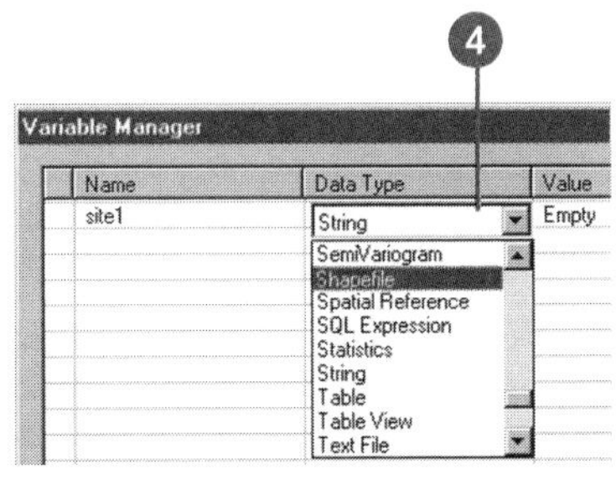

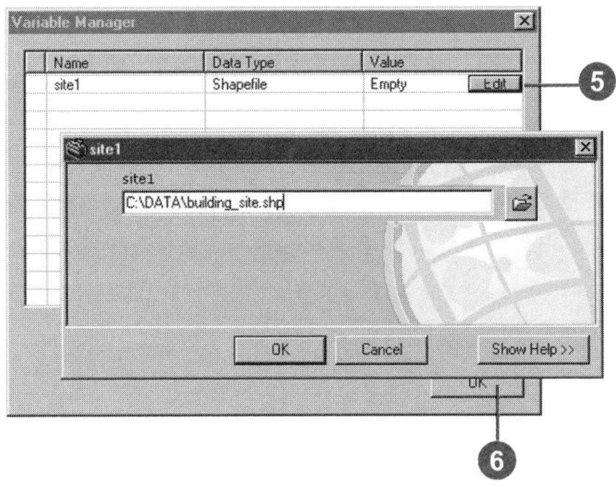

Managing variables

When you have created a variable, you may want to alter its properties. You can alter the name of the variable, its data type, or the value you have set for the variable using the Variable Manager.

For instance, you may want to drop a different field from a table other than what is defined for the value of the variable, modify the new values defined for a remap variable, or change the coordinate system defined for a coordinate system variable. ►

Editing variables

1. Right-click the command line at any time and click Variables to open the Variable Manager.
2. Click the property of the variable you want to alter—its name, data type, or value.

   To alter the name property, click the name and type a new one.

   To alter the data type property, click the data type specified, then select an alternative data type from the dropdown list.

   To alter the value specified for the variable, click the value, then click the button to the right of the value to alter the value set.
3. When you have finished editing the variable's properties, click OK.

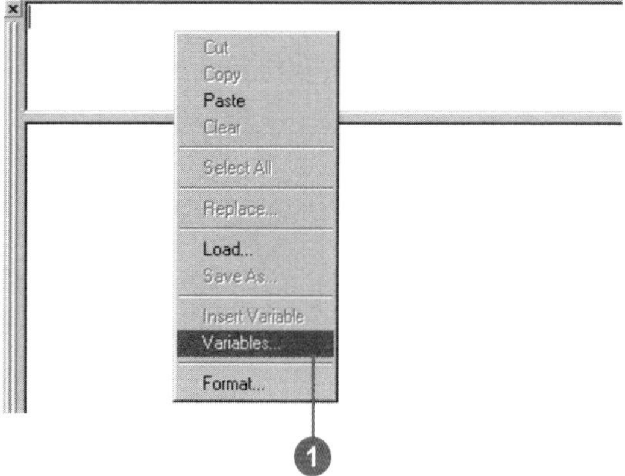

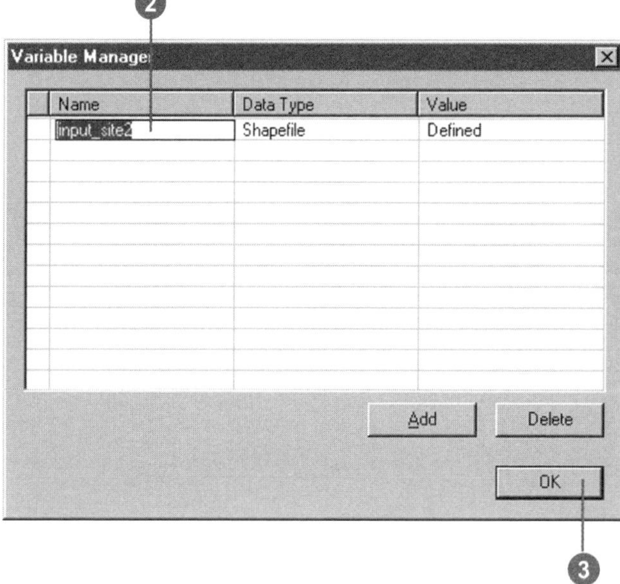

If you create multiple variables within the same application, you'll eventually want to clean up the Variable Manager and remove the variables you no longer use. Variables can be easily deleted from the Variable Manager if they are no longer required. ►

Removing variables from the Variable Manager

1. Right-click in the command line at any time and click Variables to open the Variable Manager.
2. Click to the left of the variable name to select the variable you want to delete.
3. Click Delete.
4. Click OK.

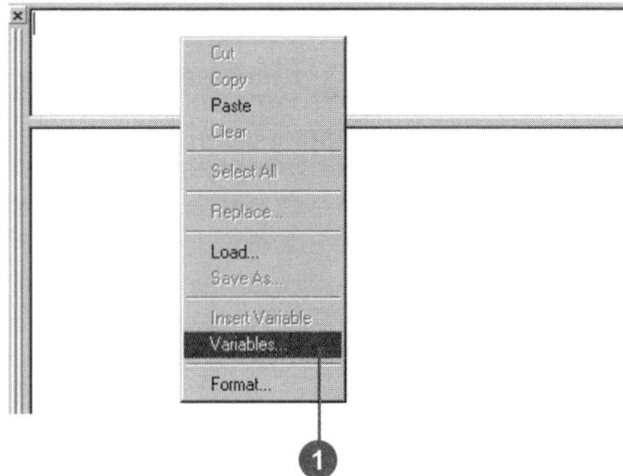

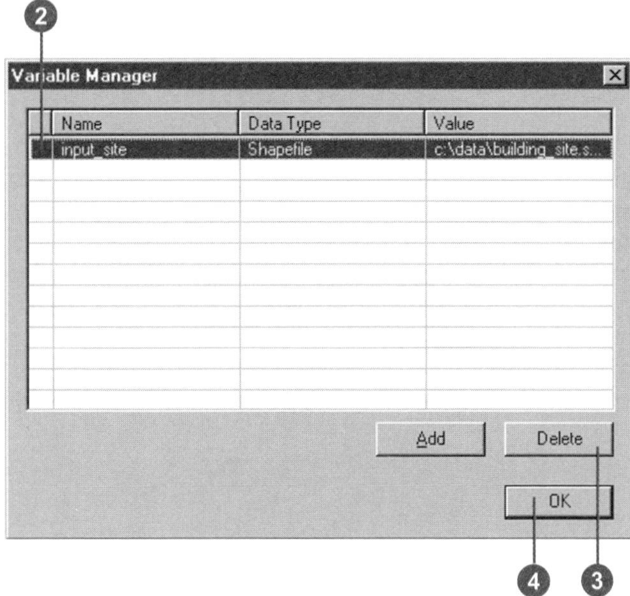

When variables are created in ArcCatalog, they are saved to the default settings when ArcCatalog is closed, or when the settings are saved as the default. All applications, when started, use the default geoprocessing settings.

When you are switching between different applications or you are working on a certain project, you may want to save the variables you have specified to a file so you can quickly load them when they are needed. Alternatively, you may save settings as the default so all other applications will use them.

The way you save variables is by saving geoprocessing settings. Along with variables, you'll also save the state of the ArcToolbox window and the state of the Environment Settings dialog box. See 'Geoprocessing settings' in ▸

Tip

Saving variables by saving the map document

If you are working in ArcMap and you save the map document, all created variables will be saved with the map document, so you do not need to save settings in this case.

Saving variables

1. Right-click the ArcToolbox window, point to Save Settings and click either To File or To Default, depending on whether you want to save settings to a file that can be loaded later, or you want to set the current settings as the default for all applications.
2. If you choose To File, click the Save in dropdown arrow and navigate to the location in which you wish to save the settings file.
3. Click the File name text box and type a name for the file with a .xml extension.
4. Click Save.

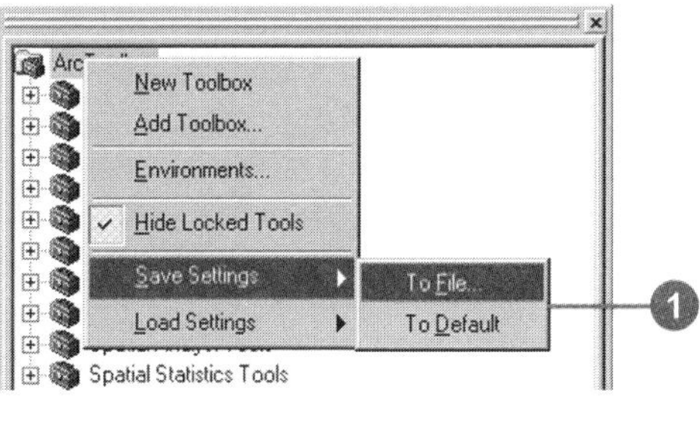

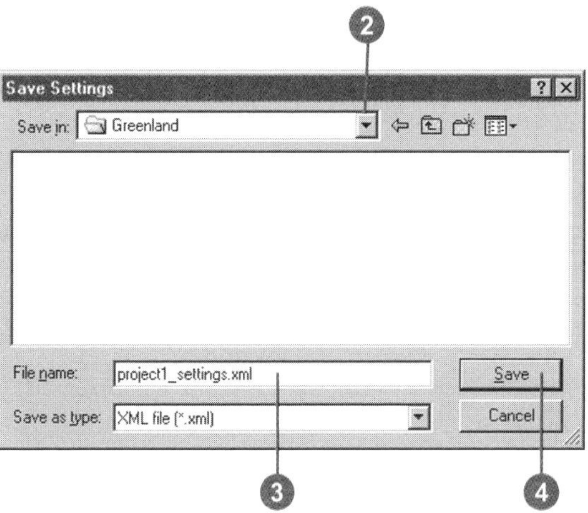

Chapter 3 to gain more of an understanding of which geoprocessing settings will be saved.

Loading variables

1. Right-click the ArcToolbox window, point to Load Settings and click either From File or From Default, depending on whether you want to load settings from a file, or you want to load the default settings for all applications.

2. If you choose From File, click the Look in dropdown arrow and navigate to the location of your saved settings XML file.

3. Click the XML file and click Open.

   Your saved variables are loaded so they can be reused at the command line.

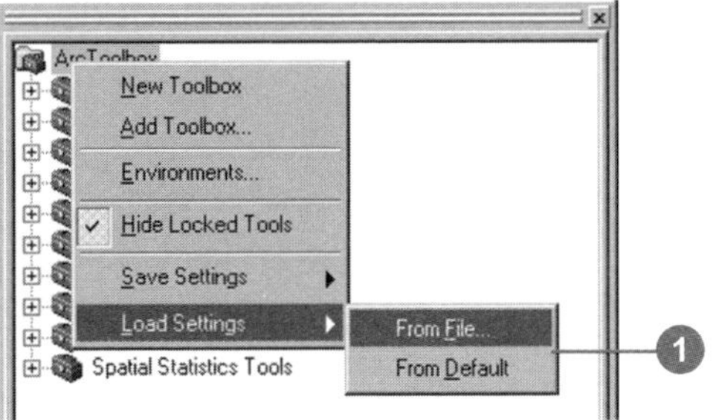

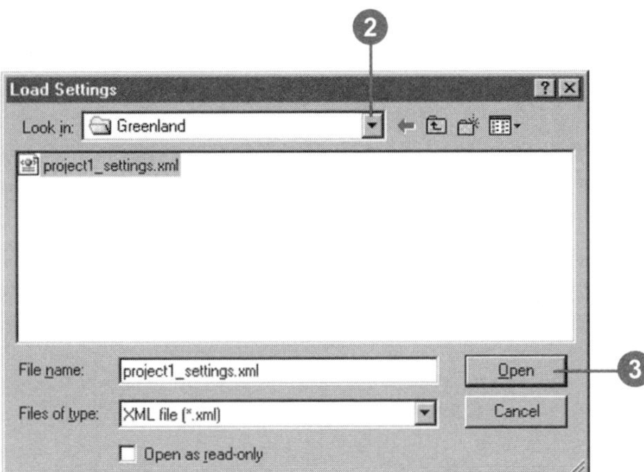

Introducing model building

8

IN THIS CHAPTER

- **What is a model?**
- **What is the ModelBuilder window?**
- **Building models**
- **A model building example**
- **Creating a new model**

The geoprocessing tools in ArcGIS make it easy to process spatial data to model aspects of the real world. However, when there are many steps involved in your geoprocessing work flow, it can be difficult to keep track of the assumptions, tools, datasets, and other parameter values you have used.

One of the easiest ways to author and automate your work flow and keep track of your geoprocessing tasks is to create a model. A model consists of one process or, more commonly, multiple processes strung together. A process consists of a tool—a system tool or a custom tool—and its parameter values. Examples of parameter values include input and output data, a cluster tolerance, and a reclassification table.

A model allows you to perform a work flow, modify it, and repeat it over and over with a single click.

This chapter introduces you to the concept of building models and the ArcGIS solution for building them. Chapter 9 explains the details of building models using the ModelBuilder window.

What is a model?

In general terms, a model is a representation of reality. A model represents only those factors that are important to your work flow and creates a simplified, manageable view of the real world.

In ArcGIS, a model is displayed as a model diagram. You automate your work flow by stringing processes together in the model diagram that will execute in sequence when the model is run.

The simple model that follows contains one process. The Buffer tool is used to create an output of buffer zones that are a certain distance around the input streams.

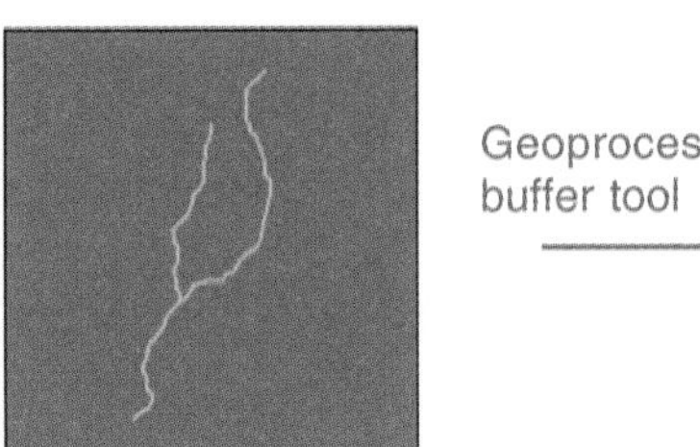

Input streams

Geoprocessing buffer tool

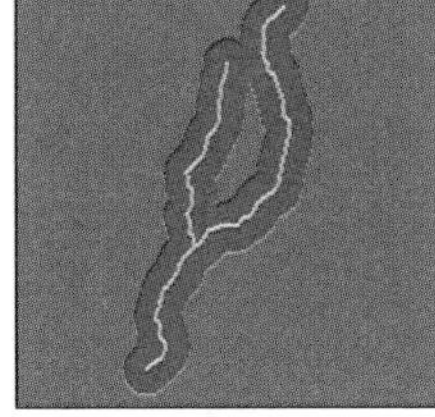

Output buffer around streams

A more complex model may combine quantitative information, for example, how far away something is or how much it costs, with qualitative information, for example, how desirable or important something is.

The model that follows considers population density and the distance to existing parks as factors in selecting a new park site. Areas of high density that are not close to existing parks on the Potential park sites map are the most suitable locations (shown in dark purple). Less suitable areas are shown in lighter shades. Gray areas mark the locations of existing parks. In this model, population density is a more influential factor, that is, has a higher weight, in site selection than distance to parks.

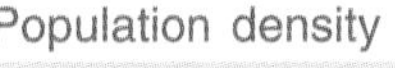

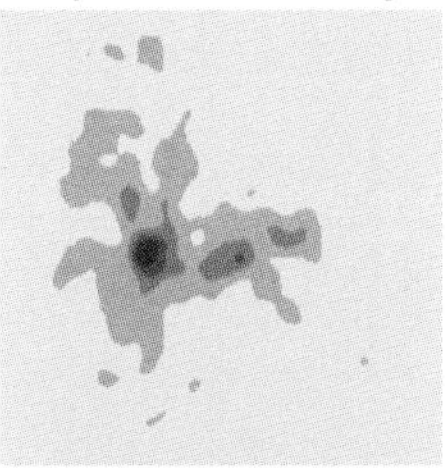

60% influence

Distance to parks

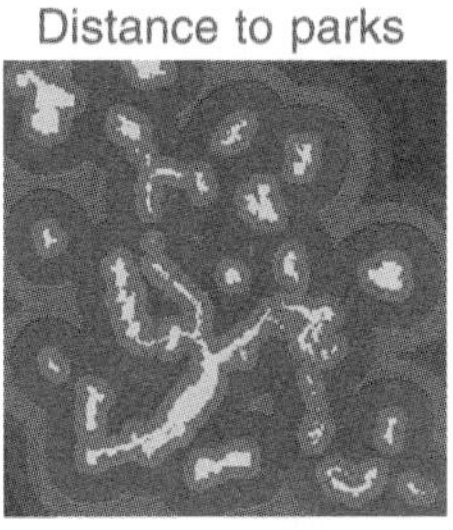

This model finds the most suitable location for a new park. The model incorporates a Weighted Overlay tool, where weights are assigned to each input based on how much influence each should have in siting a new park.

Potential park sites

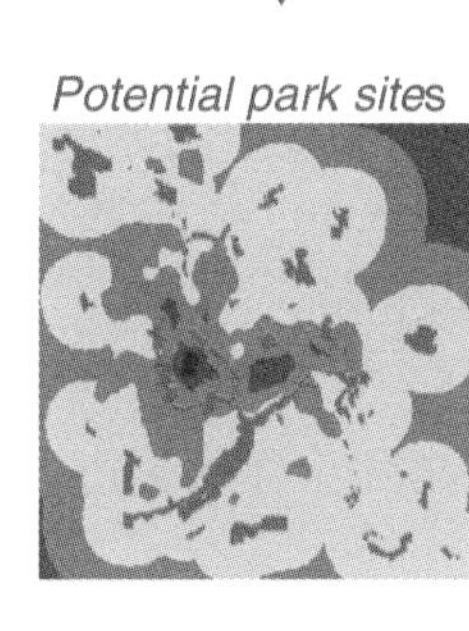

40% influence

The following model illustrates the preceding work flow.

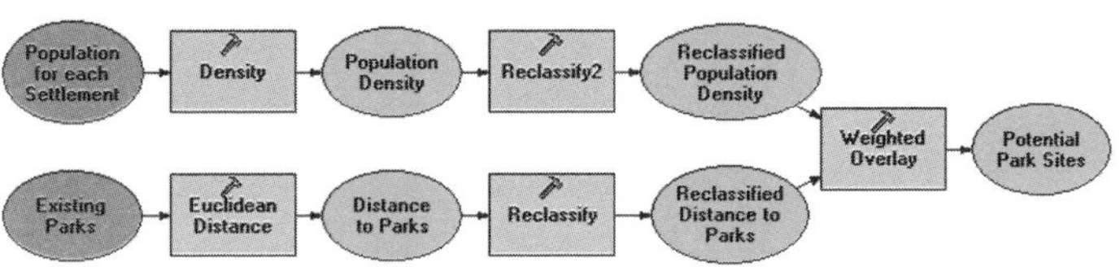

There are five processes in this model. One process calculates the population density from an input population dataset. A second process calculates the distance to parks from a dataset of existing parks. A third process reclassifies the Population Density output, and a fourth process reclassifies the Distance to Parks output. The fifth process takes the outputs from the two reclassifications and uses them as inputs to the Weighted Overlay process, where weights are applied to each input dataset based on the percent influence each should have. The output suitability map enables the decision maker to identify potential park sites.

Why build models?

Building a model helps you manage and automate your geoprocessing work flow. Managing processes and their supporting data can be difficult without the aid of a model. A sophisticated model contains a number of interrelated processes. At any time, you may add new processes, delete existing processes, or change the relationships between processes. You may also change assumptions or parameter values, for example, replace old datasets with newer ones, or consider alternative scenarios in which input factors are prioritized differently. Building a model helps you manage this complexity in a number of ways:

- It makes processes and the relationships between processes explicit, and the model you create is dynamically updated whenever a change is made.
- It lets you set values for the parameters of each tool, and it records this information, making the model output easily reproducible.
- It lets you edit the structure of the model by adding and deleting processes or changing the relationships between the processes.
- It lets you edit the parameter values defined for tools to experiment with alternative outcomes.

What is the ModelBuilder window?

The *ModelBuilder window* is the interface you use to create models in ArcGIS. A ModelBuilder window is displayed immediately when you create a new model; see Chapter 5, 'Working with toolsets and tools', for details on how to create a new model. The ModelBuilder window consists of a display window in which you build a diagram of your model, a Main menu, and a toolbar that you can use to interact with elements in your model diagram. You can run a model from within the ModelBuilder window or from its dialog box.

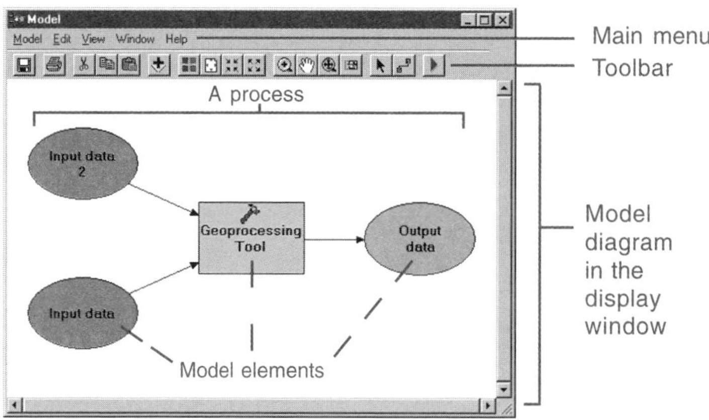

There are five pulldown menus on the Main menu. The Model menu contains options for running, validating, viewing messages, saving, printing, importing, exporting, and closing the model. You can also use this menu to delete intermediate data and set properties for the model. The Edit menu lets you cut, copy, paste, delete, and select model elements. The View menu contains an Auto Layout option that applies the settings specified in the Diagram Properties dialog box to your model. It also contains options for zooming in and out. The Window menu contains an overview window that you can use to display the entire model while you zoom in on a certain part of the model in the display window. From the Help menu, you can access the ArcGIS Desktop online Help system and the About ModelBuilder box.

The toolbar gives you quick access to much of the available functionality in the ModelBuilder menus and more.

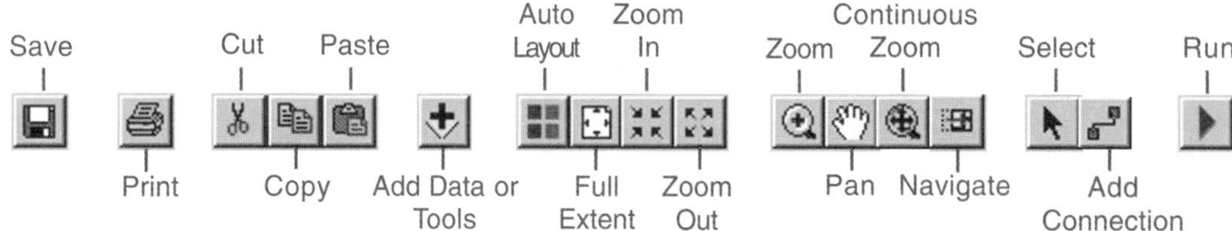

Building models

Inside a ModelBuilder window, the display window is the working area where you build a diagram of your model. The diagram you build looks like a flowchart. It consists of processes linked together that will run in sequence when the model is run.

Elements in the model diagram represent tools and their parameter values. A process consists of a tool element and its parameter values. In a typical process, elements represent the input data parameter value, the tool that operates on the input data parameter value, and the derived data parameter value. Connector lines indicate the sequence of processing. When you supply a value for an input data parameter, either within the tool's dialog box or by dragging data into the ModelBuilder window, the element created is a variable that can be shared between processes. Variables can be created for any parameter, making it easy to share any parameter value between processes.

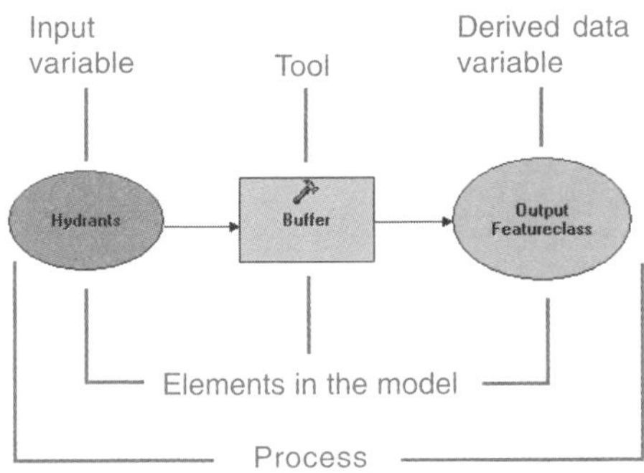

There will often be several processes in a model, and they can be chained together so that the derived data from one process becomes the input data for another process.

The diagram that follows is a conceptual overview of a model built from three processes.

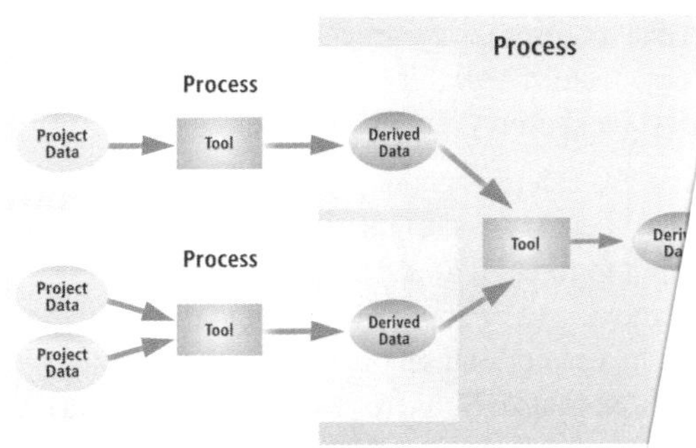

A conceptual overview of a model: A model conta processes, and a process contains a tool elemei its parameter values. In the example above, proj derived data parameter values are shared betw processes.

Each element in a model has a unique syr

Project data elements repre graphic input data that exists before the m he data referenced by these elemer input parameter values for tools roject data elements are represented l ovals and are variables by default. Varia arameter values so they can be sh processes.

Tool elements represen ıs to be performed on input dat llues. Tool elements are represent ctangles.

Derived data element output data created by a tool. Dat these elements does not exist until th The exception is when running a tool e project data, such as Add Field. I derived data is actually the project ditional field added. Derived dat ess can serve as input data for anot ived data

ments are represented by green ovals and are iables by default. Variables expose parameter values hey can be shared between processes.

Value **ue** elements reference nongeographic data meter values. Values set for these elements can be as input to tools in a model where appropriate. nples include a constant value used to multiply the alues in a raster dataset or the cluster tolerance for a tool to use. Value (or nongeographic data) nts are represented by light blue ovals.

Derived Value **d value** elements reference nongeographic data ter values that are created by running a tool. An e of a *derived value* is the output value from the Calculate Default Cluster Tolerance tool. values from one process can serve as input r other processes. Derived value elements are ed by light green ovals.

tor is a line showing the sequence of . Data elements and tool elements are together. The connector arrow shows the f processing.

In additi c model elements, there are text labels, which ar ements that place explanatory text in a model. A art of the processing sequence. The default text for e e changed, and labels can be attached to elements the model diagram.

ting text

nents

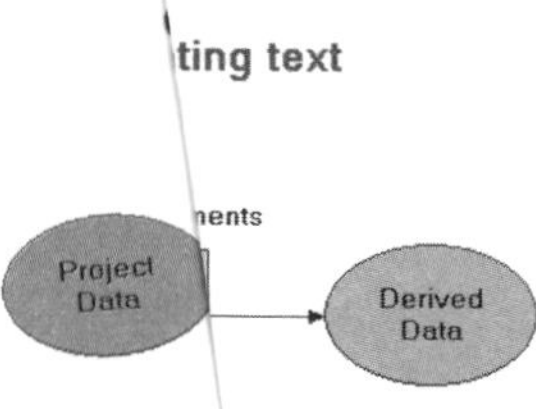

A model can be simple or complex. The simplest possible model contains a single process. In the model that follows, stream data is processed to create a dataset of buffer zones around streams. You can see the flow of project data (Input Streams) into the Buffer tool and from the Buffer tool to the derived data (Output Streams Buffer).

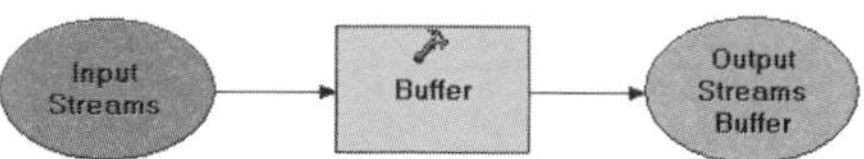

In the model that follows, the land parcels and the 100-year flood boundary are intersected to identify those land parcels that fall within the flood boundary.

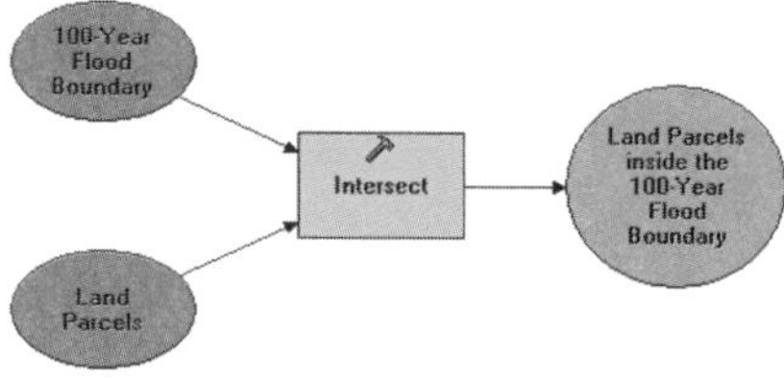

Models become more complex as processes are added. The model below has two parallel processes that share the same input data. The model generates two output datasets from one input elevation dataset.

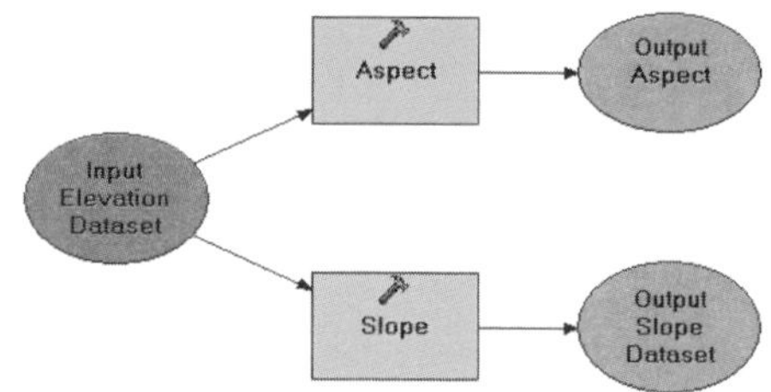

You can chain processes so that the output of one process becomes the input to subsequent processes. In the model that follows, one process interpolates a continuous elevation surface from an input dataset of measured elevation points. The derived data (Output Elevation Grid) is the input data to a second process that derives elevation contours.

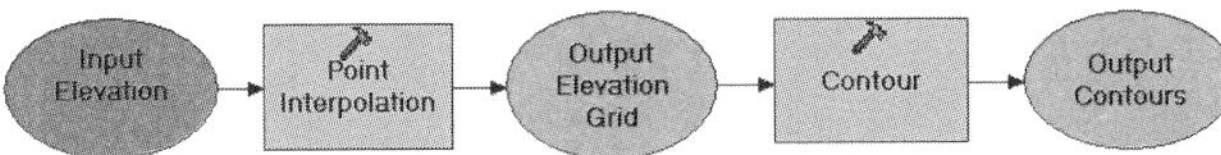

The model below finds suitable residential properties that are for sale. Suitable properties are those that are within an appropriate price range, less than five miles from a school, and outside the worst crime areas. First, buffer zones are created around schools to find residential areas close to schools. Next, crime zones, property parcels for sale, and school buffers are unioned together to produce an output dataset. The output features will have the attributes of the input features they overlap. The union output is then queried using the Make Feature Layer tool to identify the desired property parcels. The expression used in the Make Feature Layer tool will select from the Union Output attribute table those property parcels that are appropriately priced, falling inside the school buffer and outside high crime areas.

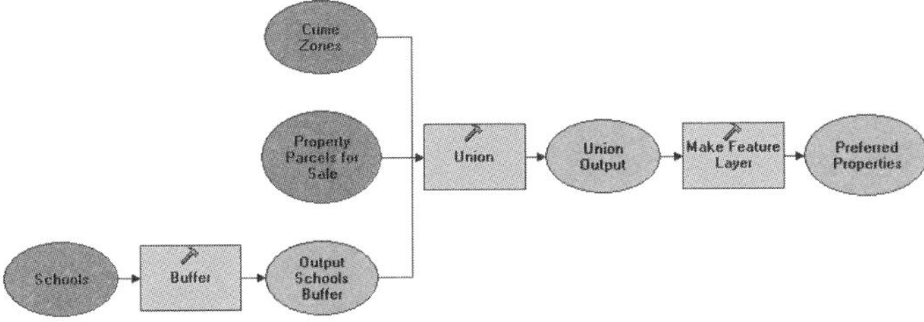

A model building example

Common spatial modeling questions involve finding a suitable location for a specific use, such as the best place to locate a business, storage site, pipeline, or park.

The goal

Before creating a model, you should ask, "What is the goal?"

In the following example, the goal is to find the easiest route through the landscape from Start to Finish.

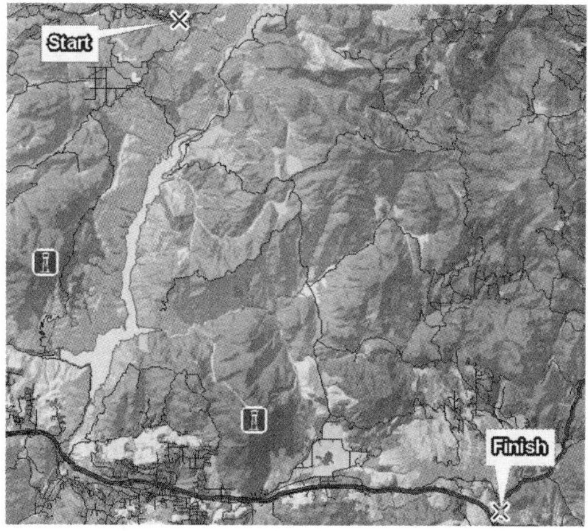

The goal: to find the easiest route from Start to Finish

Factors influencing the goal

The second question you should ask is, "What factors will influence the goal?"

In the example, the route must avoid the locations that can be seen by the people in the two observation towers, it must travel through vegetation types that are easiest to traverse, and it must travel through areas where the terrain is less steep.

Identifying input datasets

Once you have answers to the above two questions, you can decide on the input data you will need to help you achieve your goal.

In this example, the following input data is required:

- Elevation data to create a slope dataset that will enable areas of steep terrain to be identified and to create a viewshed dataset that will enable areas that can be seen from the observation towers to be identified
- Vegetation data to enable vegetation types that are easiest to travel through to be identified
- The location of the starting point for the path
- The location of the finish point for the path
- The location of the observation towers

Deriving datasets

You should decide whether any datasets need to be derived from your inputs in order to achieve your goal.

In this example, in order to reclassify a slope dataset to identify areas suitable for travel, the slope of the terrain needs to be derived from an elevation dataset using the Slope tool.

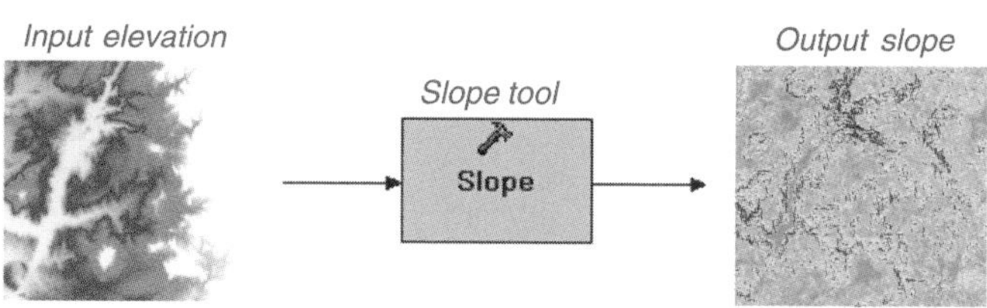

Areas shaded in red are the locations with the steepest slopes

The viewshed from the observation towers also needs to be calculated using the elevation data and the observation tower locations to identify those areas that can be seen from the observation towers.

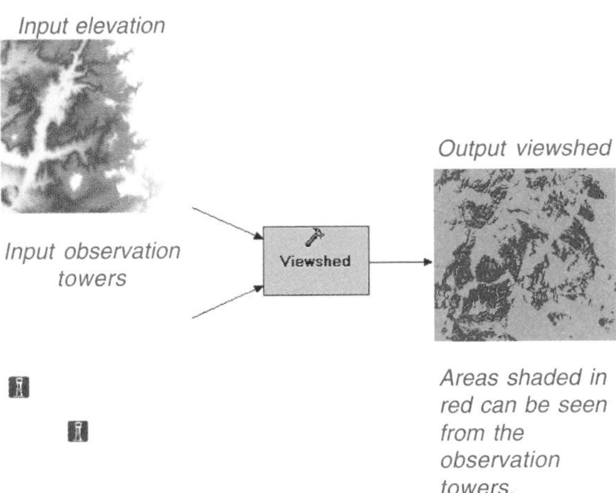

Areas shaded in red can be seen from the observation towers.

Creating a suitability map

In order to calculate a path through the landscape, the next step is to calculate how suitable each area (or cell) is to travel through, or how much it will cost to travel through each cell. This process, often referred to as weighted overlay, can be used for any kind of site selection or suitability model.

There are two steps involved in this process. First you would use the Reclassify tool to determine the categories or classes for each input factor and assign them a value representing how difficult or desirable it is to travel through a cell with those characteristics. You would then weight each input dataset by its overall importance using the Weighted Overlay tool.

Reclassifying derived data

Reclassification involves assigning higher or lower values to more suitable locations (or cell values). By examining the values within each dataset, you can determine which are more suitable. In this example, suitable values are those that represent ease of travel. Values for cells that represent grasslands in the input vegetation dataset will be given a lower value when reclassifying because they are easiest to travel through. If the range of new values to assign was from 1–10, values representing grasslands would be given a new value of 1.

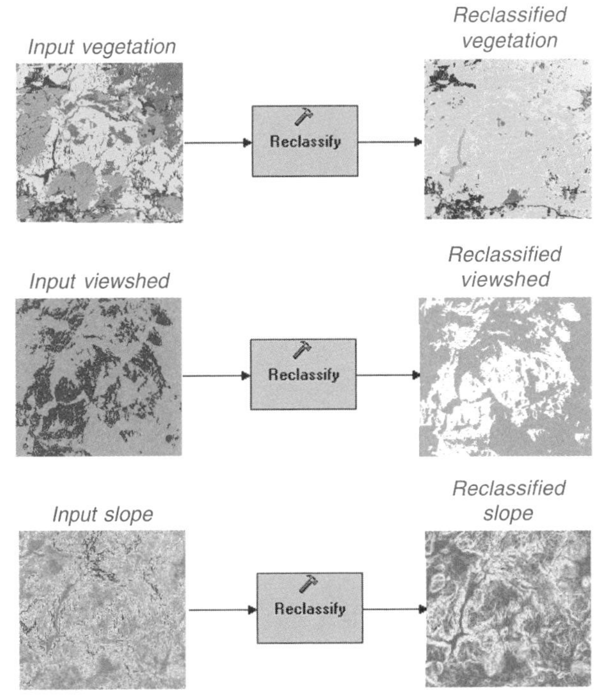

Weighting datasets

By combining the reclassified datasets using the Weighted Overlay tool, a higher influence (or weight) can be assigned to certain datasets. If all inputs share the same influence, you can assign each of them an equal percentage of influence. However, some datasets might require more influence in the weighted overlay process. For example, it might be more important to avoid steep slopes than to pass through vegetation types that are easiest to travel through.

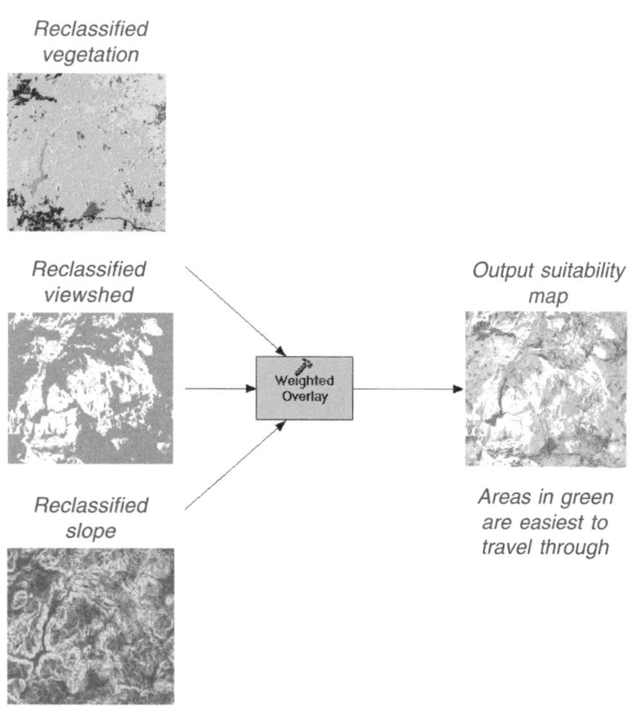

Identifying the path through the landscape

Once the suitability map (or the cost surface) has been created, you can use it to assess the suitability of a location for a particular purpose. In the output suitability map, areas that are easiest to travel through are displayed in green, and areas that are more difficult to pass through are displayed in red.

To calculate the least-cost path through the landscape, use the Cost Distance tool. It takes the cost surface created from the Weighted Overlay tool and calculates from each location (cell) the accumulated cost of traveling from any location back to the starting point. It also calculates which direction to take from any particular location back to the starting point using the least costly route.

The last process in this model uses the Cost Path tool and calculates a path through the landscape from the finish point along the least costly path back to the starting point.

You can see in the graphic that follows that the path takes the easiest route through areas where the terrain is less steep and the vegetation less harsh (areas shaded in green) and avoids those areas (shaded in dark red) that can be seen by the people in the observation towers.

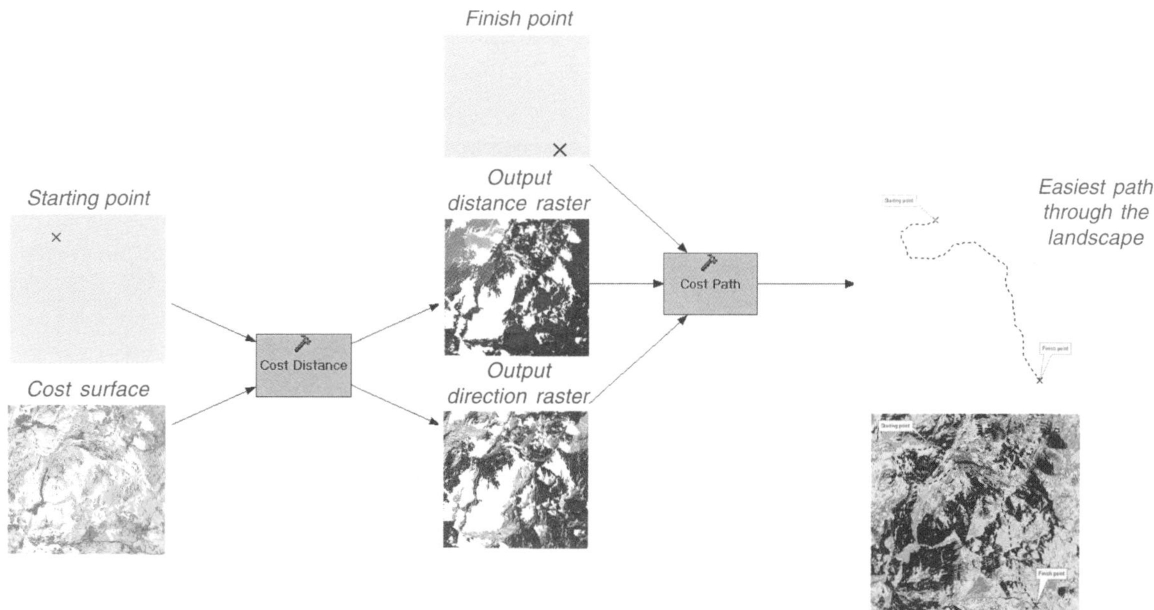

A typical geoprocessing work flow. The resultant path takes the easiest route through areas where the terrain is less steep and the vegetation less harsh (areas shaded in green) and avoids those areas (shaded in dark red) that can be seen by the people in the observation towers.

Creating a new model

Below is a summary of the series of steps you would take to create a new model.

Creating a model

When you create a new model in a toolbox in either the ArcCatalog tree or ArcToolbox window, a ModelBuilder window opens, allowing you to start building your model.

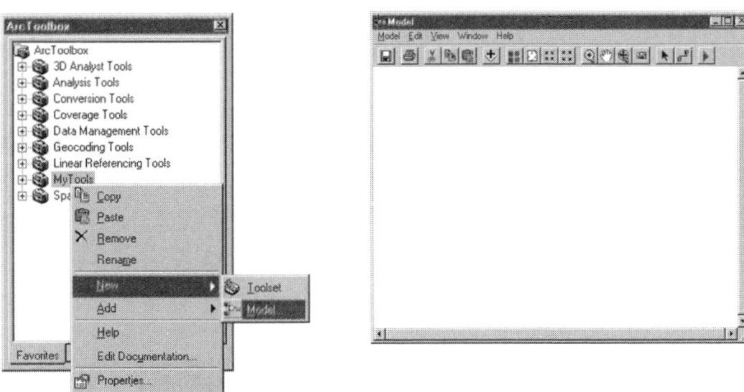

For more information on creating a new tool, see Chapter 5, 'Working with toolsets and tools'.

Building the model

There are multiple ways to add data and tools to a model. Click the Add Data or Tools button and add data, or drag input data from the ArcCatalog tree or layers from the table of contents of any ArcGIS Desktop application with a display (such as ArcMap). Alternatively, supply the input data parameter values inside the tool's dialog box.

Click the Add Data or Tools button and add tools from within toolboxes on disk, or drag a tool from the ArcCatalog tree or the ArcToolbox window.

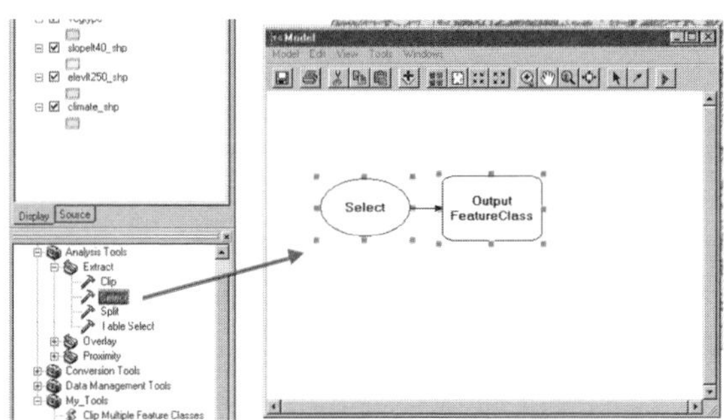

Double-click the element that represents the tool to supply the necessary parameter values for that tool. After adding a tool, you can also drag appropriate input data onto the tool, and the input data parameter value in the tool's dialog box will be supplied automatically.

Continue to add data and tools and supply values for parameters within each tool's dialog box. Connect processes until your model diagram is complete.

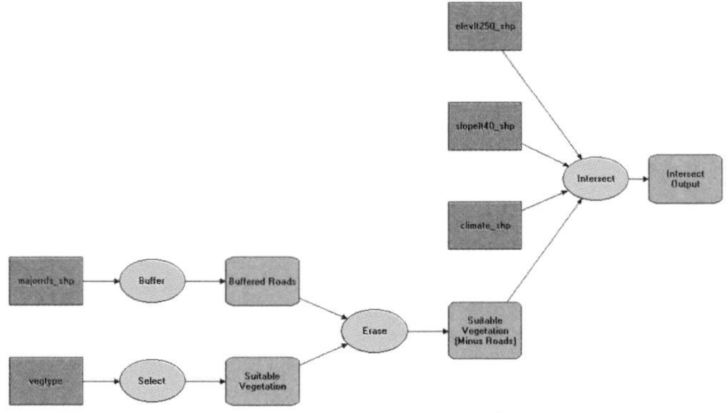

Saving and renaming the model

Once you have built your model, you should rename it to reflect its contents. You can rename the default name given to the model and its label. The label is the display name for the model. If you are referencing the model at the command line or within a script, you'll use the name.

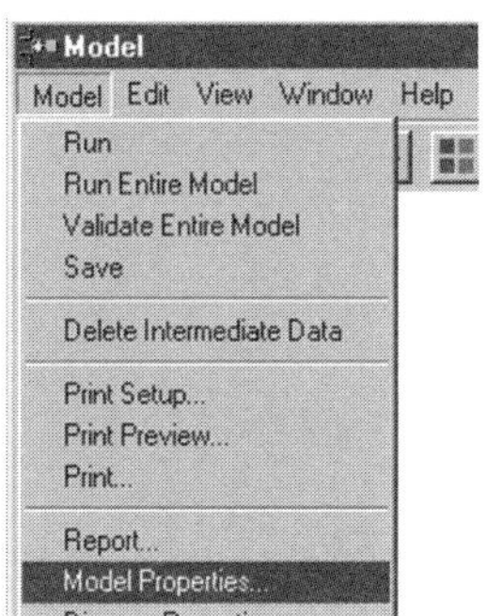

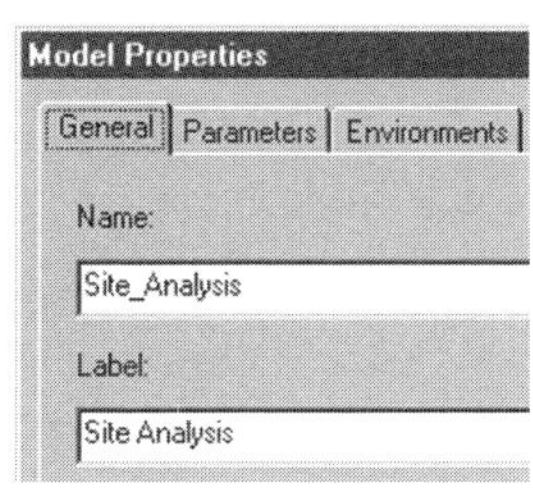

You should then save the model. Note that it is good practice to also save the model as you are building it.

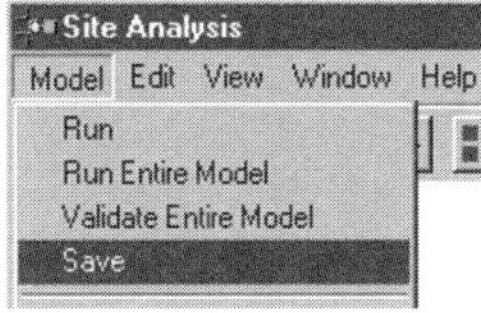

Running the model

You can either run the model from within the ModelBuilder window, or you can close the ModelBuilder window and run the model from its dialog box.

Generally, you'll run the model from within the ModelBuilder window as you are building it to ensure you are getting the results you expect from each process. You can run the entire model, only the processes that have not already run, or individual processes.

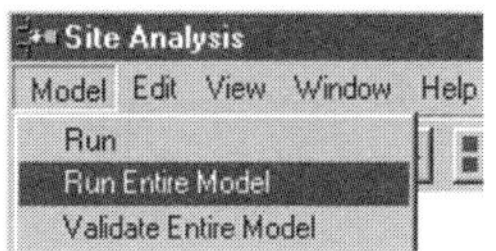

Setting model parameters

You'll likely want to set certain parameters as model parameters, so the user of your model can specify values for these parameters when running your model from its dialog box. You can do this in the ModelBuilder window or in the model's properties dialog box. By setting model parameters you can control which parameter values the user of the model can specify and which are hard coded inside the model. For instance, you might want users to be able to specify the input dataset they want to use.

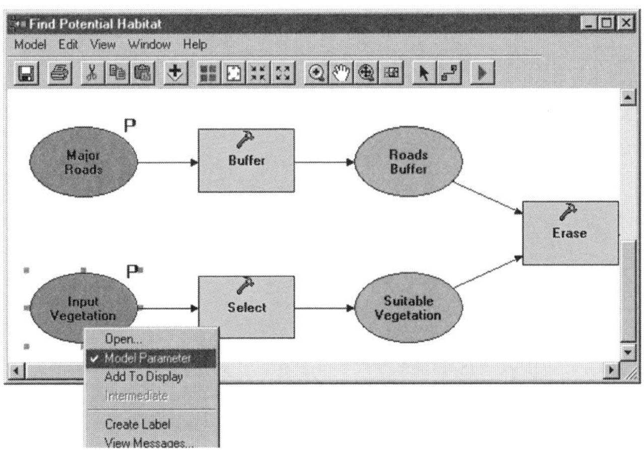

Once you are happy with the model and have set the required parameters as model parameters, save the model and open its dialog box to view the parameters that are set.

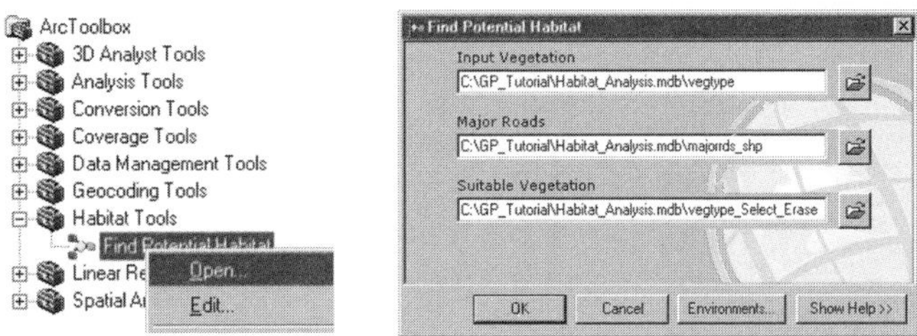

Modifying the model

You can easily modify parameter values in the model to explore "what if" scenarios and obtain different solutions, such as applying the same model to different geographic areas by changing the input data.

Documenting the model

You can create documentation for the model, including a summary of what the model does, an explanation of which input data it accepts, and a description of the expected results. You can include an illustration, parameter information, and example code for running the model at the command line or inside a script.

Documentation helps the user of your model. It provides information about what the model does and describes the parameter values that must be set.

For more information on documenting tools, see Chapter 5, 'Working with toolsets and tools'.

Sharing the model

You can share a model with others by sharing the toolbox that contains the model. However, sharing the toolbox does not share the data, scripts, Help files, or stylesheets it uses. To be sure you share all sources of information and that the path to all information sources will not have to be repaired, set relative pathnames for your model, then archive your geoprocessing work using an archiving program. See 'Sharing your geoprocessing work' in Chapter 3 for more information on distributing your work. See 'Storing relative pathnames' in Chapter 5 for information on relative paths.

Sharing a model lets you open your methodology to wider scrutiny and helps you refine and standardize modeling techniques. Models can be dragged into other models as with any system tool, allowing you to incorporate model components that have been developed by experts in various disciplines.

This chapter has introduced the concept of building models and explained the process of creating a new model. The next chapter will explain in more detail the various aspects of building models in the ModelBuilder window.

Using the ModelBuilder window

IN THIS CHAPTER

- **Building a model**
- **Element states**
- **Running a model**
- **Working with variables**
- **About intermediate data**
- **Controlling the flow of processing**
- **Using the Select Data tool**
- **Working with model elements**
- **Working with diagram properties**
- **Navigating in the model**
- **Validating and repairing a model**
- **Print setup**
- **Generating a report**
- **Documenting a process**
- **Importing and exporting a model**

The ModelBuilder window provides a graphical environment in which you can build models. The previous chapter introduced building models and the ModelBuilder window. This chapter explains in more depth how to build and work with models in the ModelBuilder window.

When you start working with the ModelBuilder window you'll find that:

- You can build a model by stringing processes together.
- You can construct processes by adding tools and setting values for the parameters of each tool.
- You can share parameter values between processes.
- You can set model parameters inside the ModelBuilder window so that the values for these parameters can be set when the model is run from its dialog box.
- You can change the default diagram properties to change the layout of the model or the symbology applied to elements.
- You can add text labels to the display window or attach labels to elements or connector lines.
- You can navigate easily in the model using the zoom or pan tools.
- You can easily repair an invalid parameter value or tool reference.
- You can print your model and generate a report.
- You can import existing models created in ArcView GIS 3, and you can export models to scripts or graphics.

Building a model

You use the ModelBuilder window to build models. When you create a new model inside a toolbox, the ModelBuilder window opens automatically so you can start to build your model.

You can edit existing models at any time. ►

See Also

See Chapter 4, 'Working with toolboxes', for information on creating toolboxes and Chapter 5, 'Working with toolsets and tools', for more information on creating a new model inside a toolbox.

See Also

See Chapter 8, 'Introducing model building', for more conceptual information about building models.

Editing a model

1. Right-click a model in a toolbox and click Edit.

   The ModelBuilder window opens so you can add or modify the processes in the model.

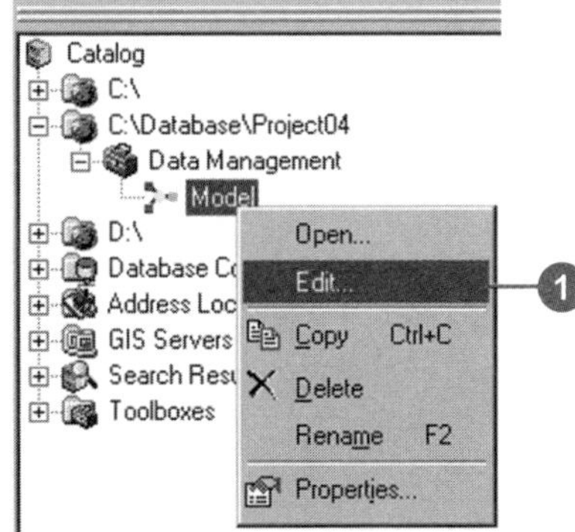

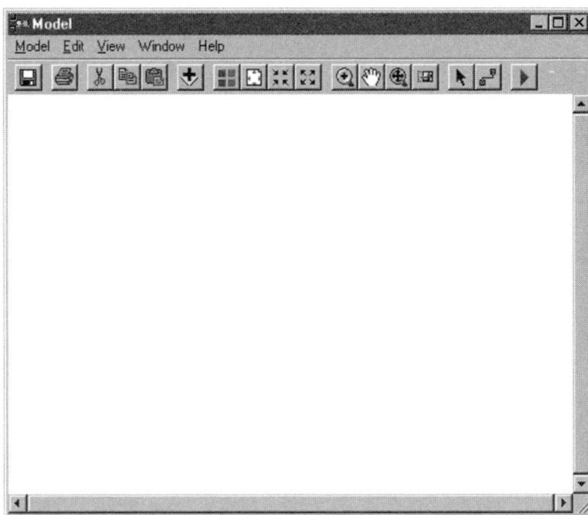

To build processes you add tools into the ModelBuilder window, then supply values for the parameters of each tool.

Both system tools and custom tools can be dragged into the ModelBuilder window. ▸

Tip

Embedding a model inside another model

When you drag a model into the ModelBuilder window, you can right-click the model and click Edit to open the model inside its ModelBuilder window. Embedding models inside other models reduces the complexity of any one model.

See Also

For more information about variables, see 'Working with variables' in this chapter.

See Also

See 'Element states' in this chapter for more information on the behavior of elements within a model.

Dragging tools into the ModelBuilder window

1. In the ArcCatalog tree or the ArcToolbox window, open the toolbox containing the tool you wish to add.
2. Click and drag the tool into the ModelBuilder window.

   Elements representing the tool and the derived data the tool will create are added to the display window.

   The derived data element is a variable that can be connected to other processes in the model.

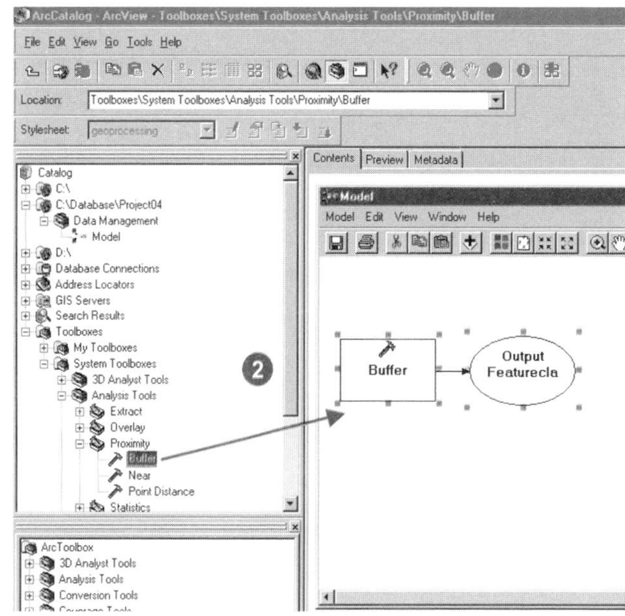

Values for parameters can be set inside the tool's dialog box.

If you are working in an ArcGIS Desktop application with a display, such as ArcMap or ArcGlobe, and you have added layers to the table of contents, appropriate layers will display in the input data parameter dropdown list inside a tool's dialog box. Alternatively, you can browse to data located on disk by clicking the Browse button next to the parameter.

Values must be supplied for required parameters. You can choose to accept the default value for optional parameters or choose to change it. You'll know if you need to supply a value for a parameter if a green circle is present to the left of the parameter. If, after supplying a parameter value, a red cross appears to the left of the parameter, the value entered is invalid. ►

See Also

See 'Working with variables' in this chapter for more information on creating and connecting variables and setting variables as model parameters that will display in the model's dialog box.

Setting values for a tool's parameters

1. After dragging a tool into a ModelBuilder window, right-click the tool and click Open.
2. Supply values for all required parameters and for optional parameters if desired.

   Optional parameters have the word (optional) written next to the name of the parameter.

   A default value is always set for output parameters. It can be modified if desired.
3. Check to be sure there are no icons to the left of any parameters.

   A green circle indicates that a parameter needs a value set.

   A red cross indicates that an invalid parameter value has been specified.
4. Click OK.

   An element is automatically created for the input data parameter, and it is connected to the tool.

   The input data element is a variable that can be shared between processes.

   The process should now be colored in, meaning it is ready to run.

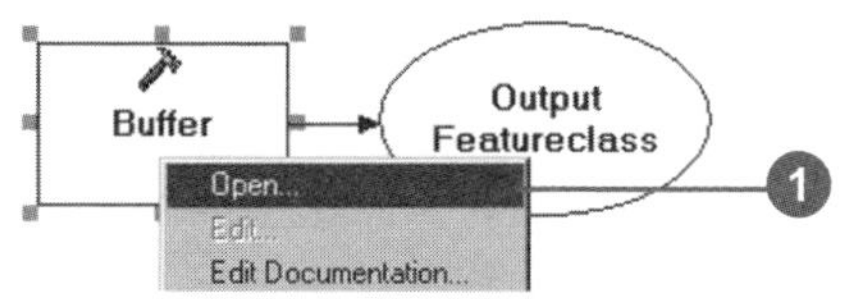

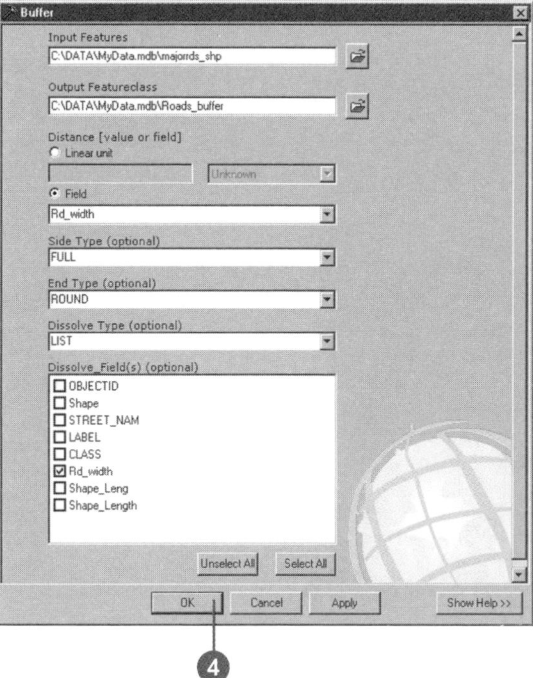

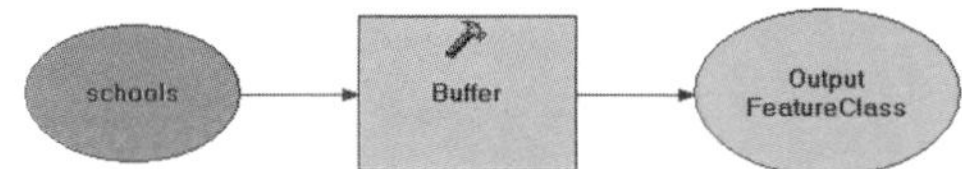

Instead of supplying the value for input parameters inside the tool's dialog box, you can drag project data from the ArcCatalog tree or from the table of contents of any other ArcGIS Desktop application, such as ArcMap, into the ModelBuilder window. The element that is added references the project data on disk and is a variable that can be connected to tools.

When you connect a project data variable to a tool, the value set for the variable will be used by the input parameter in the tool's dialog box. ►

Tip

Dragging multiple inputs into the ModelBuilder window

You can drag multiple input datasets into the ModelBuilder window from the ArcCatalog tree. Click the folder or geodatabase containing the datasets, then click the Contents tab and click multiple datasets while holding down the Ctrl or Shift key. With the datasets selected, click and drag them into the ModelBuilder window.

Tip

Adding tools and data

You can click the Add Data or Tools button in the ModelBuilder window and browse for data or tools.

Dragging and dropping project data

1. Click the project data that you wish to use and drag it into the ModelBuilder window.

   Data can be dragged from the ArcCatalog tree, or you can drag layers from the table of contents of any other ArcGIS Desktop application, such as ArcMap.

   An element representing the project data is added to the display window.

   The project data element is an input variable that can be connected to other processes in the model.

2. Click Add Connection.
3. Click the input variable.
4. Click the tool you want to connect it to.

   If the value set for the variable is of the correct input data type, the variable will connect to the tool. When all required parameter values are set, the process will be colored in.

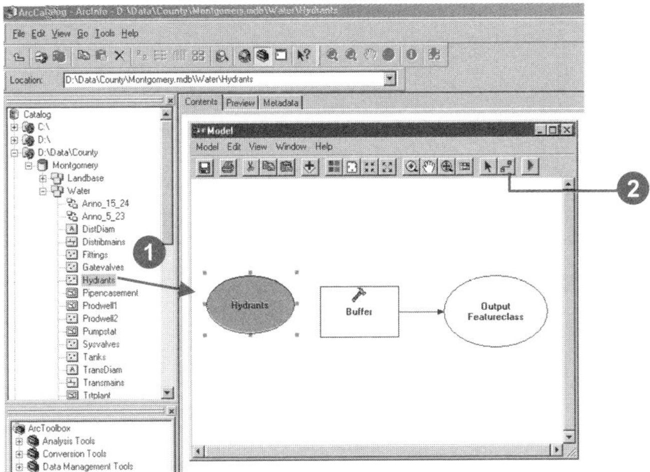

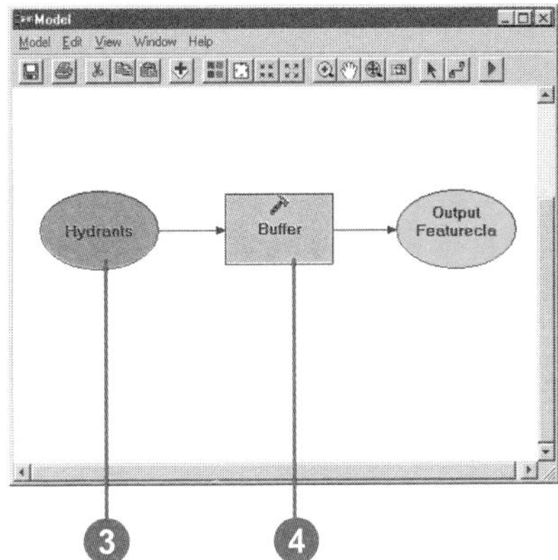

You can drag data into the model and place it directly on a tool. In doing so, a project data variable is created and attached to the tool. The value set for the variable is used by the input data parameter in the tool's dialog box. This provides a quick way to connect project data to a tool.

Project data can be dragged from the ArcCatalog tree or from the table of contents of any other ArcGIS Desktop application, such as ArcMap. ▸

Tip

Using appropriate values for parameters

If you find you can't use a particular value for a parameter, the value you are trying to use is not of the correct data type for the parameter. All inappropriate data types are filtered out of the Browse dialog box and the dropdown list for a parameter, and you cannot connect a variable containing an inappropriate value to a tool using the Add Connection tool.

See Also

You can choose to be prompted to select the parameter you want to apply a variable's value to when connecting variables to tools. See 'Displaying valid parameters when connecting variables' in this chapter for more information.

Dragging project data onto a tool

1. Click and drag the project data that you want to use directly onto a tool in the display window.

   An element representing the project data is added to the display window and is automatically connected to the tool.

   The project data element is a variable. The value set for the variable is used by the input data parameter in the tool's dialog box. The variable can be connected to other processes in the model.

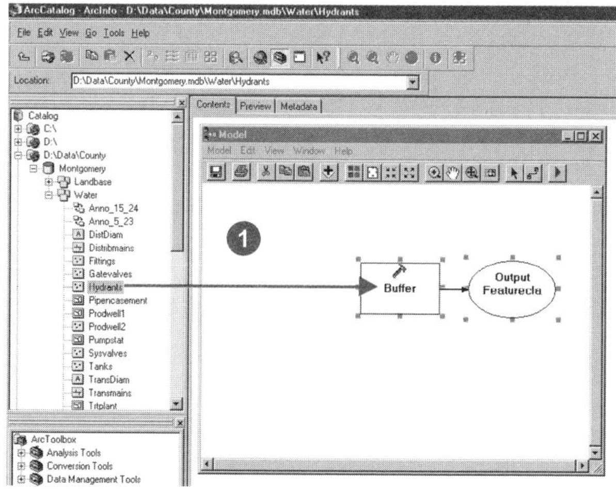

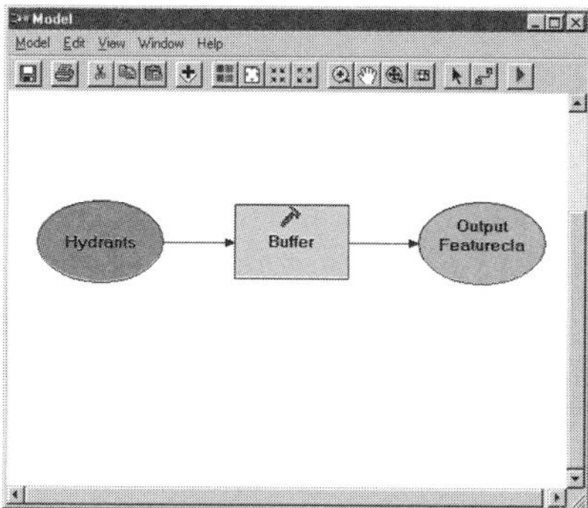

While building a model by adding processes and connecting them together, you should save your model. When you save a model, any additions or changes made to the model are saved.

You can change the display name and the label for the model via the Model Properties dialog box. It is best to change a model's default name and label to something more meaningful, so you can easily find the model the next time you want to access it.

Tip

An alternative way to rename a model

Right-click the model in its toolbox in the ArcCatalog tree or the ArcToolbox window and click Rename. Note that you should not alter the name of a model this way if you are currently editing it. When the model is saved, the current name set in the Model Properties dialog box will be applied.

Tip

Problems saving a model

If the toolbox that contains your model is read-only, you will not be able to save any changes you have made to the model. Make the toolbox writable to save your changes.

Saving a model

1. Click Model, then click Save.

   The model is saved with its existing name and label.

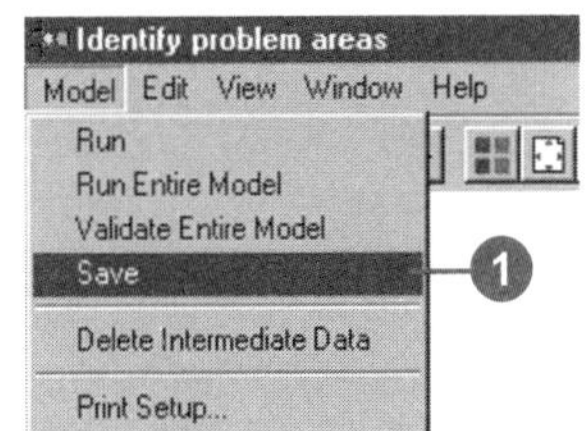

Renaming a model

1. Click Model, then click Model Properties.
2. Click the General tab.
3. Type a name for the model.

   This is the name you'll use when running the model at the command line or when adding it to a script.

4. Type a label for the model.

   This is the display name for the model. The display name appears as the name of the model in its toolbox in the ArcCatalog tree or the ArcToolbox window.

5. Click OK.

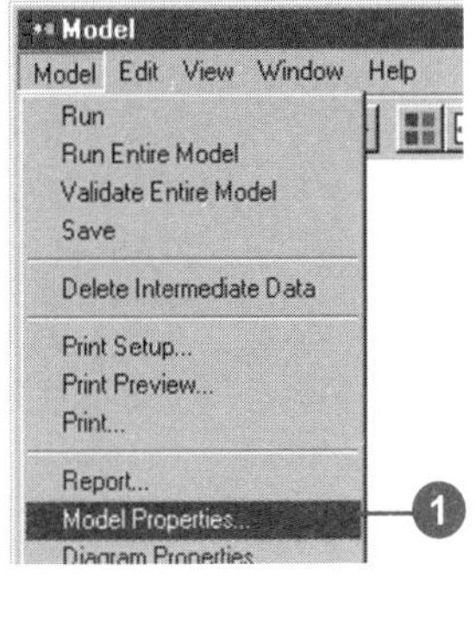

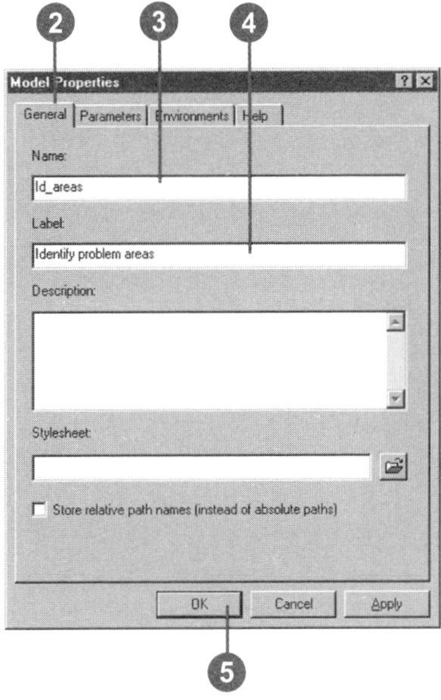

Element states

Each process in a model is in one of three states:

- Not ready to run
- Ready to run
- Has been run

The state of a process depends on the state of its elements. A process is ready to run when each of its elements is ready to run.

By default, elements that are not ready to run are symbolized in white. An element is not ready to run if the required parameter value or values for that element have not been set. When you initially drag a tool into a ModelBuilder window, the tool is in a not ready to run state because the required parameter values have not been specified, as the graphic that follows shows.

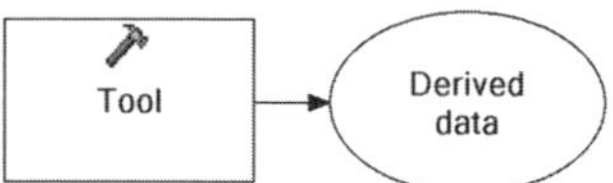

This process is not ready to run.

Elements that are ready to run are symbolized in color—input (or project) data elements are blue, tool elements are yellow, and output data (derived data) elements are green. A process is ready to run when all elements have been supplied with the required parameter values.

This process is ready to run.

When a process has run successfully, the tool and derived data elements are displayed with drop shadows, indicating that the process has run and the derived data has been generated.

This process has been run.

Running a model

There are three ways to run a model from within the ModelBuilder window:

- Run the processes that are ready to run that have not already been run.
- Run all processes including those that are ready to run and those that have already been run. ▶

See Also

In addition to running a model within a ModelBuilder window, you can run the model from its dialog box. See Chapter 5, 'Working with toolsets and tools', for more information. See 'Setting a model's parameters' in this chapter for information on setting parameters that will display on the model's dialog box.

Running only the ready-to-run processes

1. Click the Model menu and click Run.

   Alternatively, click the Run button on the ModelBuilder window toolbar.

   All processes that are in a ready-to-run state will run. In the example, only the Select and Erase tools will run. The Buffer tool has already been run, indicated by the drop shadow around the tool and derived data elements. This process will not run again.

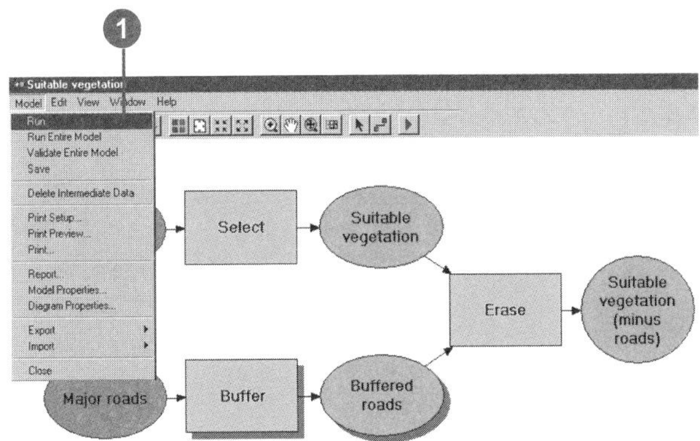

Running all processes

1. Click the Model menu and click Run Entire Model.

   All processes will run, including those that have already run, such as the Buffer tool in the example.

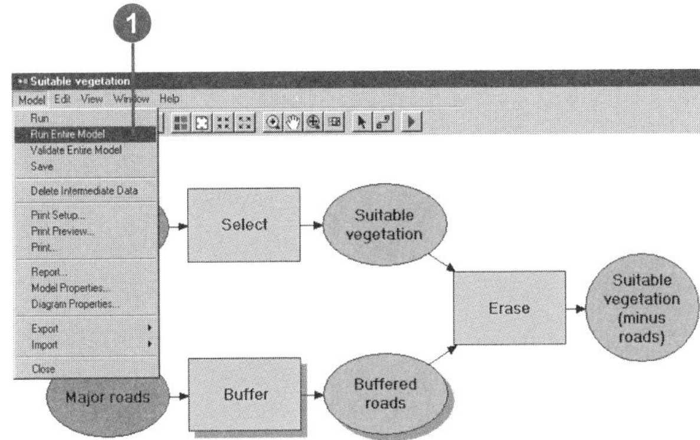

- Run a particular process in the model by itself. If the process the tool belongs to is linked to a chain of processes, earlier processes in the chain will also run if they have not already been run. Later processes in the chain of processes will not run, but if they have already run, their status will change to ready to run. ►

Running a single process

1. Right-click the tool of the process you want to run.
2. Click Run.

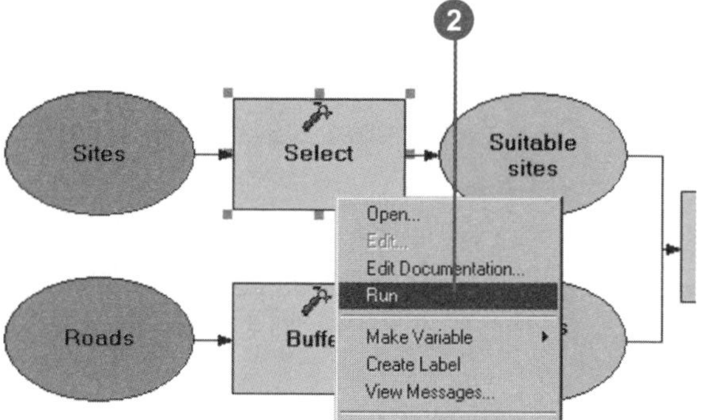

By setting the Add To Display property for derived data elements, the data referenced by these elements will be added as layers to the table of contents of the ArcGIS Desktop application you are working in after their processes have run, provided you are not working in ArcCatalog.

The layers added will honor the 'Results are temporary by default' setting in the Options dialog box, accessed via the Tools menu on the Standard toolbar.

The Add to Display option only applies to data created by running a model within the ModelBuilder window. When a model is run from its dialog box, only data referenced by derived data variables that are set as model parameters (seen in the model's dialog box) will be added to the display. The user of a model usually only cares about the final result, not the intermediate data created by the model. ►

Adding derived data to the display

1. Either before or after running a process, right-click the derived data element and check Add To Display.

   The derived data is added as a layer to the display if the process has been run.

   If the process has not yet been run, the derived data will be added to the display the next time it is run.

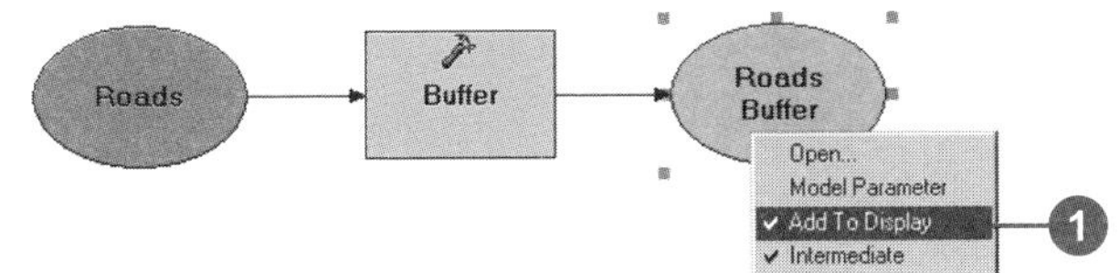

When you run a tool, status messages are displayed in the progress dialog box and also in the message section of the Command Line window (provided it is open). The status messages give the following information for each tool run: the parameter values specified, the time the tool was run, the status of the execution, and the time the tool finished executing. If errors occur when a tool is executing, these messages will also be displayed.

Status messages produced for all tools in the model can also be viewed within the ModelBuilder window. You can view messages for all tools or just the messages that relate to a particular tool.

See Also

Before running a model, you can validate it to check that all the tool references and parameter values set are valid. Any errors encountered when validating can be viewed in the model's report or a tool's Messages dialog box within the ModelBuilder window. Invalid tool references or parameter values can be repaired. See 'Validating and repairing a model' in this chapter for more information.

Viewing all status messages

1. After running tools inside the ModelBuilder window, click the Model menu and click Report.
2. Check to display a report in a temporary window or to save a report to a file.

   If saving a report to a file, click the Browse button and navigate to the location on disk into which you want to save the file. Type a name for the file and click Save. Alternatively, type the path to a location and type a name for the XML file (with a .xml extension).
3. In the temporary window, or after navigating to the location of the XML file you created on disk and opening the XML file using Internet Explorer, scroll to the Processes section and expand the contents of one of the tools.
4. Expand the Messages section to see messages for the tool.
5. Click OK.

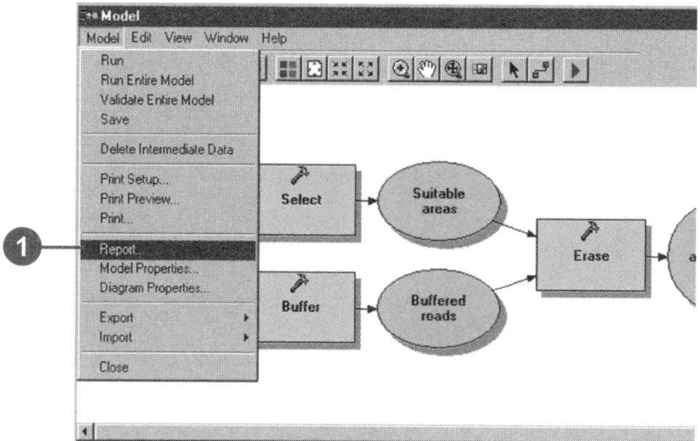

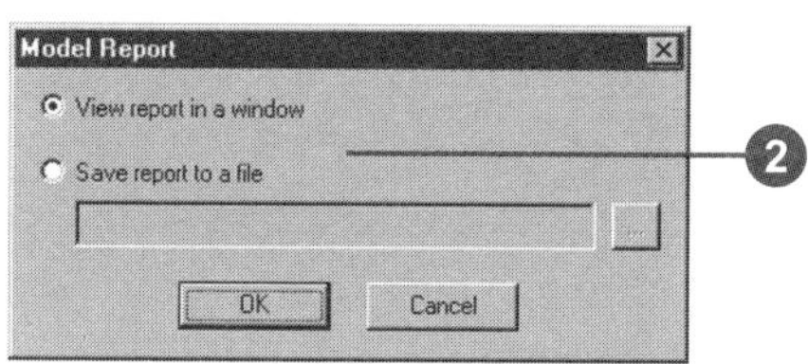

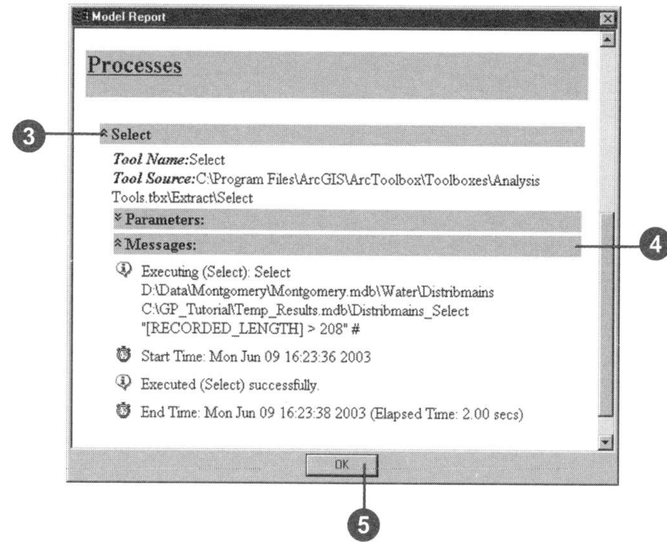

Viewing status messages for a single tool

1. After running a tool inside the ModelBuilder window, right-click the tool for which you wish to see status messages and click View messages.

   Status messages are displayed in a Messages dialog box.

2. Click OK to close the Messages dialog box.

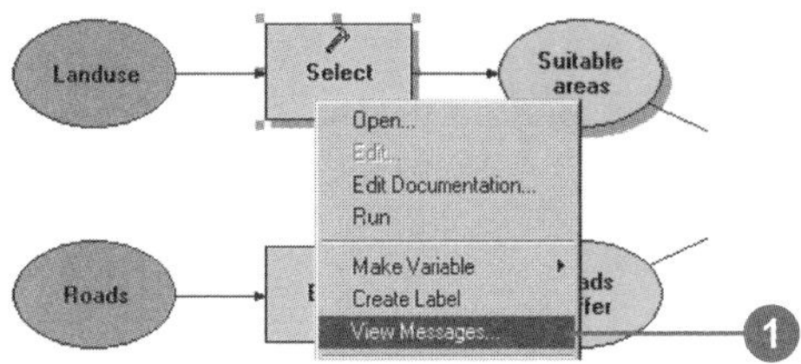

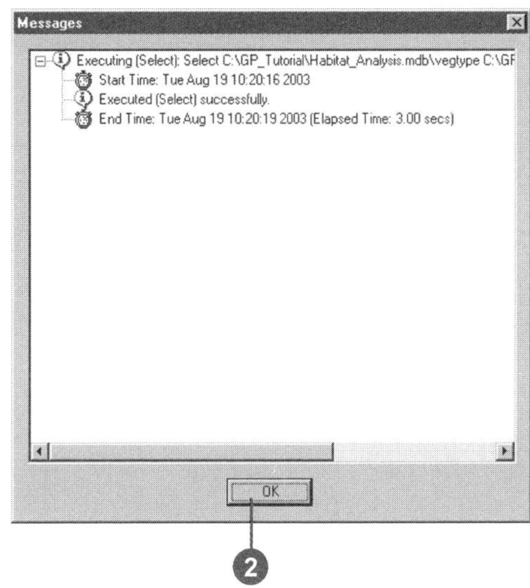

Using nonexistent output data

After adding a tool to a ModelBuilder window and supplying its parameter values, the tool usually knows the properties of the derived data it will create, so you can connect the derived data as input for the next tool without having to first run the initial tool.

In the following example, the Add Join tool accepts a layer as the input to which other layers, feature classes, or tables can be joined. The Make Feature Layer tool knows the properties of the output it will create—a layer that is based on the input feature class on disk and the parameter values specified within the tool. As the properties of the derived data 'Roads Layer' are known, the Add Join tool can use this information to provide choices for its input parameter values, such as available fields.

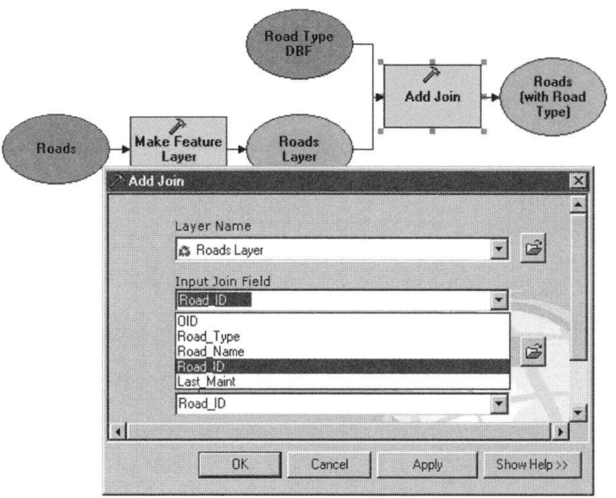

In some cases, if the derived data does not exist (in other words, the preceding tool has not been run), the succeeding tool cannot extract the information it needs in order to populate parameters with available values. You'll encounter this when creating a model. You might come across a parameter for which you expect to see a choice list displaying available values, such as a list of fields. Examples of such cases include when connecting the derived data from a process that runs a script as input for another tool or when connecting the derived data from a data conversion tool, such as the Import From Interchange File tool, as input for another tool.

In the example that follows, the Import from Interchange File tool is used to create a coverage from an E00 file of soil types. The Dissolve tool takes the output from the Import From Interchange File tool and dissolves the boundaries between soil polygons of the same type. As the Import From Interchange File tool has not yet been run, the Dissolve Item parameter in this tool's dialog box does not know what the available fields are that could be used as the Dissolve field.

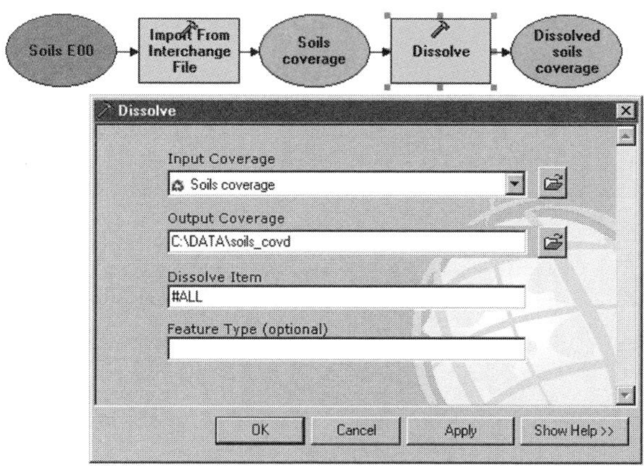

If a tool in your model is not able to set its output properties without being run, this will propagate through all succeeding tools in your model. In the example below, the Index Item parameter on the Index Item tool has no items available because the Import From Interchange File tool has not been run.

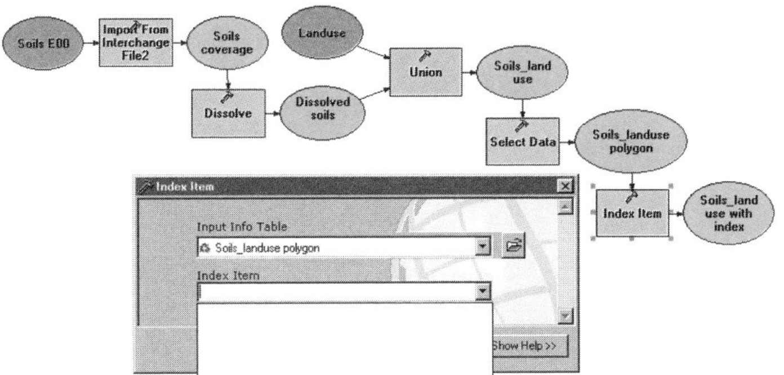

When you encounter such a situation, you have two options:

1. If you know the behavior of the tools in your model and know what derived data they will create, you can type the values for parameters in the tool's dialog box and continue to build your model. In some tool dialog boxes, you can click to add items to lists (such as using the Add Field button), then enter the text for the item.

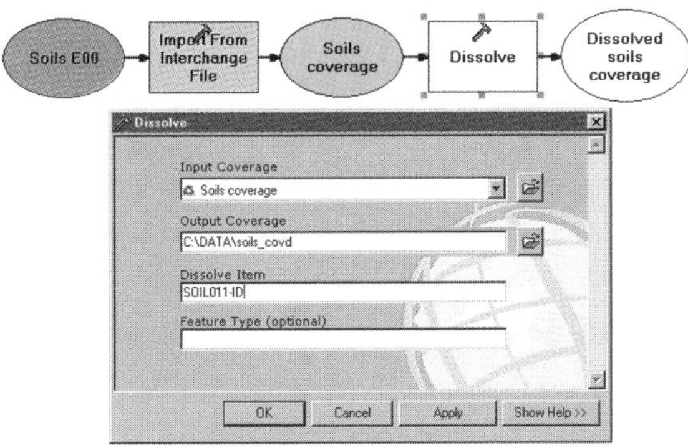

If values are known, type values for parameters if available parameter values are not presented for you in a tool's dialog box. In the example, the name of the field in the coverage that will be used for the dissolve is entered (SOIL011-ID).

2. If you are unsure of a tool's behavior and the next tool in your model requires information from this tool in order to provide choices for its parameter values, simply run the tool. Values will appear for parameters in the succeeding tool's dialog box. This happens because the input parameters are known once the preceding tool has been run.

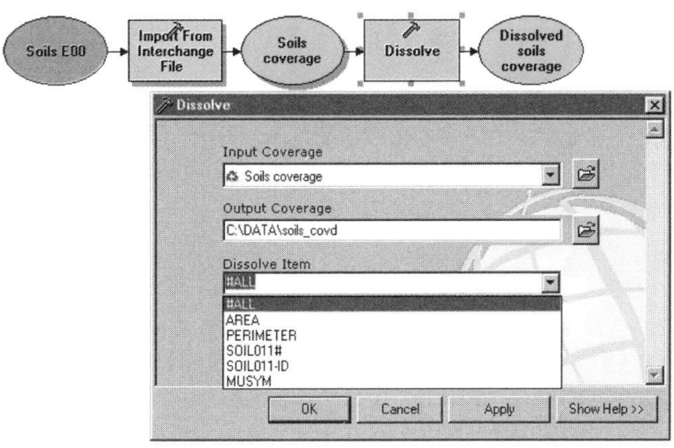

Run the preceding tool (Import From Interchange File in the example) to populate the available parameter values in the succeeding tool's dialog box (Dissolve in the example).

Working with variables

You can think of a variable as a 'container' that holds a value that can be changed. In the context of a model, a variable can be created and its value used in place of a parameter value.

Reasons to create variables:

- You may want to share the values set for variables between processes. To keep you from having to supply a value for a parameter each time it is required in a tool's dialog box, you can simply use the value set for the variable each time it is needed.
- You may want to create empty variables simply to construct a model as a plan of the work flow you will follow.
- You may want to set variables as model parameters that will display in the model's dialog box. ►

Tip

Selecting variables

A variable is identified by its icon in a parameter's list of values.

Connecting variables using the Add Connection tool

1. Click the Add Connection tool.
2. Click the variable containing the value that you want to share between processes.
3. Click a tool to which you want to connect the variable.
4. Click the variable again.
5. Click the next tool to which you want to connect the variable.

   Continue steps 2 and 3 until you have shared the variable with all desired tools.

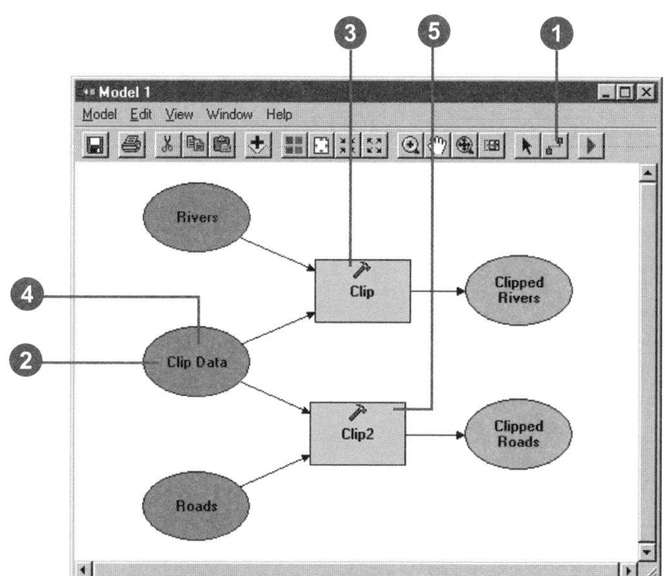

Connecting variables using the parameter's dropdown list

1. Double-click a tool.

   The tool's dialog box opens.
2. Click the dropdown list for the parameter and select the variable.

   The variable's value is used by the parameter.
3. Click OK.

   Continue steps 1–3 until you have shared the variable with all desired tools.

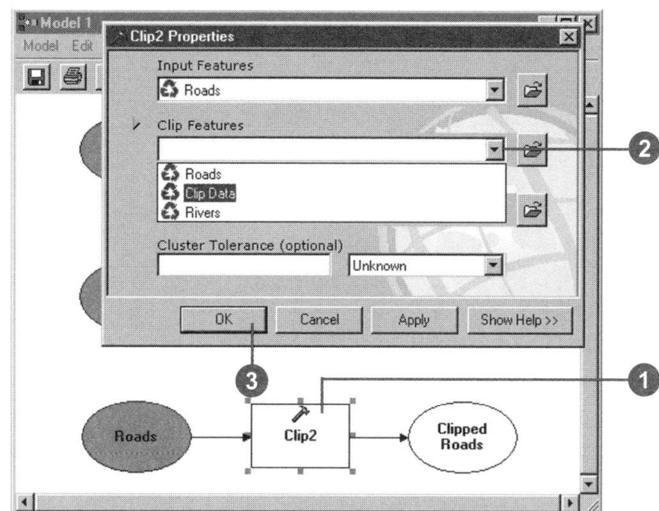

If you connect a variable to a tool and the variable's value is accepted by more than one possible parameter, you may find that you'll have to open the tool's dialog box and alter the parameter values set to apply the value set for the variable to the correct parameter.

You can choose to be prompted to select the correct parameter when connecting variables to tools.

For instance, the Clip tool has two input parameters: an Input Features parameter and a Clip Features parameter. You can choose to be prompted to apply the value set for the variable to the correct input parameter for the Clip tool. ▸

Displaying valid parameters when connecting variables

1. Click the Tools menu and click Options.
2. Click the Geoprocessing tab.
3. Check When connecting elements, display valid parameters when more than one is available.
4. Click OK. ▸

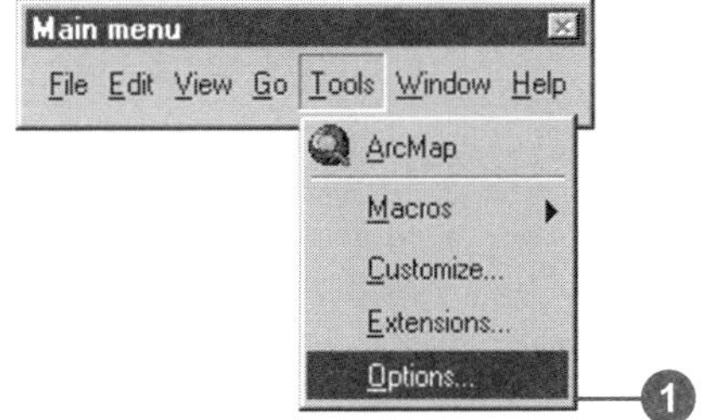

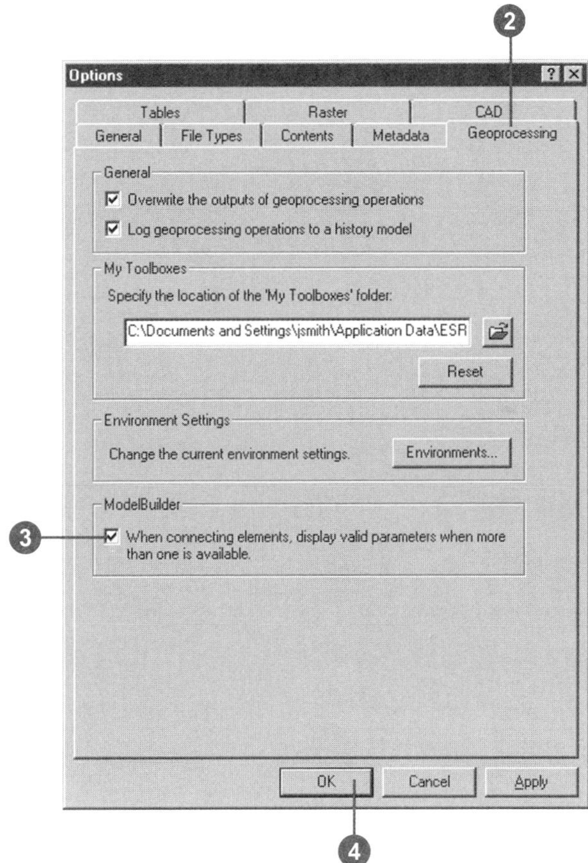

5. Inside your model, connect the input variable to a tool.

   If the value set for the variable could be applied to multiple parameters, the Select Parameter dialog box will display.

6. Click the parameter to which you want the variable's value to be applied, then click OK.

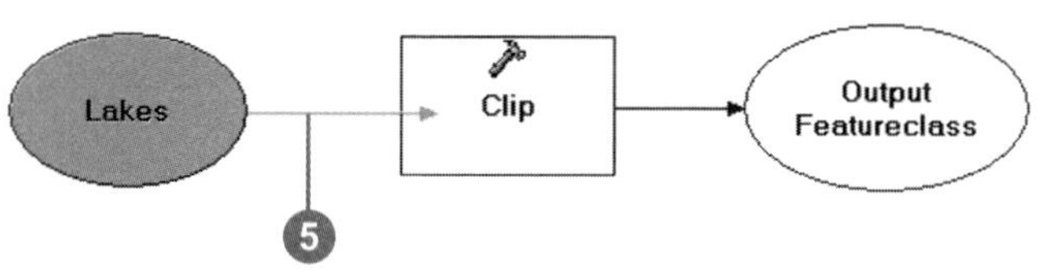

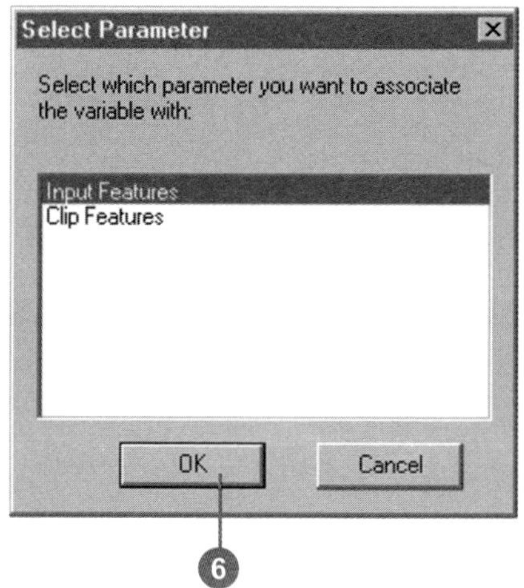

A variable is created when you drag data from the ArcCatalog tree onto the ModelBuilder window or supply a value for input dataset parameters inside a tool's dialog box. The value set for the variable is the input data on disk. Variables are automatically created for input and output data because you usually want to connect processes in your model together.

All other parameters for a tool must be set as variables before they can be shared between processes. ▸

See Also

See 'Building a model using empty variables' in this chapter for an alternative way to create variables.

Creating a variable for a parameter

1. Right-click the tool and point to Make Variable.
2. Click From Parameter and click the parameter for which you want to make a variable.

   The variable is added as an element in your model and is automatically connected to the tool.
3. Right-click the variable and click Open.
4. Specify or alter the value for the variable and click OK.

   The variable is colored in, meaning it has a value set and is ready to run.

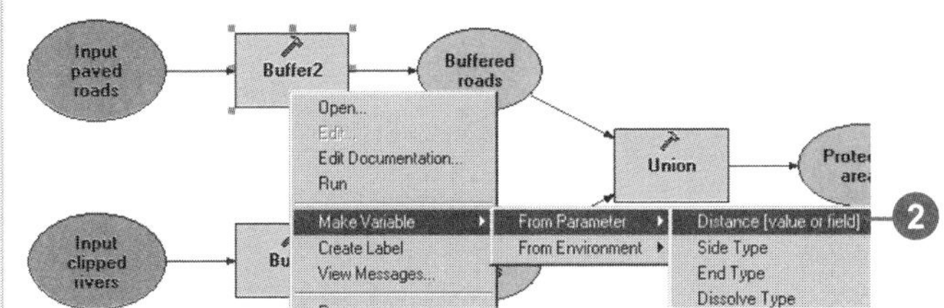

The buffer distance parameter is set as a variable to expose the parameter, so the value set for the parameter can be shared between processes.

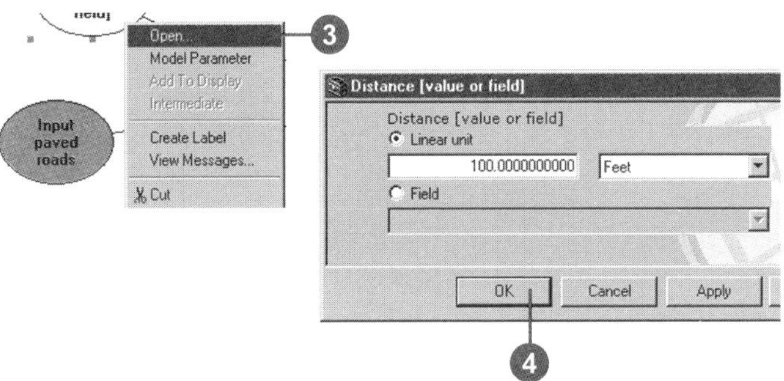

Right-click the created variable, click Open, and set a value for the variable.

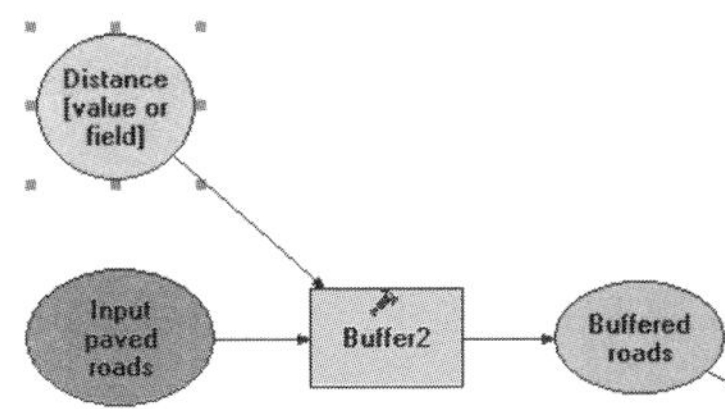

The variable is colored in, meaning it is ready to run.

You can make a variable from an environment setting and have the value set for the variable used automatically by the tool. For instance, you may want to use a different workspace for the results of some of your processes.

Rather than setting this environment setting for all processes—by right-clicking each tool, clicking Properties, then clicking the Environments tab to change the default value set—you can make a variable from an environment setting, give it a value, then connect it to all desired tools. ▸

See Also

See 'Building a model using empty variables' in this chapter for an alternative way to create variables.

See Also

See Chapter 6, 'Specifying environment settings', for information on the hierarchical nature of environment settings.

Creating a variable for an environment setting

1. Right-click the tool, point to Make Variable, then click From Environment and click the section of environment settings that contains the setting that you want to make into a variable.
2. Click the environment setting.

   A variable is created for the environment setting, and it is automatically attached to the tool. If a value for the environment setting has been specified in the Environment Settings dialog box, it will be used by the variable.
3. Right-click the variable and click Open, or double-click the variable, and specify or alter its value.
4. Click OK.

   The variable is colored in, meaning it has a value set and is ready to run.
5. Click the Tools menu and click Options.
6. Click the Geoprocessing tab.
7. Check When connecting elements, display valid parameters when more than one is available.
8. Click OK. ▸

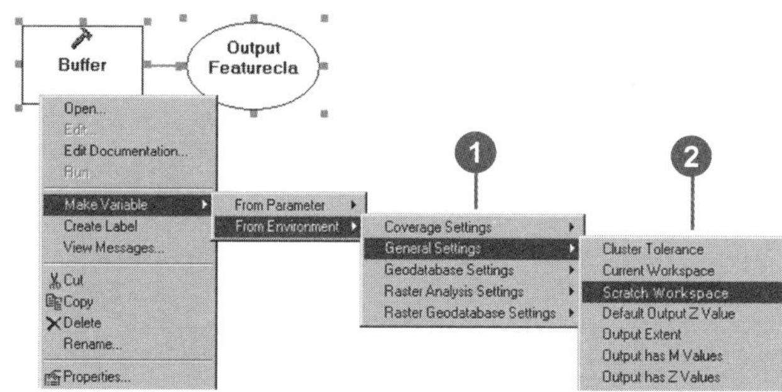

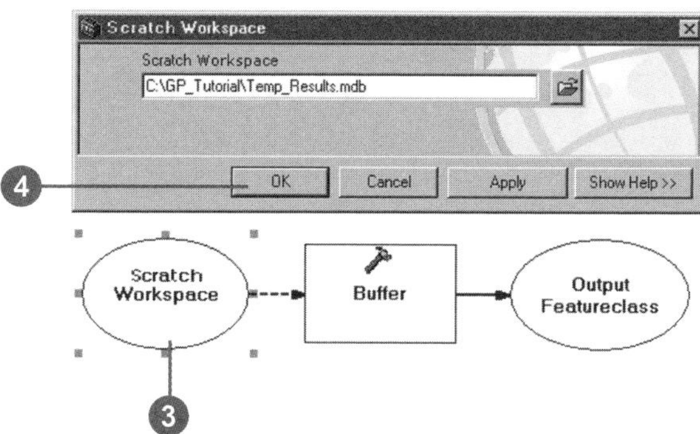

9. Click the Add Connection tool, click the environment setting variable, then click the tool.
10. Click the environment setting that you want to associate the variable with, then click OK to connect the environment setting variable to another tool in the model.
11. Follow steps 9 and 10 until the environment setting variable has been shared with all desired tools.

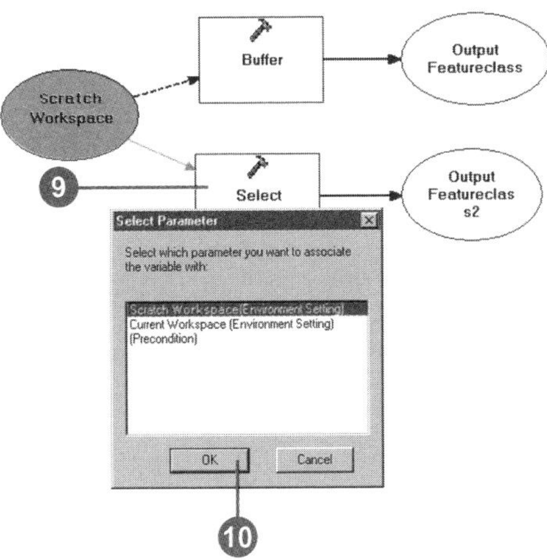

If you make a variable from an environment setting by right-clicking the tool and clicking Make Variable, the variable is automatically connected to the tool.

There are two ways you can work with environment setting variables created by right-clicking the display window and clicking Create Variable.

You can connect the variable using the Environment Settings dialog box for a tool added to the ModelBuilder window, or you can connect the variable directly to a tool using the Add Connection tool. By checking the option to display valid parameters when connecting variables (on the Geoprocessing tab of the Options dialog box, accessed via the Tools menu on the Main menu of the application you are using), you will be prompted to select the appropriate environment setting for which you want to use the value set for the variable, when you connect the variable to a tool.

Connecting variables for environment settings via the Environment Settings dialog box

1. Add a tool into the ModelBuilder window either by dragging the tool from the ArcCatalog tree or ArcToolbox window or via the Add Data or Tools button.
2. Right-click the display window and click Create Variable.
3. Click the data type you want to use for the variable.
4. Click OK.

   An empty variable is created.
5. Double-click the variable to give it a value, then click OK. ▸

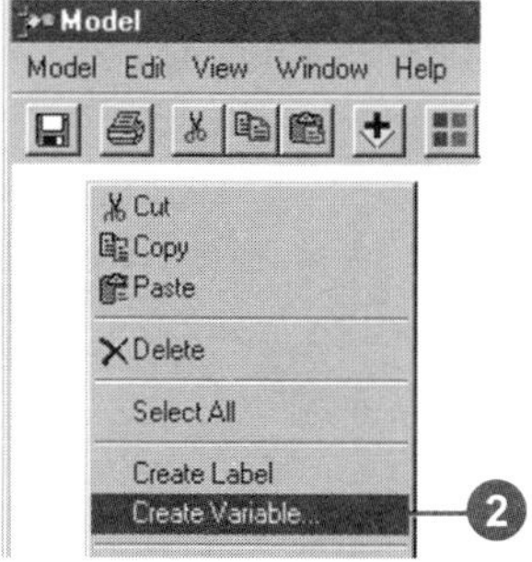

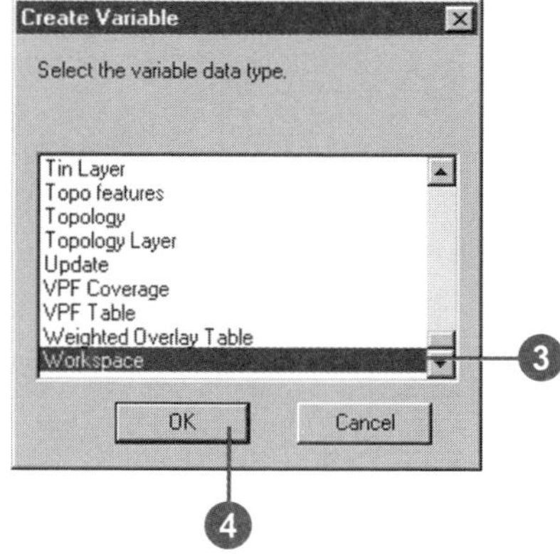

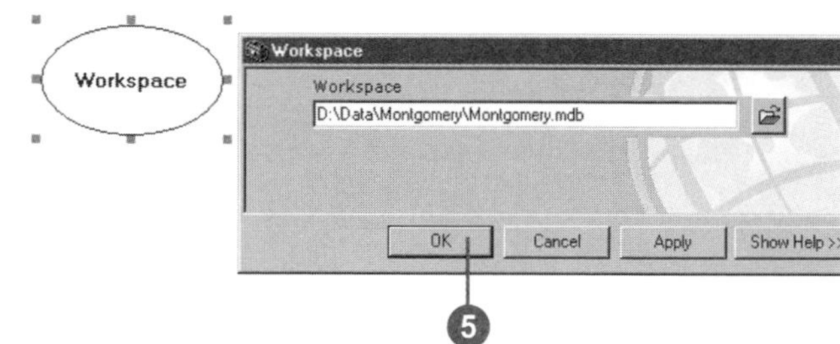

6. Right-click the tool and click Properties.
7. Click the Environments tab.
8. Navigate to the environment setting for which you want to use the variable's value.
9. Check the environment setting, then click Values. ▸

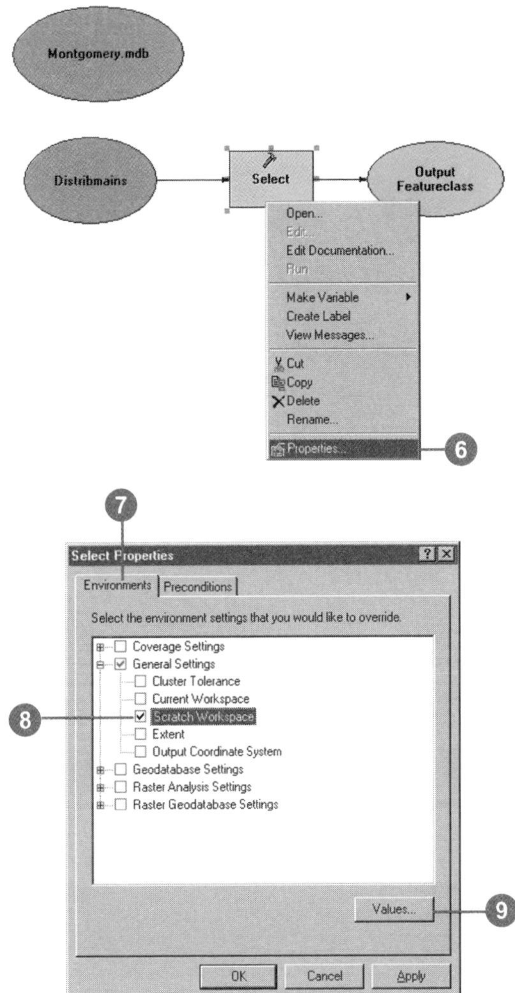

10. Click the dropdown arrow for the environment setting and click the variable.
11. Click OK.
12. Click OK on the Properties dialog box.

    The variable will be connected to the tool, and its value will be used by the tool.

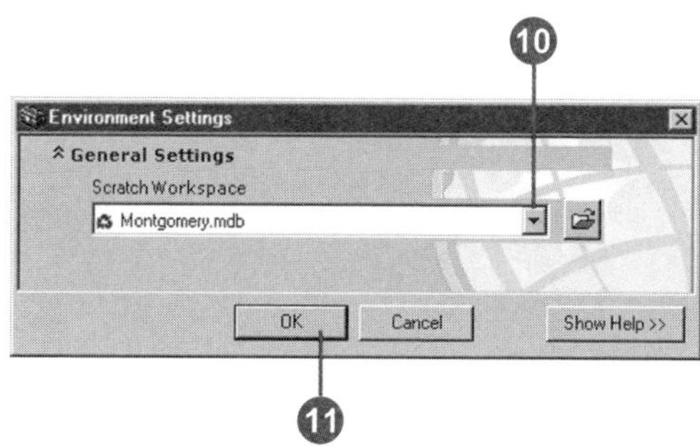

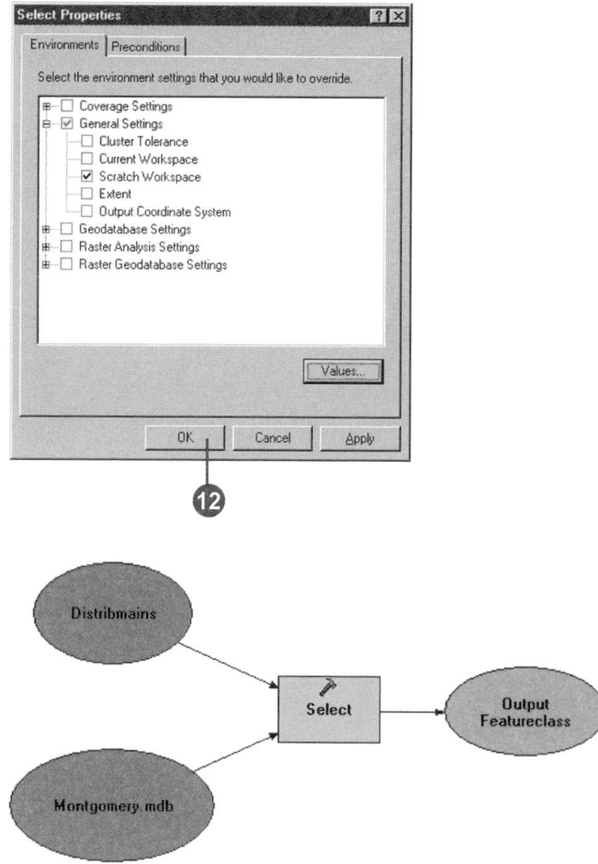

Connecting variables for environment settings using the Add Connection tool

1. Click the Tools menu and click Options.
2. Click the Geoprocessing tab.
3. Check When connecting elements, display valid parameters when more than one is available.
4. Click OK. ▸

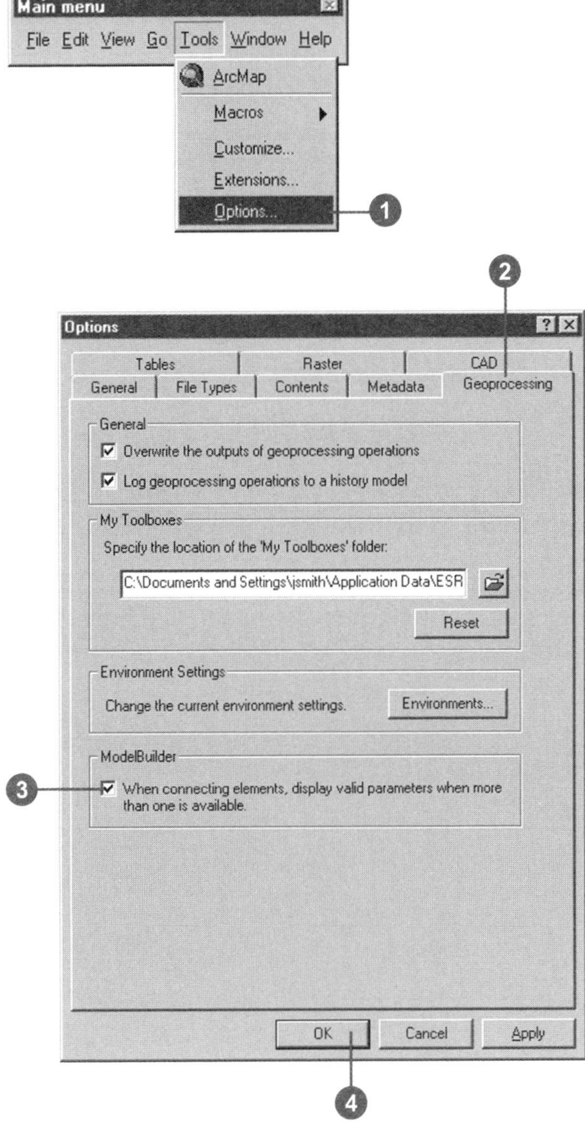

See Also

Follow steps 1–5 of the task 'Connecting variables for environment settings via the Environment Settings dialog box' earlier in this chapter to add a tool to the ModelBuilder window and to create a variable.

5. Click the Add Connection tool.
6. Click the variable created for an environment setting.
7. Click the tool.

   A dialog box displays valid environment settings.
8. Click the desired environment setting and click OK.

   The value set for the variable is used as the environment setting for the tool.

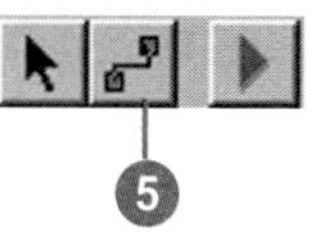

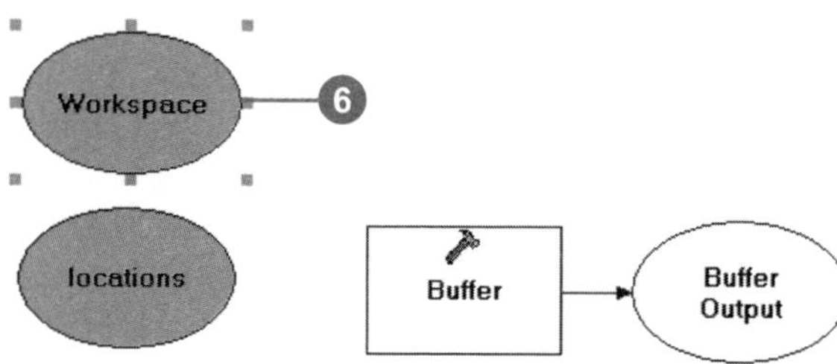

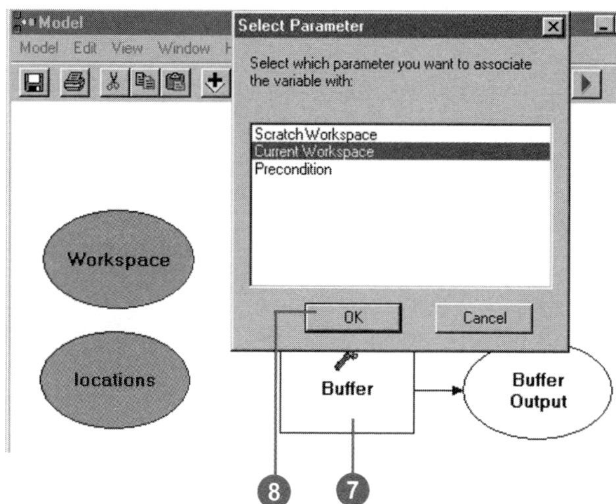

Variables you have created for environment settings do not have to be attached to tools. After their creation, they can be set in the model's Environment Settings dialog box so their value applies to all processes within the model. This is an alternative to right-clicking the model in its toolbox and clicking Properties, then setting Environment Settings. The difference is that by creating a variable, you can then make it into a model parameter so users of your tool can set the value for the environment setting when they run your model from its dialog box. The value supplied will be used by all processes in the model. ▸

See Also

See 'Setting a model's parameters' in this chapter for more information on setting model parameters.

Setting a variable for an environment setting as a model parameter

1. Right-click a toolbox, point to New, and click Model. The ModelBuilder window is opened.

   Alternatively, if a model has already been created, right-click the model and click Edit.
2. Right-click the ModelBuilder display window and click Create Variable.
3. Scroll through the list of data types and select the data type, such as Workspace, for the variable you want to create.
4. Click OK.
5. If you want a default value to display, double-click the variable and enter a value, then click OK. ▸

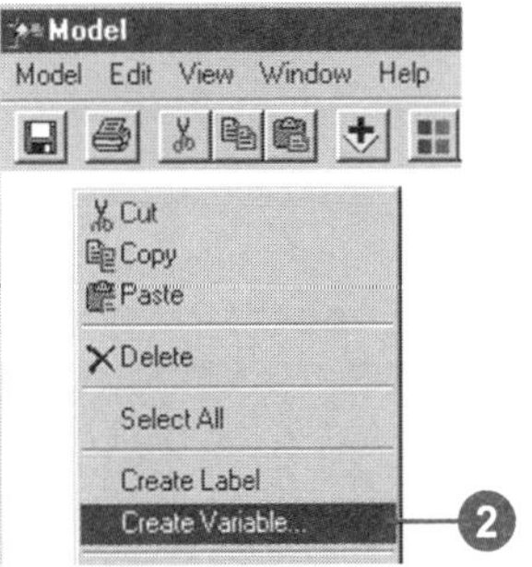

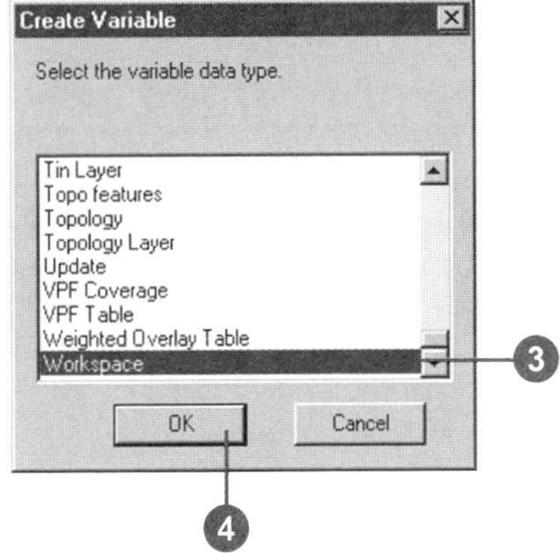

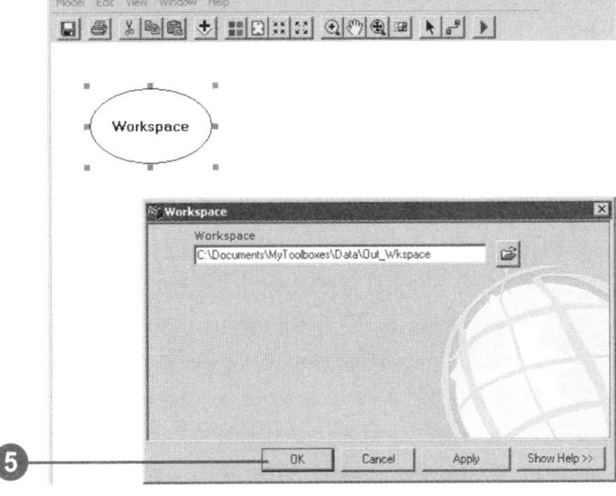

6. Click the Model menu and click Model Properties.
7. Click the Environments tab.
8. Expand the section containing the environment setting for which you want to use the value set for the variable, such as the General Settings section.
9. Check the environment setting you want to apply the variable's value to, such as Scratch Workspace.
10. Click Values.
11. Click the dropdown arrow for the section.
12. Click the dropdown arrow for the environment setting and click the variable.
13. Click OK, then click OK on the Model Properties dialog box. ▸

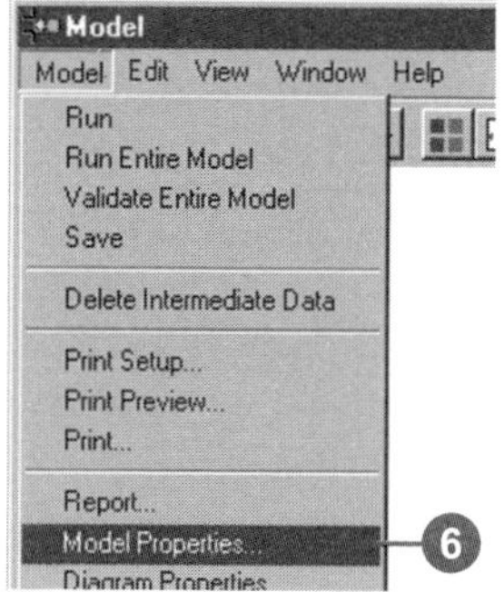

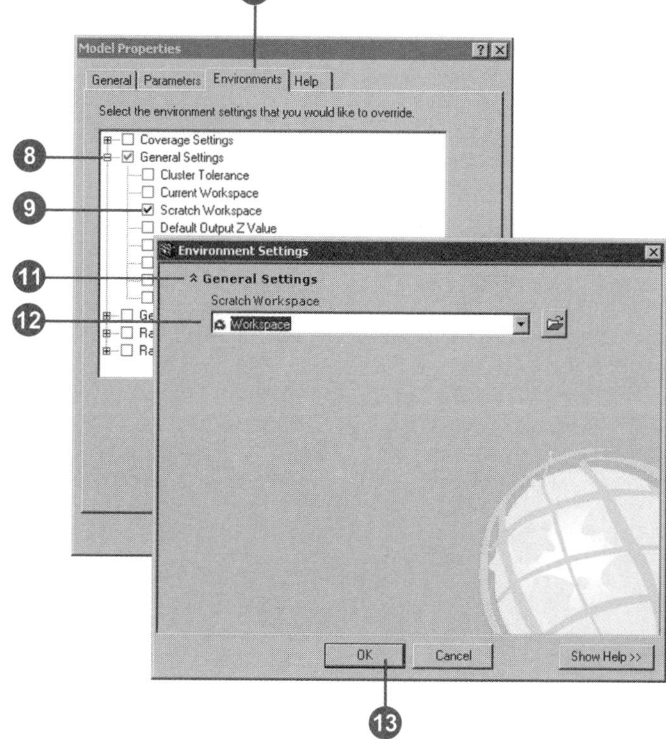

14. Right-click the variable and click Model Parameter.
15. Click the Model menu and click Save.
16. Click the Model menu again and click Close.
17. Right-click the model in its toolbox and click Open to view the created parameter.

    The value entered will be used by all processes in the model. It can always be altered by the user of the tool.
18. Click OK.

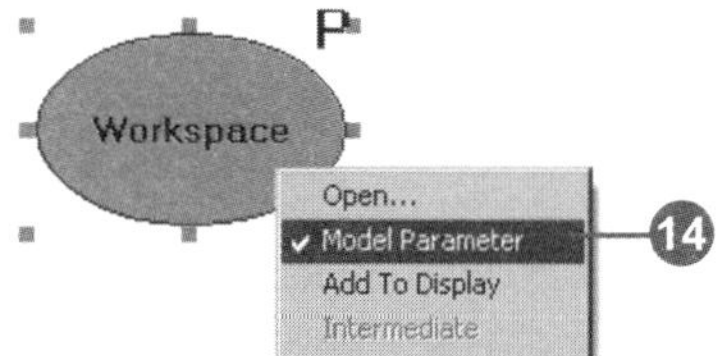

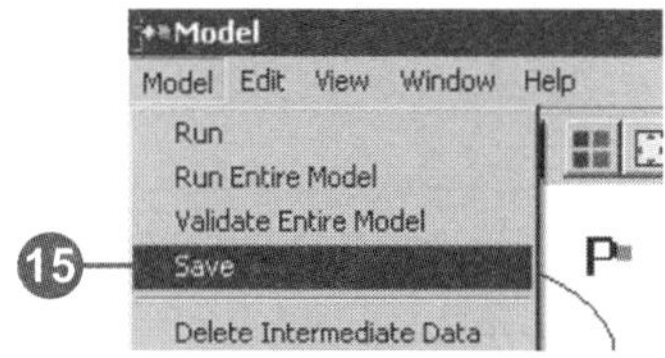

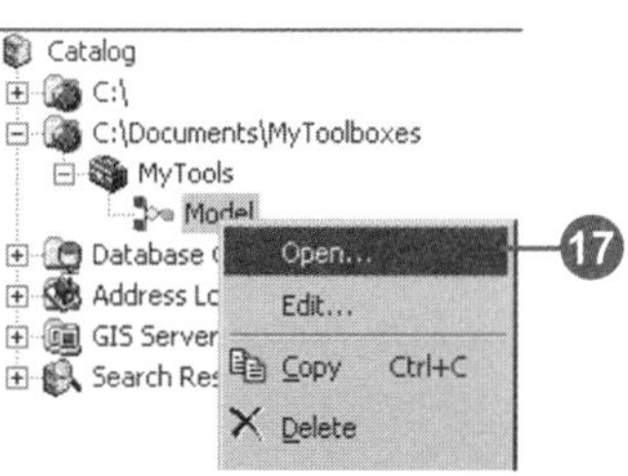

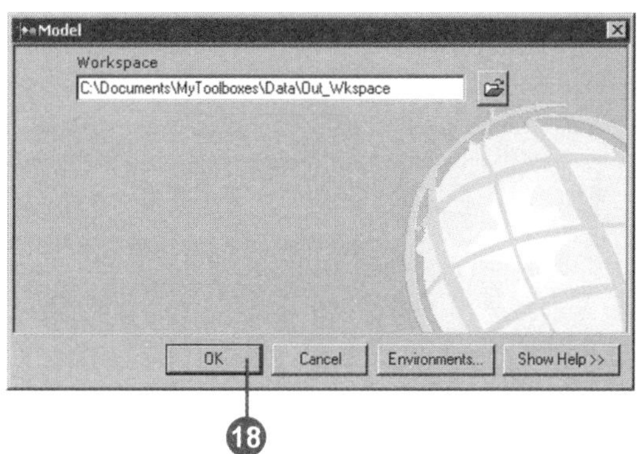

You do not need data to create a model in the ModelBuilder window. Many times you want to generate a plan—a work flow of the tasks involved—before you consider where your data will come from or what other parameter values you'll set.

You can build a model to use as a plan of your work flow by creating empty variables and attaching them to tools. You can fill in the values for parameters and environment settings later. ►

Tip

Setting values for variables

Right-click any variable and click Open to set its value.

Building a model using empty variables

1. Right-click a toolbox, point to New, and click Model. The ModelBuilder window is opened.

   Alternatively, if a model has already been created, right-click the model and click Edit.
2. Right-click the ModelBuilder display window and click Create Variable.
3. Scroll through the list of data types and select an appropriate data type for the variable you want to create.
4. Click OK.
5. Drag a tool onto the display window from the ArcCatalog tree or ArcToolbox window.
6. Click Add Connection.
7. Click the input data variable, then click the tool.

   The input data variable will be connected to the tool.
8. Continue creating variables, adding tools, and connecting them together to create processes. Connect the processes together until your model is complete.

   You now have a plan of your work flow. You can supply the values for parameters and environment settings later.

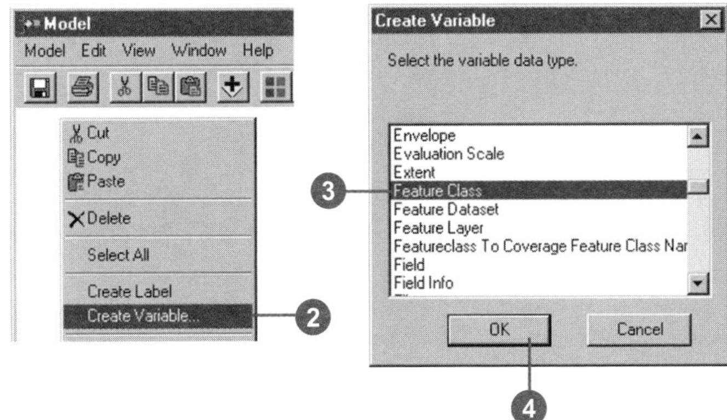

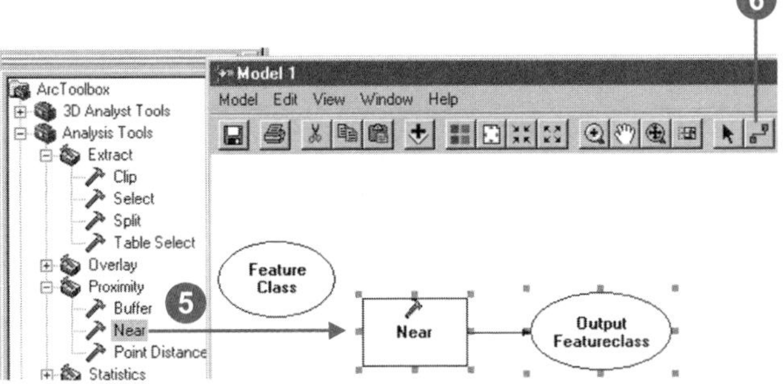

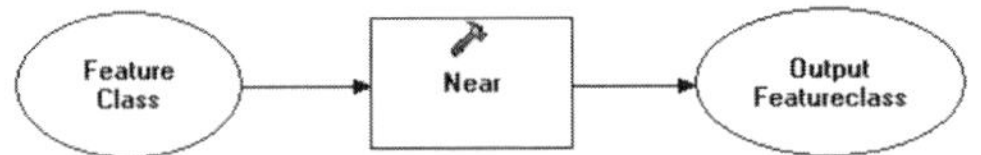

When you open the dialog box of a model in the ArcCatalog tree or ArcToolbox window, you'll see that the dialog box is blank. There are no parameters that the user can specify values for. Parameters in each process in the model can be exposed as model parameters, so the user of your model can supply the values for these parameters.

To expose model parameters, the parameters in the relevant processes must first be set as variables.

Input and output parameters are, by default, variables, so their values can be shared between processes. Other parameters can be set as variables. For more information, see the tasks on connecting variables and 'Creating a variable for a parameter' earlier in this chapter.

After setting a process's parameter as a variable in your model, you can expose the variable as a model parameter by right-clicking the variable and clicking Model Parameter. Model parameters will display in the dialog box of the model.

See Also

See Chapter 5, 'Working with toolsets and tools', for an alternative way to set model parameters and to change their order.

Setting a model's parameters

1. Right-click the variable and click Model Parameter.

   To set more than one variable as a model parameter, select the desired variables first.

   Notice that variables set as model parameters are labeled with a 'P'.

2. Click the Model menu and click Save, then close the ModelBuilder window.

3. Right-click the model in the ArcCatalog tree or ArcToolbox window and click Open.

   The user of your model can now specify values for the exposed parameters before running your model.

   Note: If you don't want a default value to display for a parameter or an environment setting, don't specify a value for the variable in the model.

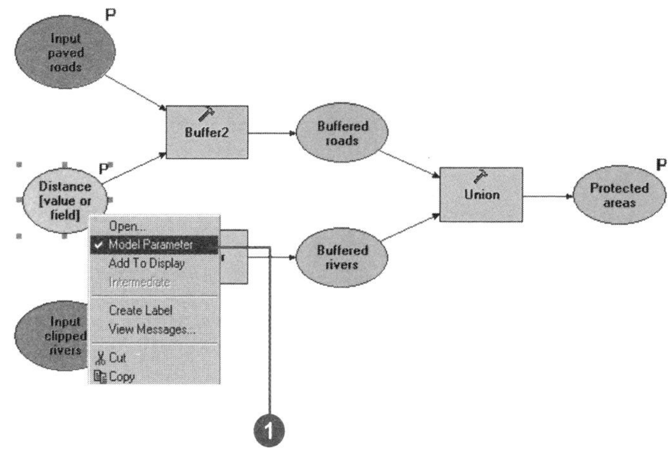

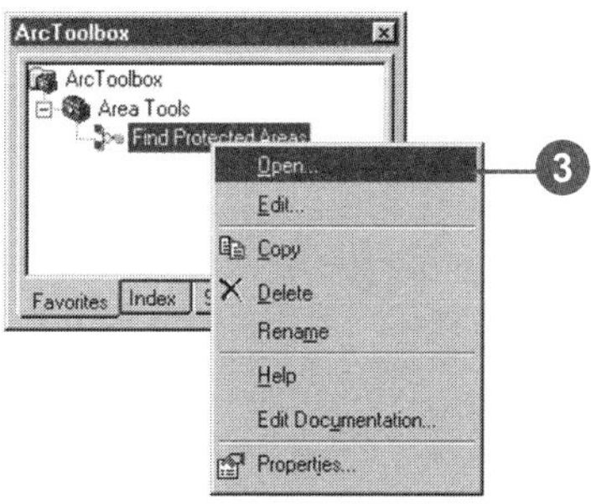

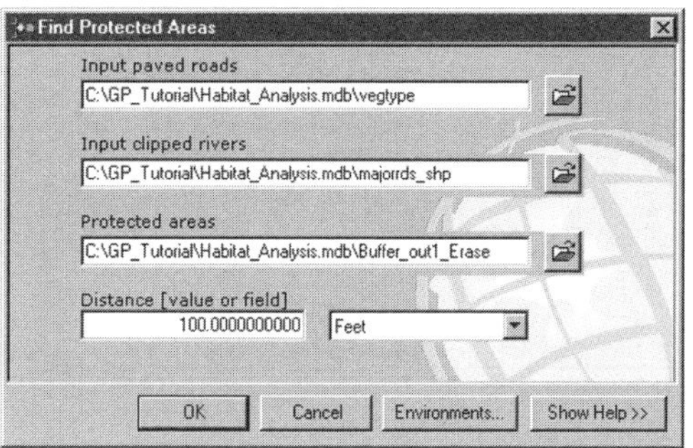

About intermediate data

When you run a model, output data is generally created from each process in the model. The derived data variables that reference this output data are, by default, flagged as *intermediate data*.

All variables created for derived data parameters reference data that is set by default as intermediate. Note that input data is always needed to run the model, so it cannot be set as intermediate.

What output data cannot be intermediate?

Variables that are set as model parameters and variables that reference data that is a derived output (where the output updates the input) cannot be set as intermediate.

When a derived data variable is set as a model parameter, the intermediate option on the variable's context menu is disabled. By setting a derived data variable as a model parameter, the user of your model will supply a value (a path and a name) for this parameter in the model's dialog box, and a result will be produced and saved on disk in the location specified.

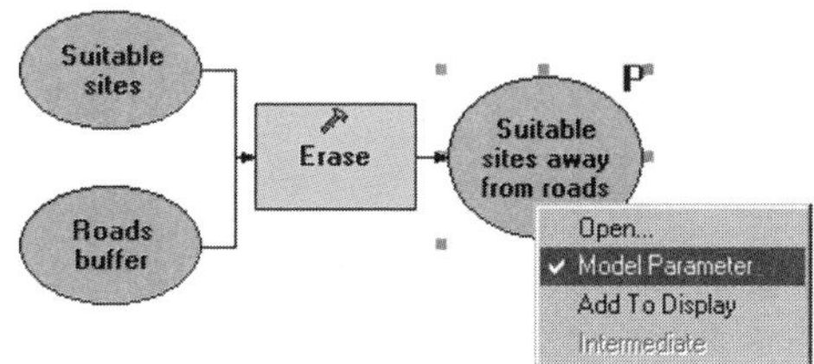

Setting a variable as a Model Parameter disables the Intermediate option. You'll generally want to set the final variable in your model as a Model Parameter so the user of your model can set the value for this parameter in the model's dialog box.

For more information on model parameters, see 'Setting a model's parameters' earlier in this chapter.

Tools that have no output parameter create derived outputs for the purpose of linking processes in a model. The output from such a tool, for example, Add Field, actually updates the input. It is not desirable for the output from such a tool to be deleted after the execution of the tool or after clicking Delete Intermediate Data, as it is the updated input that will be deleted. In such cases, the Intermediate option on the derived data variable context menu is unchecked and unavailable, so the output will not be deleted.

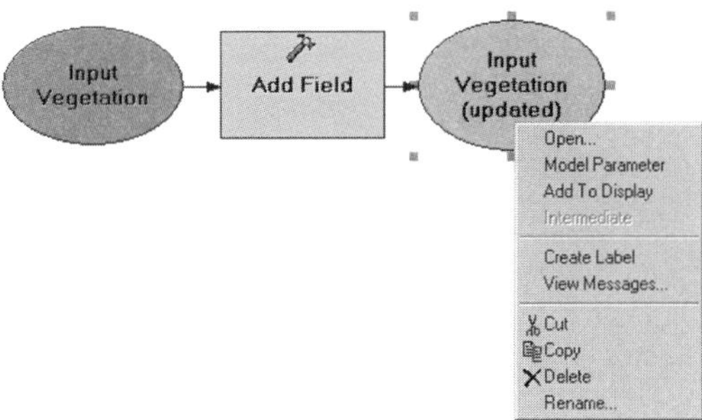

The Intermediate option is unchecked and unavailable for all variables that reference derived outputs (the input data for a tool).

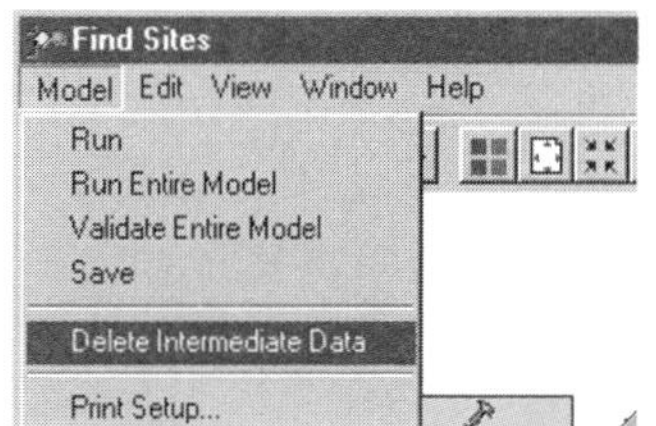

The data referenced by derived data variables that have the Intermediate option flagged is deleted by clicking Delete Intermediate Data.

How is intermediate data handled?

After running a model within the ModelBuilder window, all output data remains on disk regardless of whether the Intermediate option is checked and even after closing the ModelBuilder window. This allows the already run state of the model to be saved between sessions so that each time you open a previously run model in its ModelBuilder window, you will not need to rerun the entire model. The Delete Intermediate Data option allows you to delete all output data that is flagged as intermediate. If you want to retain a particular output, be sure to first right-click the derived data variable that references the output data and uncheck Intermediate so this output is not deleted when you click Delete Intermediate Data.

When you run a model from its dialog box or the command line, all derived data flagged as intermediate will automatically be deleted at the end of the execution of the tool.

If you want intermediate data to be created and saved after the model has executed from its dialog box, you can either set derived data variables inside the model as model parameters or uncheck the Intermediate option on the context menu of the derived data variable. It is important to uncheck Intermediate if you have not set your final variable to be a model parameter, so that the final output from running your model from its dialog box or from the command line is not deleted.

See 'Working with intermediate data' in the section that follows for directions on how to set variables as intermediate and how to remove intermediate data.

Working with intermediate data

By default, the Intermediate option is checked on the context menus of all derived data variables. When you run a model from its dialog box or the command line, or you click Delete Intermediate Data within the ModelBuilder window, output data referenced by derived data variables with the Intermediate option checked will be deleted.

The Intermediate option is unchecked and disabled for derived data variables set as model parameters and for derived data variables that reference output data that is derived output (where the tool updates the input data).

Intermediate data can be deleted. In doing so, all output data referenced by derived data variables with the Intermediate option checked will be deleted. All affected processes within the model will return to a ready-to-run state.

See Also

For more information on intermediate data, see 'About intermediate data' earlier in this chapter.

Setting output data to intermediate

1. Right-click the derived data variable and check Intermediate.

   To make multiple variables intermediate, select them first.

   When you run the model from its dialog box or if you click Delete Intermediate Data within the ModelBuilder window, output data referenced by the derived data variable will be deleted.

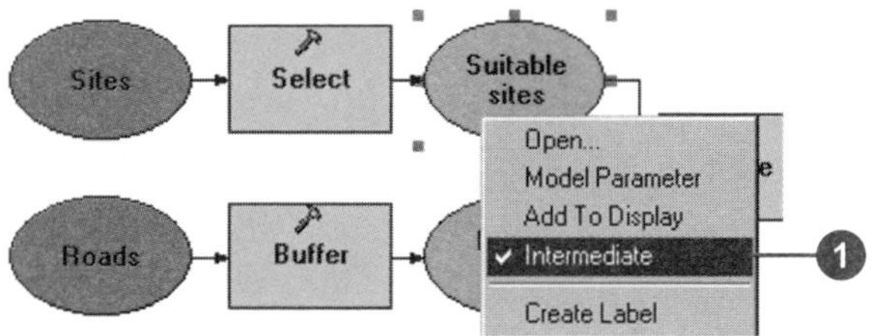

Deleting intermediate data

1. After running a model inside the ModelBuilder window, click the Model menu and click Delete Intermediate Data.

   This will delete from disk all output data referenced by derived data variables that are set as intermediate.

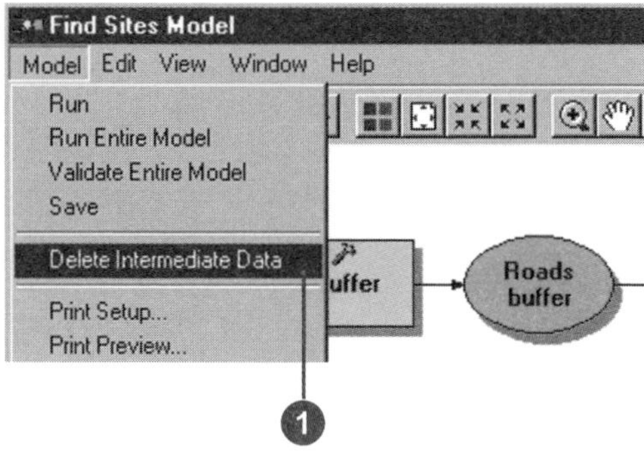

Controlling the flow of processing

There may be times when you want to control the flow of processing in a model. You may want a certain process in your model to run only if a condition is met.

In the task to the right, 'Checking for a condition', an example is followed. There are two routes that can be taken. If the required field doesn't exist in the input feature class's attribute table, the branch of the model is run that creates a layer from the input feature class, then adds a field so a selection can be made using the attributes in the field. If the field exists, the branch of the model is run that creates the selection using the attributes within the field.

See Also

The script used in the example can be copied from the ArcGIS Desktop online Help system. In the Search tab, type "controlling the flow of processing" and double-click this title in the Select topic box. Expand the section 'How to control the flow of processing' and expand the task 'Checking for a condition'. Click 'View an illustration' under the first step.

Checking for a condition

1. Create the script that will determine which branch of the model will be taken.

   View the comment lines in green in the script to the right to gain an understanding of the workings of the example script.

2. Right-click a toolbox, click Add, then click Script. ▸

```
# Description: Controling the flow of processing: Check if a field exists
# then take the appropriate route in the model.
# Created by: ESRI
#*********************************************************************
from win32com.client import Dispatch
import sys
# Create the GeoProcessing object.
gp = Dispatch("esriCore.GpDispatch.1")
# Set the value for the input variable to be entered by the user.
InFC= sys.argv[1]
# Declare a boolean variable to determine if the field exists. It's value
# is set to false as it is assumed that the field does not exist until th
# input data is checked.
bFieldExists = 0
# Get the fields from the input feature class.
flds = gp.ListFields(InFC, "*", "ALL")
fld = flds.next()
# Check to see if the field exists or not.
while fld:
    fldname = fld.name
    if (fldname == "HABITAT"):
        bFieldExists = 1
    fld = flds.next()
# If the field exists, the value for the first derived output parameter
# (Exists) is set to true. The branch of the model that runs Select is
# executed.
if bFieldExists == 1:
    gp.SetParameterAsText(1, "True")
    gp.SetParameterAsText(2, "False")
# If the field doesn't exist, the value for the second derived output
# parameter (Does Not Exist) is set to true. The branch of the model that
# runs Make Layer, Add Join, then Select is executed.
else:
    gp.SetParameterAsText(1, "False")
    gp.SetParameterAsText(2, "True")
```

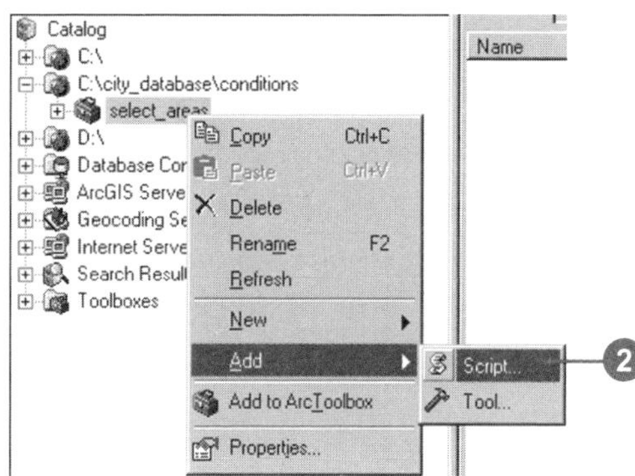

3. Type a name and a label for the script.
4. If desired, type a description for what the script does.
5. Optionally, check Store relative path names so that if the script is moved, the path to the script will not have to be repaired.
6. Click Next.
7. Browse or type the path to the script.
8. Click Next. ▸

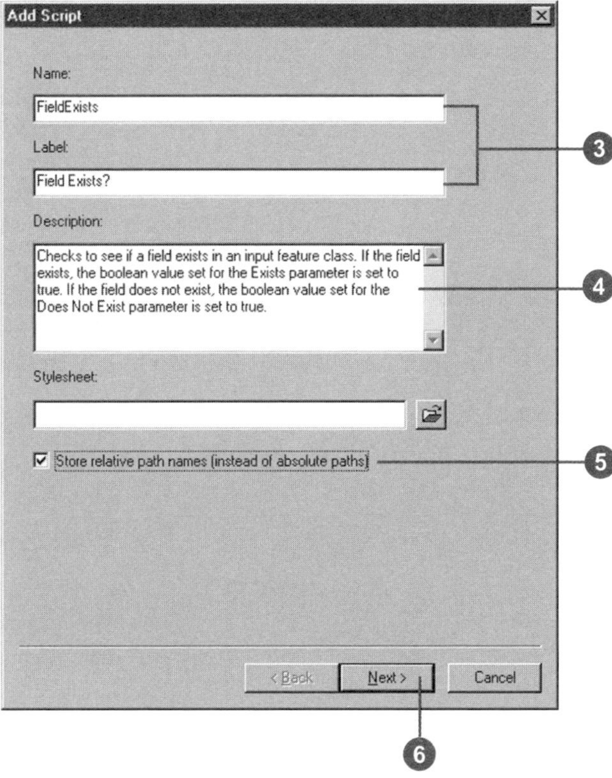

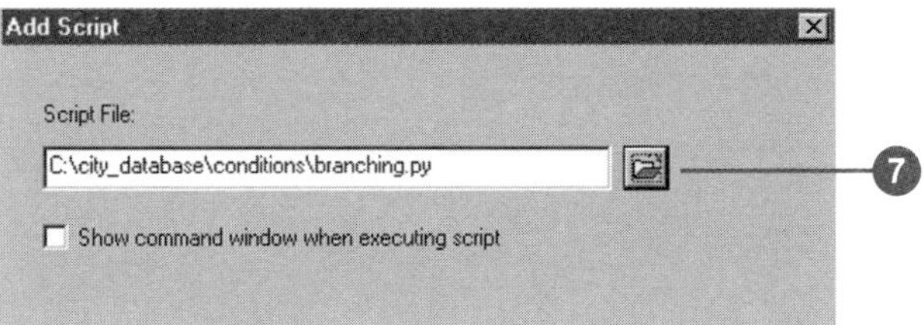

9. Set up the script parameters and their properties that correspond to the variables in the script.

   In the example, three parameters need to be set. The input feature class parameter needs to be set so the tool's user can input the feature class to use. This parameter relates to the InFC variable that is set in the script to equal a user-defined value (sys.argv [1]).

   The properties for this parameter are that it is a required input.

   Two Boolean parameters also need to be set in the example: Exists and Does Not Exist.

   The properties for both these parameters are that they are both derived output. This means that the tool's user will not be able to supply a value for these parameters in the tool's dialog box. The value for these parameters will be set by the tool. Both Boolean parameters are set to be outputs, and a default value of false has been given.

10. Click OK on the script's dialog box. ▸

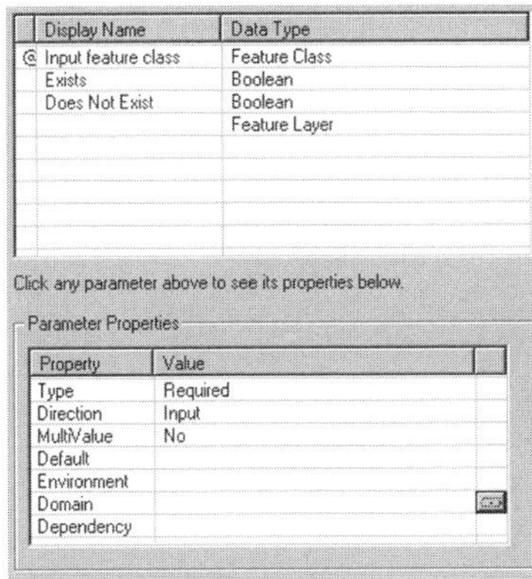

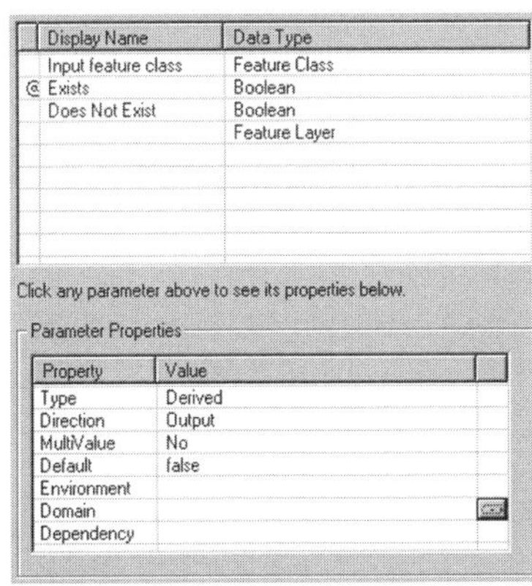

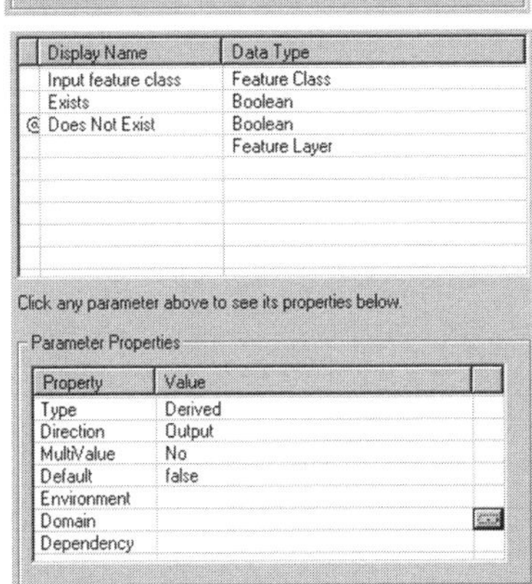

11. Right-click a toolbox, point to New, and click Model.
12. Click the created script and drag it into the new model.

    In the example, the Field Exists? script is dragged into the new model. ▸

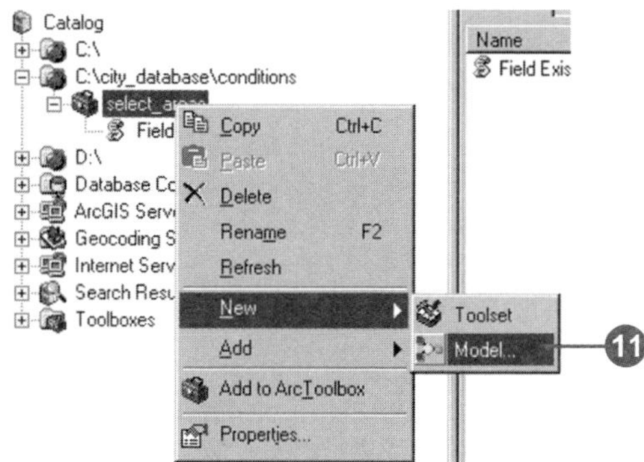

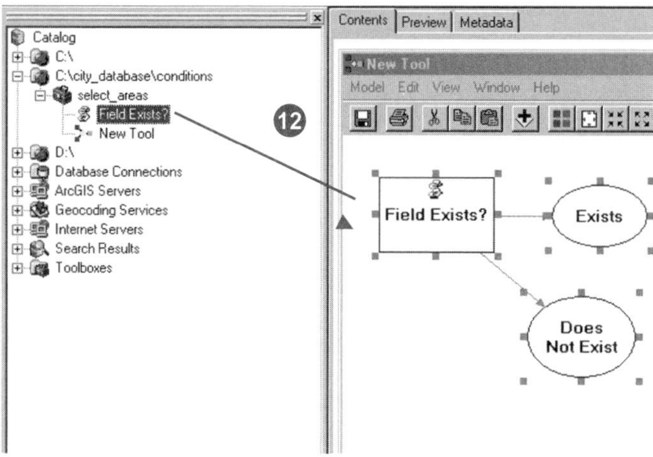

Tip

An alternative way to set preconditions

Click Tools on the Main menu in the ArcGIS Desktop application you are working in and click Options. Click the Geoprocessing tab and check When connecting elements, display valid parameters when more than one is available. You can then use the Add Connection tool within the ModelBuilder window to set the preconditions.

13. Click the tool that should be run if the condition is met and drag it into the ModelBuilder window.

    In the example, if the condition is met—that is, if the field exists—the Select tool will run.

14. Right-click the tool that will run if the condition is met and click Properties.

    In the example, this is the Select tool.

15. Click the Preconditions tab.

16. Select the variable to use for the precondition.

    In the example, this is the Exists variable. If the field exists in the attribute table of the feature class, the Select tool will run using this field.

17. Click OK.

18. Click the tool that should be run if the condition is not met and drag it into the ModelBuilder window.

    In the example, if the condition is not met, that is, if the field does not exist, the Make Layer, Add Join, and Select tools will run.

19. Right-click the tool that will run if the condition is not met and click Properties. ▸

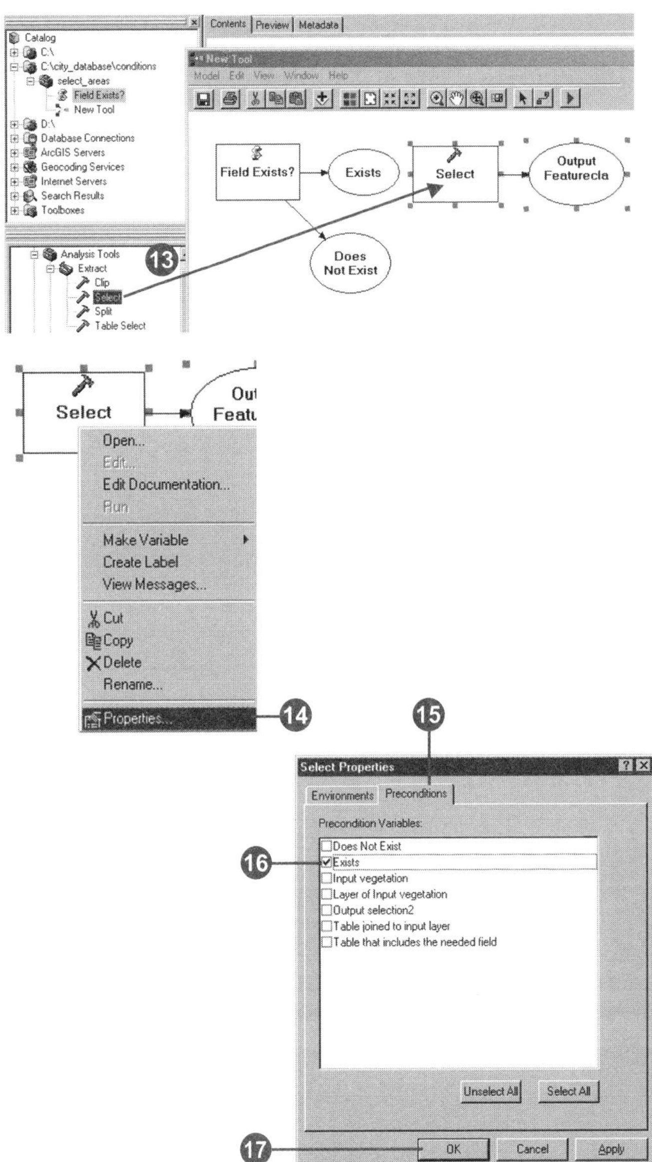

20. Click the Preconditions tab.
21. Select the variable to use for the precondition.

    In the example, this is the Does Not Exist variable. If the field does not exist in the attribute table of the feature class, the Make Layer tool will run, then the Add Join tool will run to join the required field to the layer. The Select tool will select the required features using the added field to select features of a certain attribute.
22. When all preconditions have been set, right-click the script and click Open.
23. Supply a value for any necessary parameters, then click OK.

    In the example, a value for the Input feature class parameter needs to be supplied. This parameter corresponds to the InFC variable in the script and the Input feature class parameter defined in the script properties dialog box.

    When the model is run, the script determines which precondition has been met and gives a value of True or False to each precondition accordingly. Based on which precondition is set to True, a certain branch in the model will execute. ▸

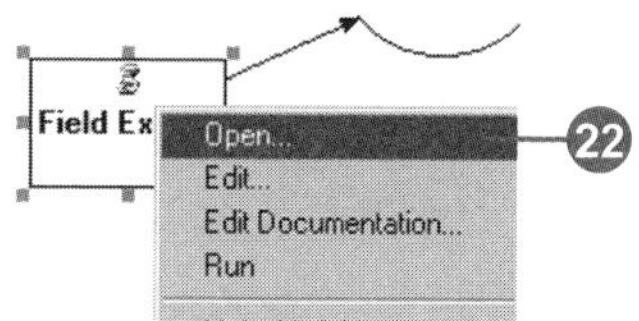

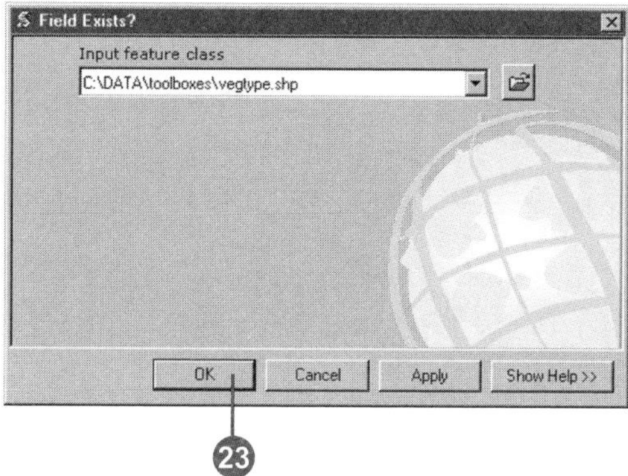

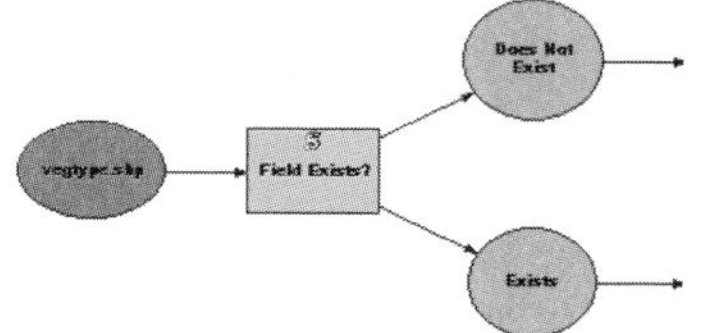

24. Click the Add Connection tool.
25. Link the input data to the tools that will run using the input data depending on whether the condition is met.

    In the example, if the field exists in the input feature class, the Select tool runs. If the field does not exist, a layer is created and the required field is added to the layer. The Select tool then runs on the layer containing the required field.

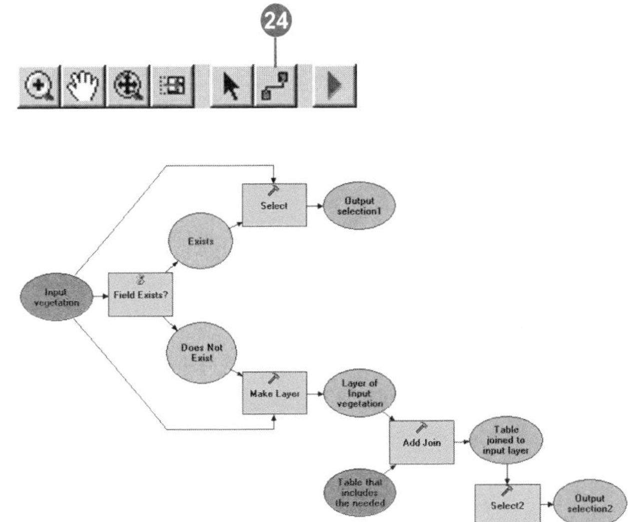

When you need to run certain processes before others, you can control the sequence of processing in your model. For example, you might have a process in your model that creates a personal geodatabase and a subsequent process that adds a feature dataset to that geodatabase. You might have a process that clips a feature class and places the resulting clipped feature class in the created feature dataset. To place the result from the Clip tool in the created feature dataset, you must first run the process that creates the feature dataset.

Tip

An alternative way to set the precondition

Click Tools on the Main menu of the application you are working in and click Options. Click the Geoprocessing tab and check When connecting elements, display valid parameter values when more than one is available. Click the Add Connection tool in the ModelBuilder window and click the output element of the process you want to run first. Click the tool of the process you want to run second. Click Precondition in the Select Parameter dialog box, then click OK. The second process will only run if the first process has run.

Running certain processes before others

1. Create and string together the processes that will run first.
2. Create and string together the processes that will run only after the first set of processes have run.
3. Right-click the tool of the process that should run only after the first process has run and click Properties.
4. Click the Preconditions tab.
5. Check the variable to use for the precondition.

   In the example, the Clip tool will not execute until the output from the Create Feature Dataset tool is created.
6. Click OK.

   The derived output variable of the process or string of processes that should run first is connected to the tool of the process or string of processes that should run second.

   In the example, the creation of the output feature class is dependent on the output feature dataset already being present, so the feature class can be created inside it.

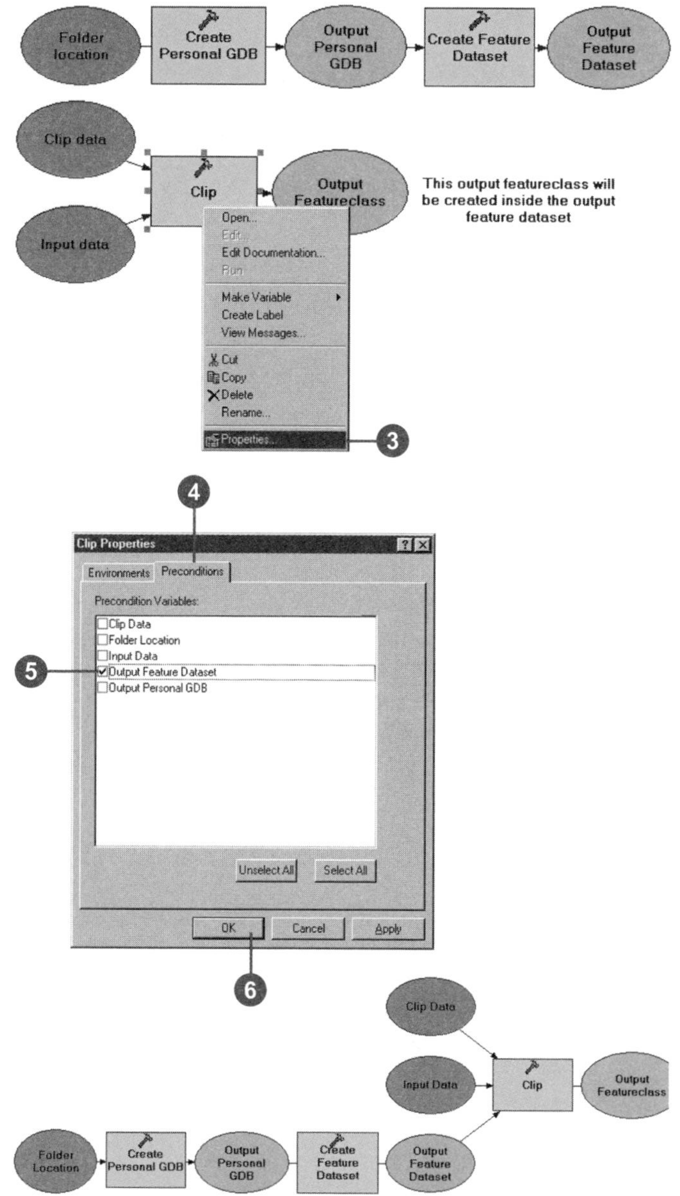

Using the Select Data tool

When building models, the output from some tools can be a folder, a geodatabase, a feature dataset, or a coverage into which results are placed, such as feature classes or tables.

To access this data so it can be used in future processing in the model, you can use the Select Data tool to extract the data you want to use. This tool is only intended for use when building models.

The input to the Select Data tool can be referred to as the parent, as it may contain other datasets, such as tables or feature classes. These other datasets, or children of the parent, are listed in the tool's dropdown list. If the input is a feature dataset, all the feature classes within the feature dataset are listed in the dropdown list. The tool's output always includes the full path to the child dataset.

Selecting the child from the parent using the Select Data tool enables you to continue processing your data after performing a task where the output data is a container such as a feature dataset and the next tool in the model requires a feature class.

1. Run the tool that places data in a workspace.
2. In the ArcCatalog tree, examine the data that is placed in the workspace and make note of the names of any data you want to use in future processing.
3. In the ArcToolbox window, expand the Data Management Tools toolbox, then expand the General toolset.
4. Click and drag the Select Data tool into the ModelBuilder window.
5. Click the Add Connection tool and click the output variable of the tool run in step 1.
6. Click the Select Data tool.

   The output from the tool run in step 1 is used as the input data for the Select Data tool.
7. Double-click the Select Data tool to open it.
8. Type the name of the data contained within the workspace that you want to process using another tool, or click the dropdown arrow to select the data.
9. Click OK on the Select Data dialog box. ▸

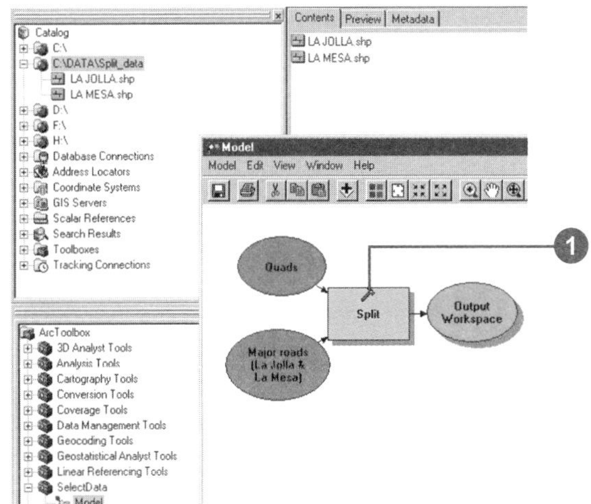

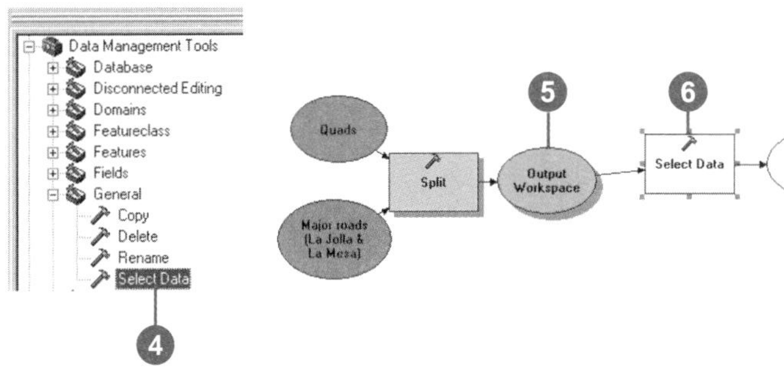

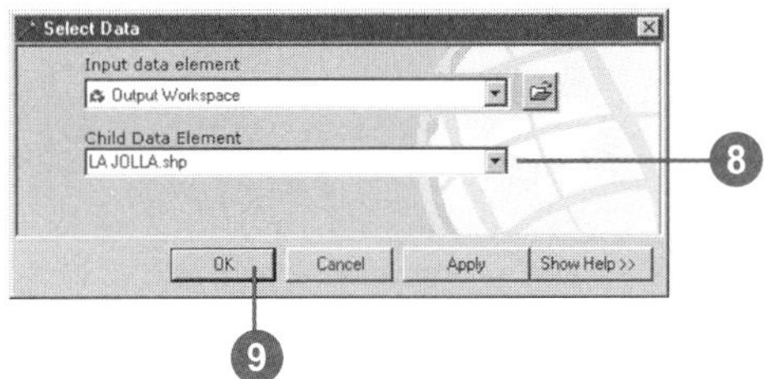

10. Follow steps 4–9 until you have selected all desired data from the workspace.
11. Click and drag into the ModelBuilder window the tool or tools for which you want to use the data selected from the workspace as input.
12. Click the Add Connection tool.
13. Click the derived data element of the Select Data tool.
14. Click the tool for which you want to use the derived data as input.

    The derived data element is connected to the tool, and the data it references is used as input for the tool.
15. Double-click the tool to supply other parameter values if necessary, then click OK.
16. Follow steps 12–15 to connect all outputs from the Select Data tools as inputs for other tools.

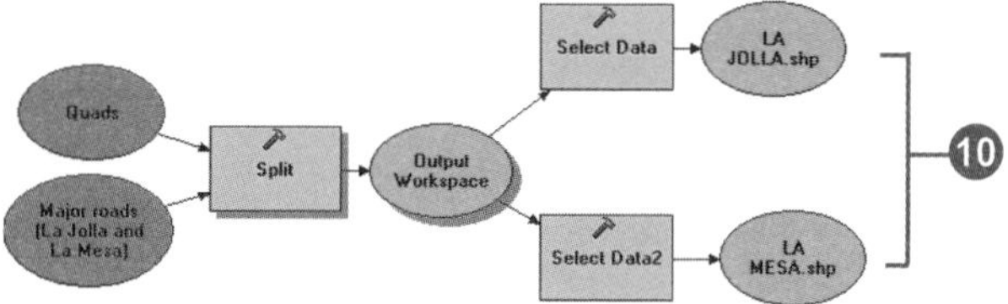

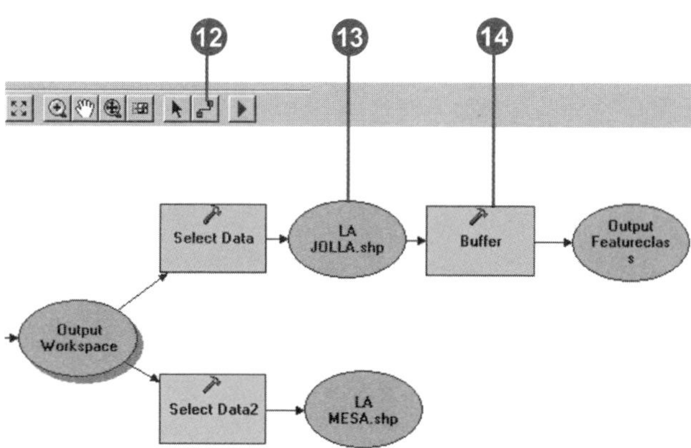

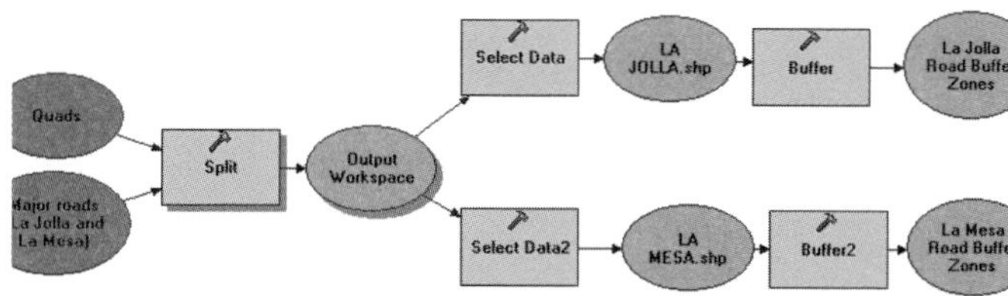

The use of the Select Data tool in a model. Use it to select data from a workspace in order to continue building a model using the data from the workspace as input to other tools in the model.

Working with model elements

All elements can be cut, copied, pasted, deleted, renamed, and disconnected from other elements.

To select an element, click the Select tool and click the element. To select multiple elements, drag a box around them with the Select tool or hold the Shift key and click them one by one. To select all elements, choose Select All from the Edit menu.

To deselect an element, hold the Ctrl key and click the element. To deselect all elements, click the display window. You can also deselect one element this way if there are no other elements you want to keep selected. ►

Tip

Alternative ways to copy, cut, or paste elements

You can also access Cut, Copy, Paste, and Delete from the Edit menu or from the display window context menu. Click the Select tool, then click the element or elements you want to cut, copy, or delete. Right-click the display window and click Cut, Copy, or Delete. You can paste a cut or copied element by clicking Paste on the context menu.

Copying and pasting elements

1. Click the Select tool.
2. Click the element or elements you want to copy.
3. Click the Copy button.
4. Click the Paste button.

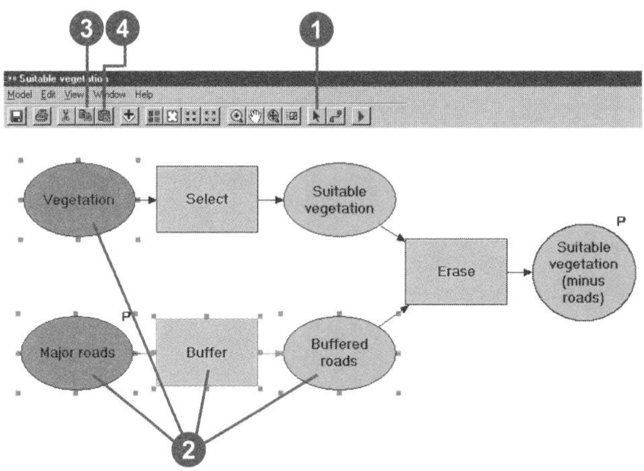

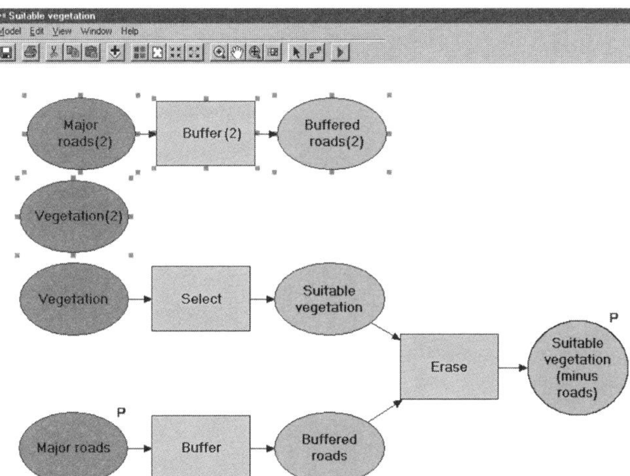

Elements can be copied and pasted or cut within the same model or into another model.

The first time you copy and paste an element within the same model, it receives the name of the element copied, plus the value (2), because it is the second copy. Each subsequently pasted element receives the name of the element copied, plus the value incremented by 1.

If you copy and paste a tool or derived data element that is connected to input data and in a ready-to-run state, the tool and derived data elements are duplicated and connected to the same input data.

If you delete, or cut, an element from a process that is ready to run or has already run, the remaining elements change to not ready to run. If you delete, or cut, a process on which other processes depend, the dependent processes become not ready to run.

If a tool or a derived data element is deleted, the tool or the derived data element it is connected to is also deleted. ▸

Deleting elements

1. Click the Select tool.
2. Click the element or elements you want to delete; press the Shift key to select multiple elements.

   Selected elements have blue handles.
3. Click the Edit menu and click Delete.

   Alternatively, right-click the display window and click Delete.

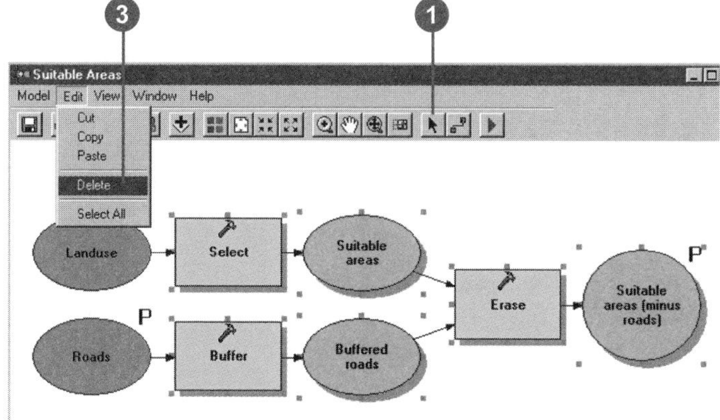

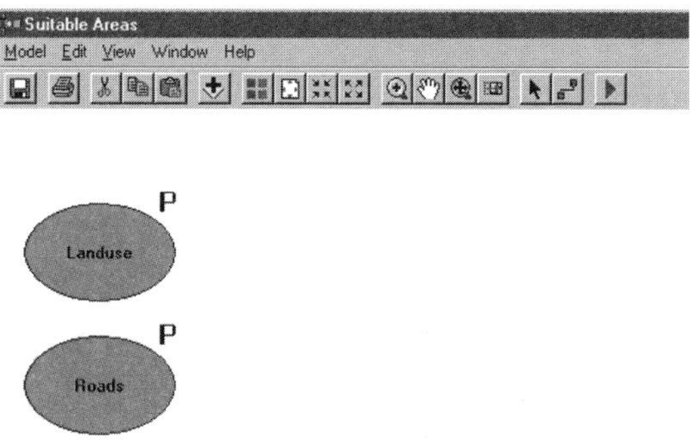

All elements can be renamed, so you can give them a more meaningful name than the one provided by default.

When you change the name for input or derived data elements, you do not change the name of the dataset on disk, only the label of the element.

You should rename the default element names given, especially for large, complex models. When you rename the elements to something more meaningful, anyone who views the model can understand its overall flow.

For variables, a meaningful name makes it is easier to choose the correct variable from a parameter's dropdown list in a tool's dialog box.

All variables, with the exception of derived data variables, can be disconnected from a tool. If the tool is deleted, the derived data variable becomes an input data variable. ▶

Renaming elements

1. Right-click the element and click Rename.
2. Type a new name for the element.
3. Click OK.

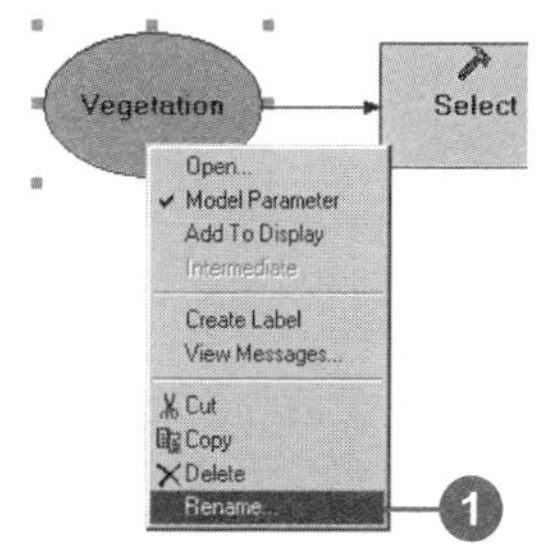

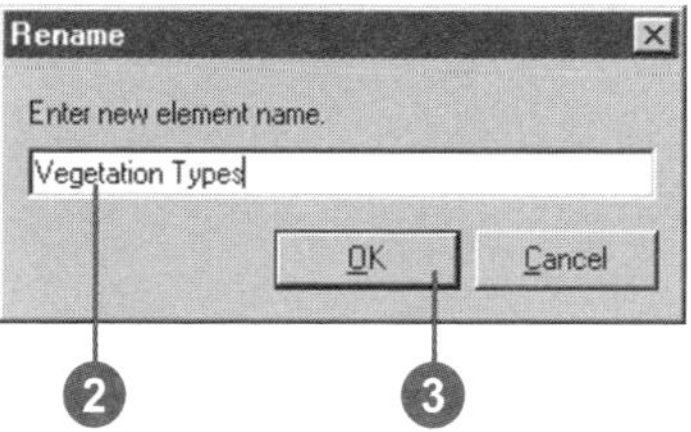

Disconnecting variable elements

1. Click the Select tool.
2. Click the connector arrow to select it, then right-click the connector arrow and click Delete.

   The variable becomes disconnected from the tool. The tool and derived data variables may turn back to a not-ready-to-run state if they were ready to run, as a required parameter value may now be missing.

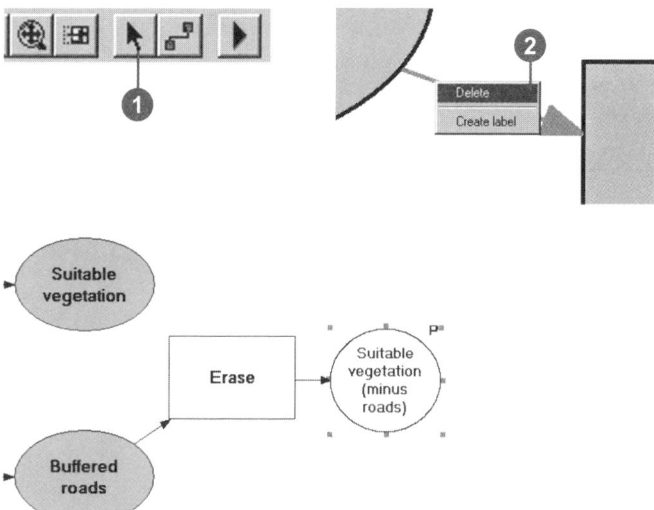

You can add text labels as free-floating text, or you can attach them to elements or connector lines in your model.

When a label is created, it uses default properties. The default text for the label reads "Label", and it is symbolized with a default font (10 point black, bold, Microsoft Sans Serif).

Default properties can be altered in the label's properties dialog box. Commonly changed properties include:

Name: The text for the label.

ToolTip: The text that appears when the mouse pointer is placed over the label. ToolTips are useful when you are zoomed out on a larger model.

Font: Click the cell next to the Font property, then click the Font button that appears to change various font properties.

Background Color: Click the cell next to the Background Color property, then click the Color button that appears to change the label's background color.

Transparent: Set this property to True, and the label will not obscure items that are behind it.

Border Color: This property sets the color of the border around the label and works in ▸

Adding free-floating labels

1. Right-click in the ModelBuilder display window and click Create Label.

   A label with a default name is added to the display window.

2. Double-click the label to enter text.

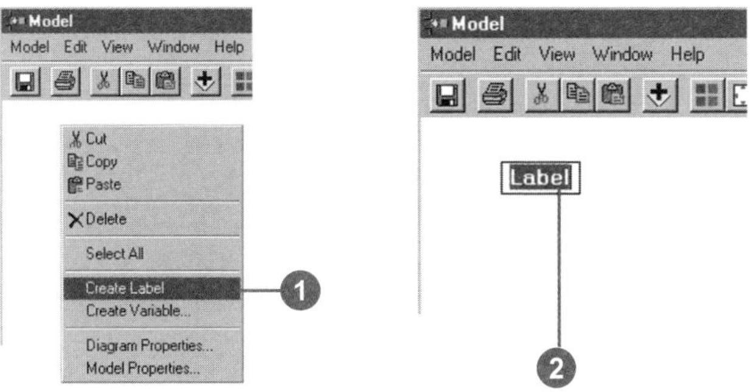

Adding labels to elements or connectors

1. Right-click the element or connector line in the display window and click Create Label.

   A label with a default name is added to the display window. It is attached to the element or connector line, so if the element or connector line is moved, the associated label will move correspondingly.

2. Double-click the label to enter text.

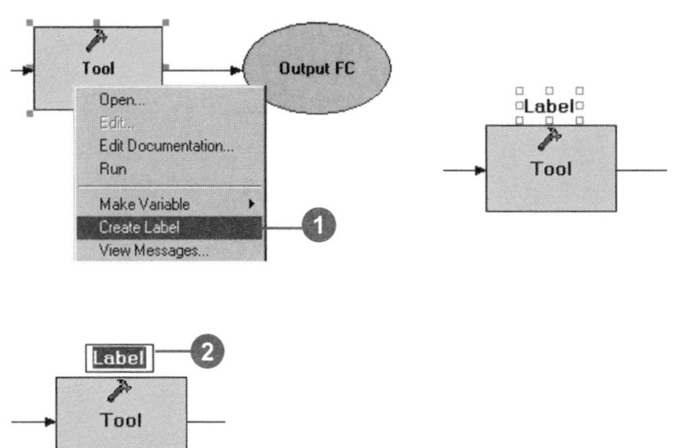

conjunction with the Border Width property.

Border Width: Click the cell next to the Border Width property, then click the dropdown list that appears to select the border width.

Fit To Name: This property specifies how tightly the background property fits to the Name property. No Fit lets you alter the Width and Height properties for the background. Tight Fit restricts the background to fit tightly around the text.

Text Justification: This property changes the way the text is justified in relation to the background—left, center, or right.

Width: This property changes the width of the background.

Height: This property changes the height of the background.

Changing the properties of a label

1. Right-click the label and click Display properties.

   A properties dialog box opens so you can change the properties associated with the label.

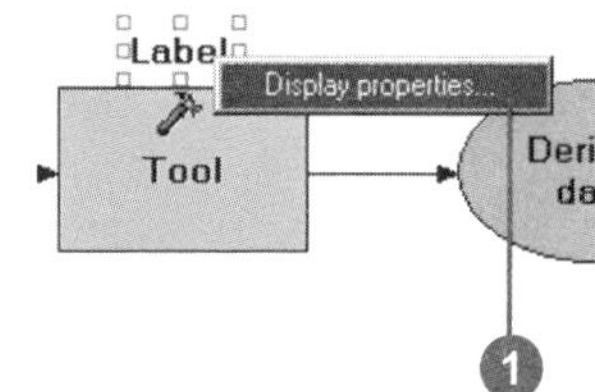

Object Properties	
Name	Label
Tool Tip	
Font	MS Shell Dlg
Background Color	
Transparent	True
Border Color	
Border Width	
Fit To Name	Tight Fit
Text Justification	Center
Width	39
Height	16
X Center	220
Y Center	82
Orientation	Outside
Region	Don't Care
X Absolute	0
Y Absolute	14
X Per Mil	-20
Y Per Mil	1000
Repositioned	True

Running text to the next line

1. Right-click the label and click Display properties.
2. Click in the cell to the right of the Name property and change the name for the label.
3. Click in the cell to the right of the Fit To Name property, click the dropdown list, and click No Fit.
4. Click the Select tool and click the label, then click the blue handles to resize the label.

   Making the label smaller in width will force the text to run onto the next line.

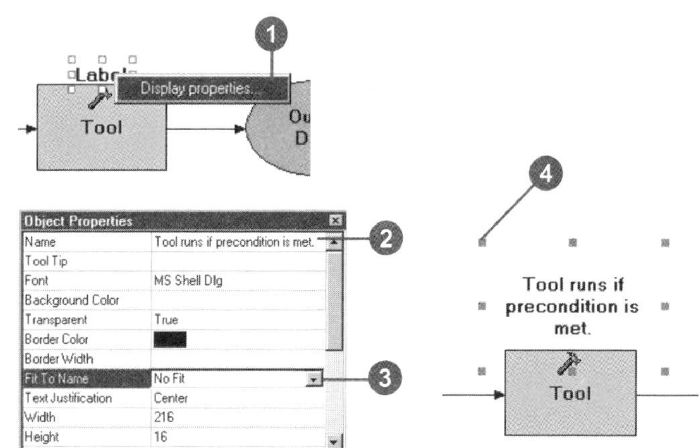

Working with diagram properties

Diagram properties determine the layout of elements and connector lines in the ModelBuilder window and the symbology of elements—their size and color. Changes made to the diagram properties are applied whenever you click Apply on the Diagram Properties dialog box or when you click the Auto Layout button on the ModelBuilder window toolbar. The Diagram Properties dialog box has three tabs: General, Layout, and Symbology.

General tab options

You can set the Layout Mode to Automatic or Manual. In automatic layout mode (the default), the placement of elements and their alignment is managed automatically. In manual mode, you can make these decisions.

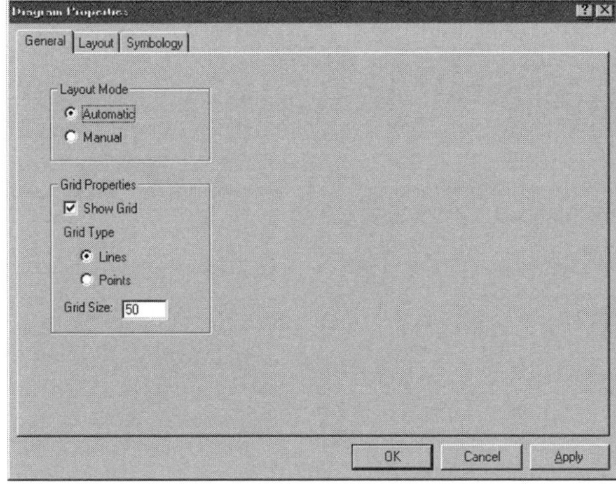

The General tab of the Diagram Properties dialog box

In automatic mode, elements are snapped to an invisible grid so they can be aligned horizontally and vertically. ModelBuilder vertically aligns project data elements with other project data elements, tool elements with tool elements, and derived data elements with derived data elements. Horizontal and vertical spacing between elements is constant. Elements are arranged to minimize the number of crossed connector lines.

The effects of automatic layout are applied whenever you click the Auto Layout button on the ModelBuilder toolbar.

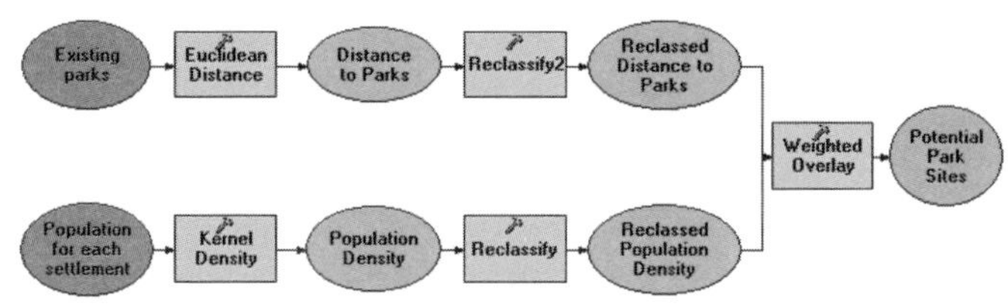

This model was created in automatic layout mode.

In manual mode, you can arrange elements in any pattern you like. This gives you freedom but requires care; layout in manual mode can quickly become jumbled. The Auto Layout tool on the ModelBuilder toolbar is disabled when you use manual mode.

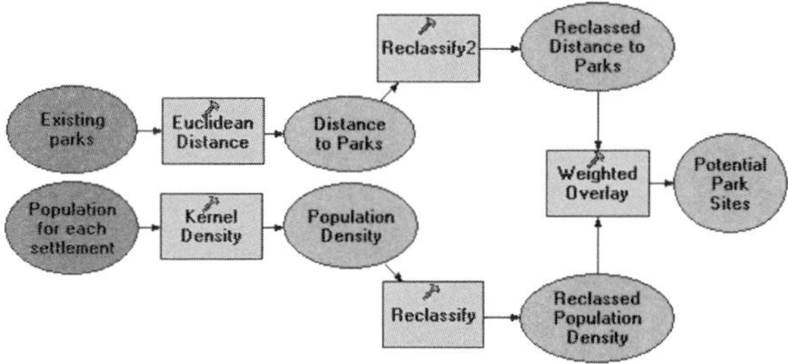

This model was created in manual layout mode.

You can start laying out a model in one mode and switch to the other at any time. You cannot, however, return to a particular manual layout once you have switched to automatic layout mode and changed the layout.

You can check to display a grid in the ModelBuilder window. If you are in automatic mode, this is the grid to which elements will be snapped. If you are in manual mode, you can use the grid to line up your elements.

Layout tab options

You control specific layout properties with the Layout tab. All layout options work only in automatic layout mode.

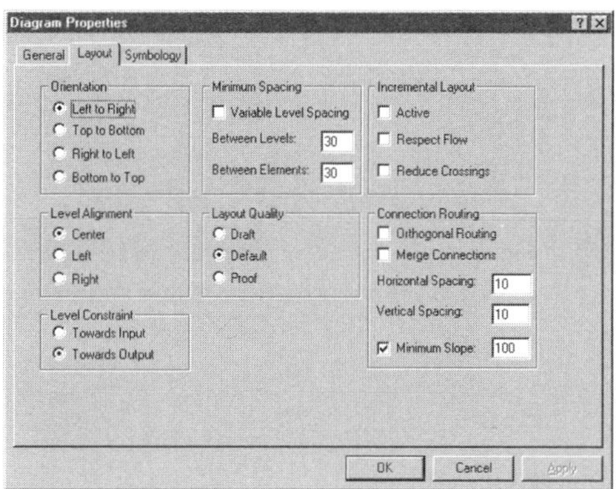

The Layout tab of the Diagram Properties dialog box

Orientation

Orientation determines the direction of processing in the model. Orientation can be from left to right (the default), top to bottom, right to left, or bottom to top.

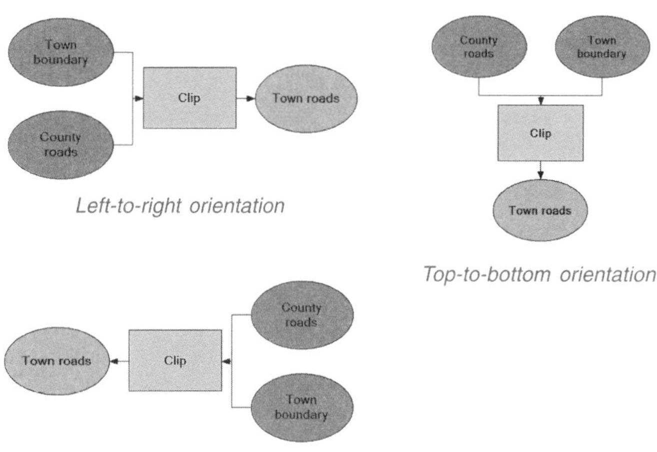

Left-to-right orientation

Top-to-bottom orientation

Right-to-left orientation

Incremental layout

With incremental layout active, a kind of compromise is achieved between automatic and manual layout. Once an element or connector has been placed, it will be moved as little as possible, minimizing the amount of reorganization that occurs within the ModelBuilder window.

Incremental layout can be turned on and off while you work. The effects of incremental layout are applied whenever you click the Auto Layout button on the ModelBuilder window toolbar.

When the option for respecting the flow is checked, the algorithm assigns elements to levels incrementally as much as possible, while respecting the flow of the connector lines. When the option to reduce crossings is set, the maintenance of the order of elements within levels is overridden when necessary. The algorithm assigns elements to levels incrementally and preserves their ordering within levels as much as possible, while trying to reduce the number of crossings.

Minimum spacing

You can set the spacing to change spacing between levels and elements. The default is 30 diagram units for each.

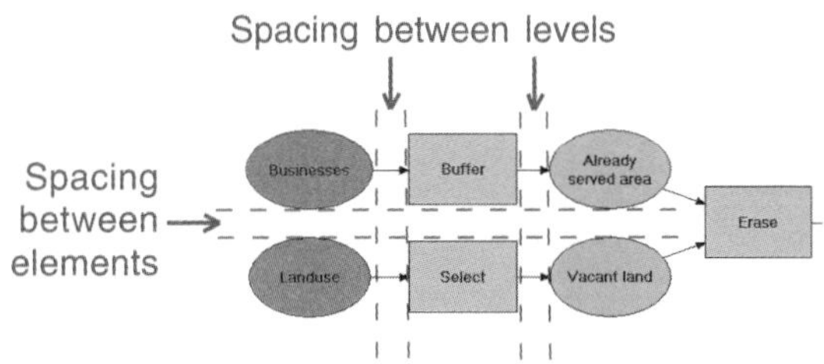

Check Variable Level Spacing to adjust the spacing between levels. In models that contain crossed connector lines, the spacing between levels is distributed so that levels with crossed lines are given more space.

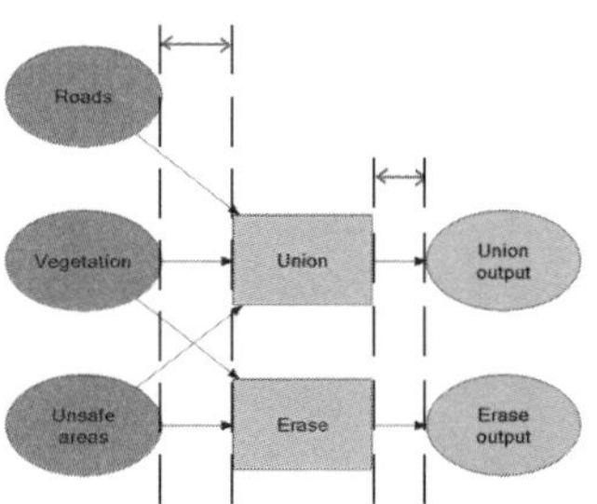

Variable level spacing is checked on. The additional space between levels connected by crossed lines improves the model's appearance.

Connection routing

Connection routing determines whether or not diagonal, or orthogonal, connector lines can be used. When Orthogonal Routing is checked off (the default), diagonal connector lines can be used. When Orthogonal Routing is checked on, all connector lines must be straight lines. When the Orthogonal Routing option is checked on, you can use the Merge Edge Channels option to merge the connector lines that connect to the same input element, so all vertical lines meet at the same point. You can also set the vertical and horizontal spacing between connectors.

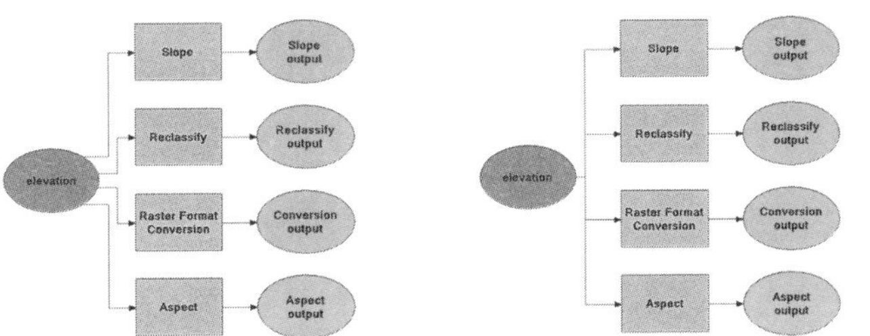

In the model on the left, Orthogonal Routing is turned on, so all connector lines are straight lines rather than diagonal (the default). In the model on the right, Orthogonal Routing and Merge Edge Channels are turned on, so vertical lines meet at the same point.

The Minimum Slope option sets the lowest slope value that any diagonal connector line can have. When the model orientation is vertical, slope is measured by the angle of the connector line from an imaginary horizontal line. When the model orientation is horizontal, slope is measured by the angle of the connector line from an imaginary vertical line.

Slope is measured in percentages: zero percent is flat, and the value of a perpendicular slope approaches infinity. In practice, increasing the minimum slope by approximately 300 percent has no noticeable effect.

Layout quality

The layout quality controls how rigorously the layout adheres to the layout rules. In draft mode there is the least adherence to the layout rules—a simpler computation is used—making the layout of elements in the display window quicker. In proof mode, the maximum amount of adherence to the layout rules is attained. The default is a compromise between draft and proof.

Level constraint

The Level Constraint option determines how the elements in different chained processes are aligned. Processes can be aligned toward the input element or toward the output element.

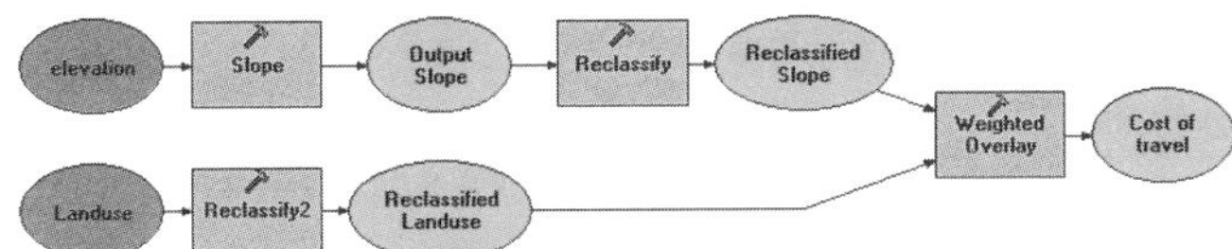

The level constraint is toward the input (the default).

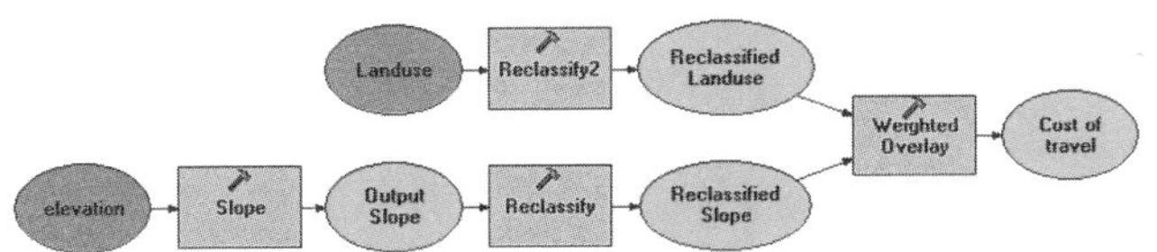

The level constraint is toward the output.

Level alignment

Level alignment determines how elements align within a level. The effects of changing the level alignment are only visible in the model if you have elements of different sizes.

The Auto Layout tool

The Auto Layout tool on the ModelBuilder toolbar is only enabled if you are working in automatic layout mode.

When the Auto Layout tool is clicked, it applies the current layout preferences to the model.

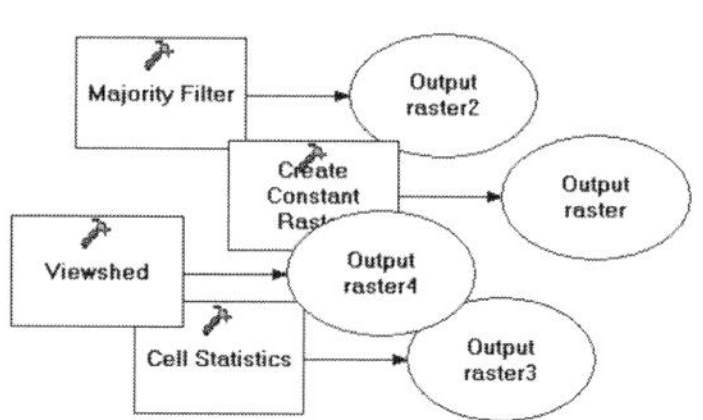

By dragging and dropping tools, they can be placed haphazardly in the display window.

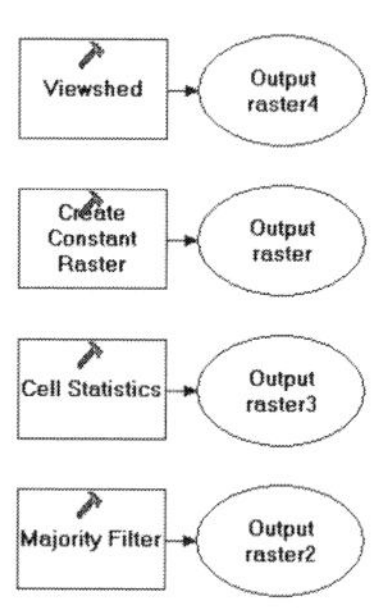

When the Auto Layout tool is clicked, the settings specified in the Layout tab of the Diagram Properties dialog box are applied.

Symbology tab options

Options on the Symbology tab allow you to control the default color of elements and the default font properties of element text, and they give you control over how elements are displayed. Tool elements are displayed as rectangles by default, and other elements are displayed as ovals.

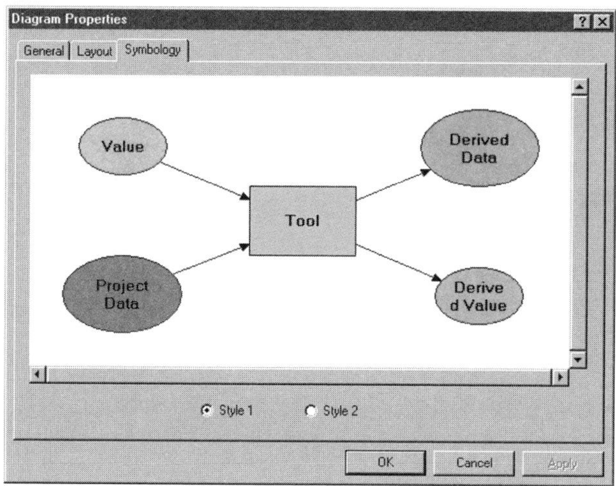

The default symbology applied to elements

You can switch the default setting around so tool elements display as ovals and all other elements display as rectangles.

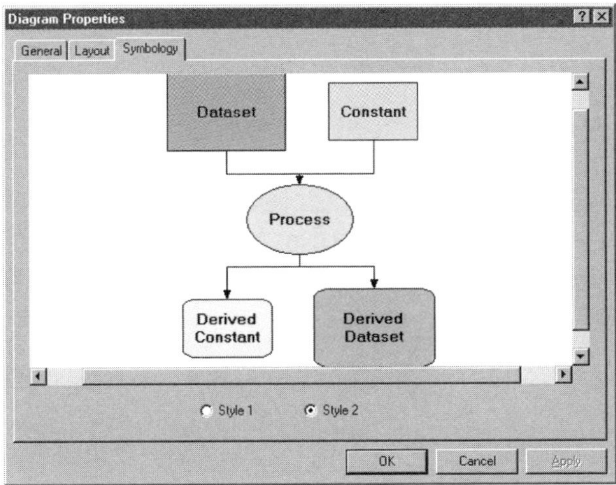

The default symbology has been altered so tool elements will be displayed as ovals and all other elements will be displayed as rectangles. The color for elements has been altered also. All elements of the same type will display in the colors specified by default.

Changing diagram properties

Diagram properties affect the appearance of the display window and its elements.

By default, the layout mode is set to automatic. This is the simplest mode to work in because the element alignment is taken care of for you. Manual layout mode, however, offers more flexibility.

By default, a grid is not shown in the display window. Elements are snapped to the grid when you are working in manual mode. ▸

Changing the layout mode

1. Click the Model menu and click Diagram Properties.
2. Click the General tab and click either Automatic or Manual.
3. Optionally, click Apply to see your changes in the display window without closing the dialog box.
4. Click OK on the Diagram Properties dialog box.

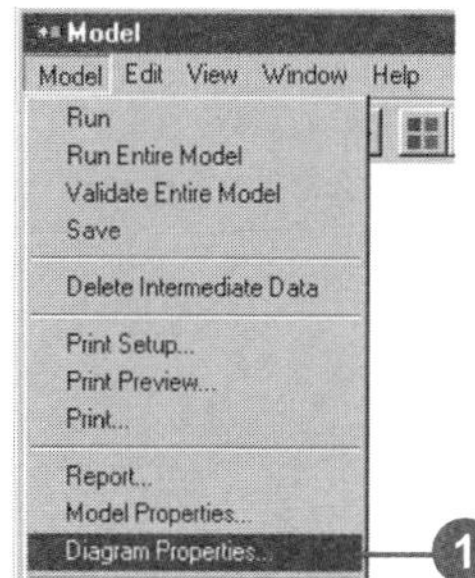

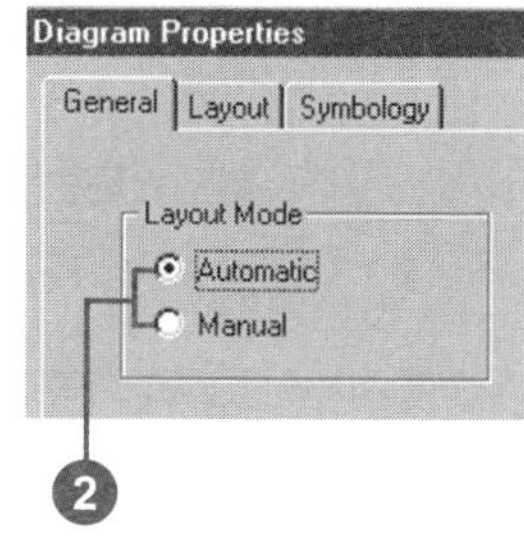

Displaying a grid in the display window

1. Click the Model menu and click Diagram Properties.
2. Click the General tab and check Show Grid.
3. Check the grid type you want to display (either Lines or Points).
4. Type a size for the grid.
5. Optionally, click Apply to see your changes in the display window without closing the dialog box.
6. Click OK on the Diagram Properties dialog box.

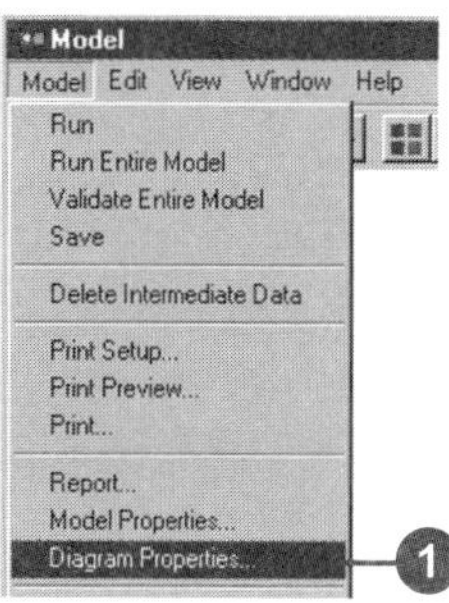

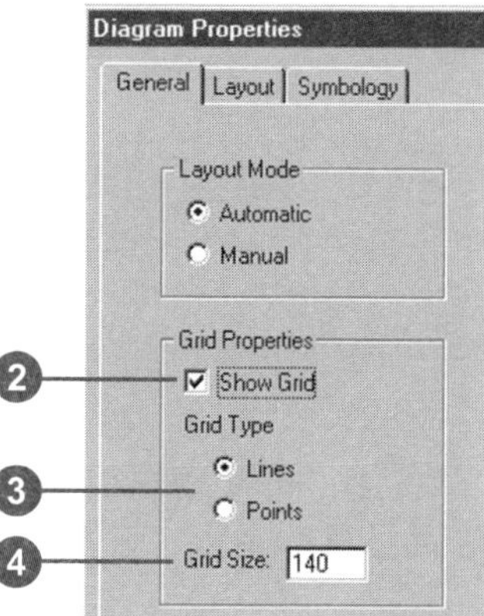

When you're in automatic layout mode, which is set on the General tab of the Diagram Properties dialog box, the default layout of the model in the display window can be altered in various ways. For instance, you may want the model to flow from right to left instead of the default flow from left to right, or you might want to change the default spacing between levels and between elements. After setting your preferred layout options, click Apply to see the changes; or you can click OK, then click the Auto Layout button on the ModelBuilder toolbar.

Changing the default layout

1. Click the Model menu and click Diagram Properties.
2. Click the Layout tab. ▸

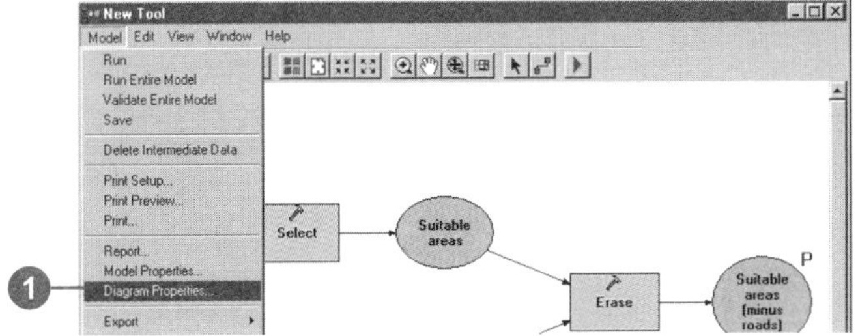

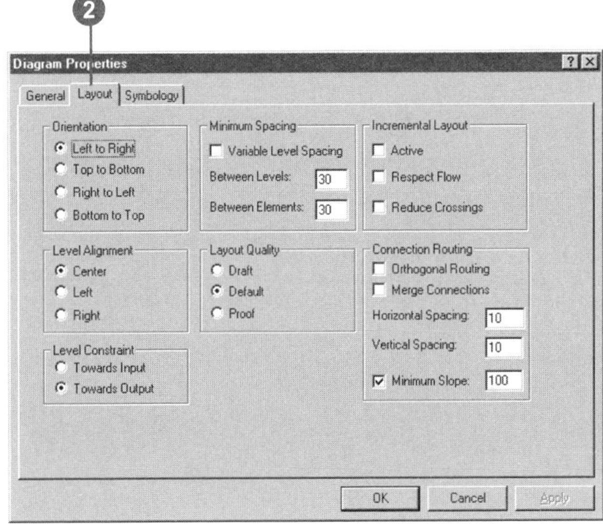

See Also

See 'Working with diagram properties' earlier in this chapter for information on each layout option.

3. Change the options as desired.

   Examples of changes that could be applied are shown in the diagram to the right.

4. Optionally, click Apply to see your changes in the ModelBuilder window without closing the dialog box.

5. Click OK.

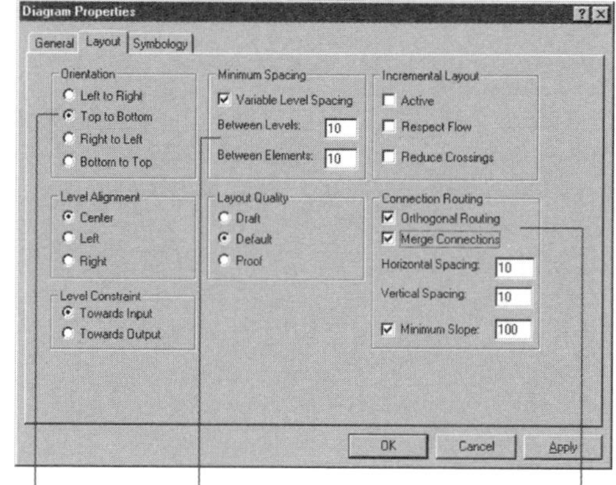

The orientation has been changed to Top to Bottom.

The spacing between levels and elements has been decreased.

Orthogonal Routing and Merge Connections have been turned on.

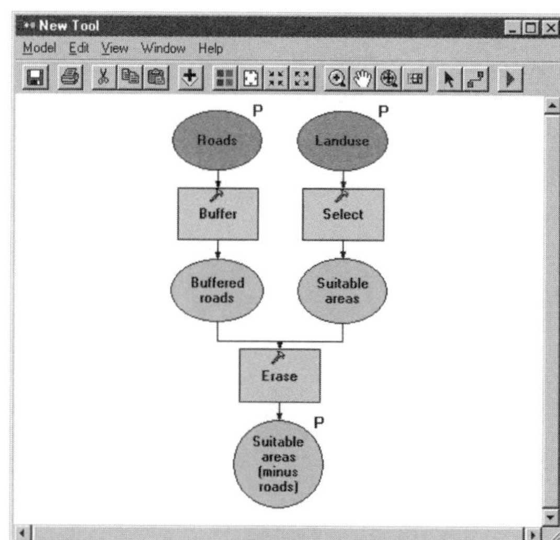

The model after the new layout options have been applied

You can change the symbology that is applied to elements by default. You can change the shape and the color of elements and the font used for element text.

Tip

Resetting changes back to the default

If you make changes and you wish to return to the default settings, simply check Style 1 or Style 2, depending on how you want the elements displayed, then click OK on the Diagram Properties dialog box.

Changing the default symbology

1. Click the Model menu and click Diagram Properties.
2. Click the Symbology tab.
3. Check the style for displaying elements.

   Check Style 1 to make tool elements rectangular and all other elements oval. Check Style 2 to make all tool elements oval and all other elements rectangular.
4. Right-click an element and click Color to change the default color for all elements of that type. ►

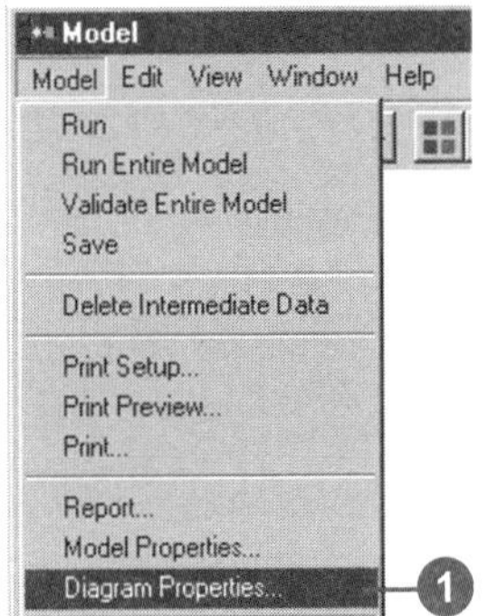

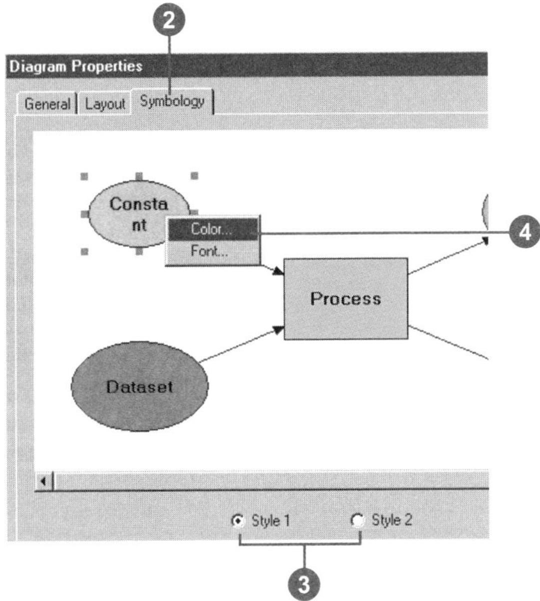

5. Click a color square from the palette or define a custom color to use for the element.
6. Click OK.
7. Right-click an element and click Font to change the default font for all elements of that type. ▸

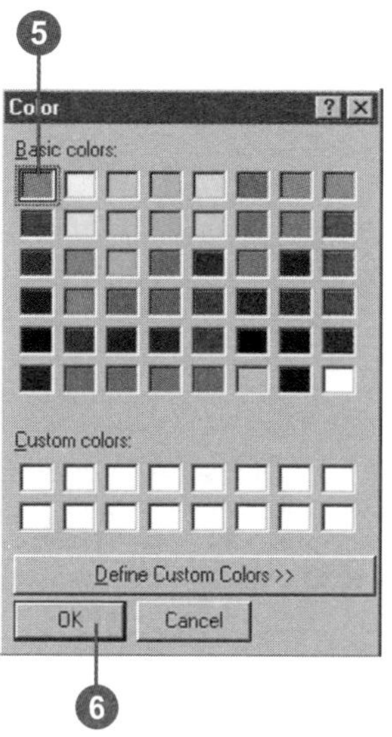

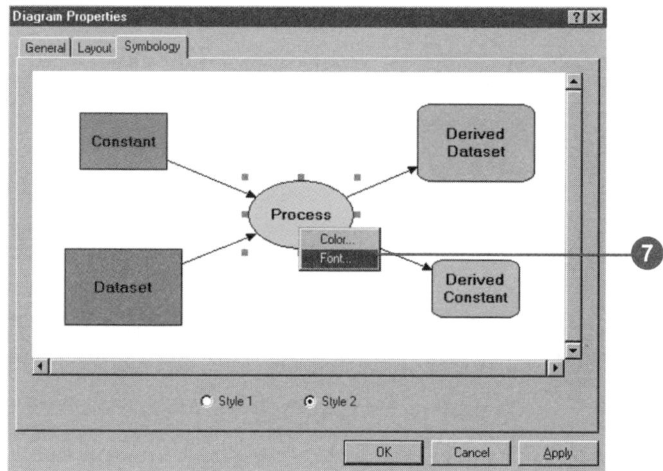

8. Make appropriate changes to the font.
9. Click OK on the Font dialog box.
10. Optionally, click Apply on the Diagram Properties dialog box to see your changes in the display window without closing the dialog box.
11. Click OK on the Diagram Properties dialog box.

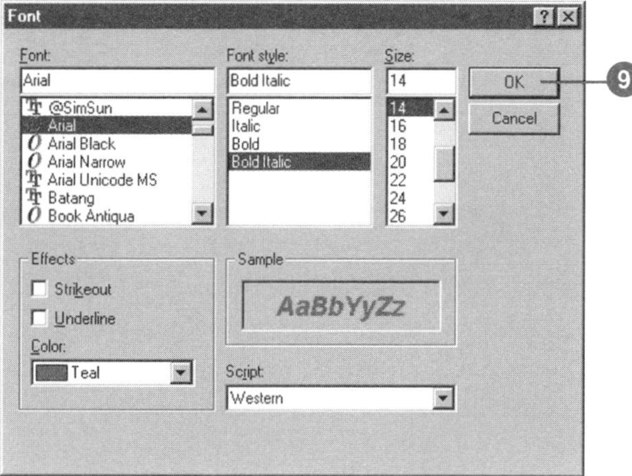

Navigating in the model

There are many ways to navigate in the model, including zooming and panning, using an overview window, and navigating through the model using the Navigate tool.

There are several zoom controls:

- The Full Extent tool zooms to the boundaries of the model.
- The Fixed Zoom In and Fixed Zoom Out tools zoom the view by constant increments.
- The Zoom tool zooms to the area of a rectangle you draw.
- The Continuous Zoom tool zooms continuously in or out when the left mouse button is pressed and the mouse is dragged forward or backward.
- The Custom option on the View menu lets you zoom by a custom percentage. ►

Tip

An alternative way to pan

Move the vertical or horizontal scrollbars—or click the arrows—to move to different parts of the model.

Zooming and panning

1. To see part of a larger model more clearly, click the Continuous Zoom tool.
2. Click and hold the left mouse button on the display window.
3. Drag the mouse down to zoom in.
4. Click the Pan tool.
5. Click the display window, hold down the left mouse button, and move the model until you reach the part you want to view.
6. To see the entire model, click the Full Extent tool.

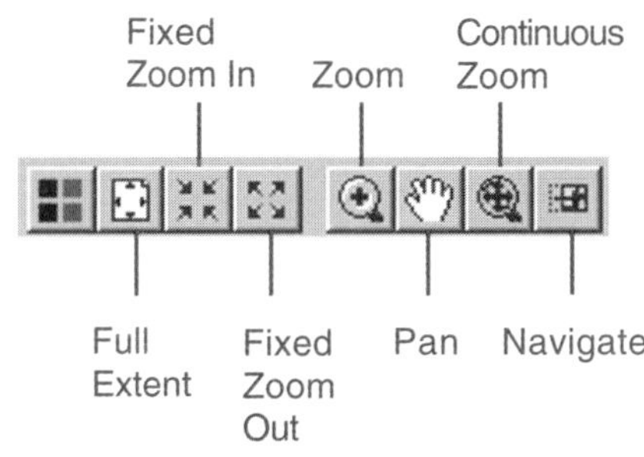

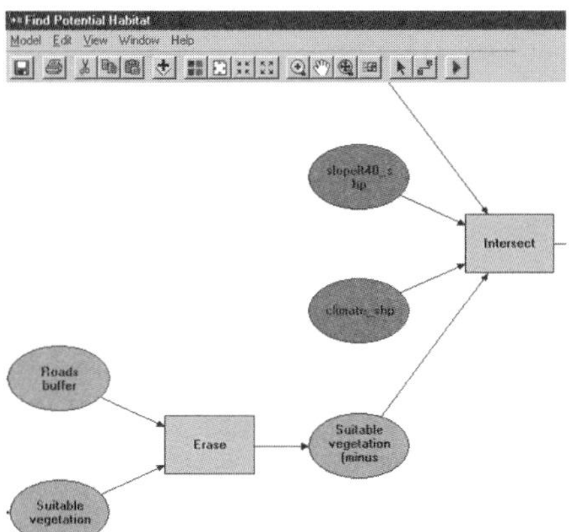

Using any of the zoom in tools, then clicking the Pan tool, you can move to different parts of the model.

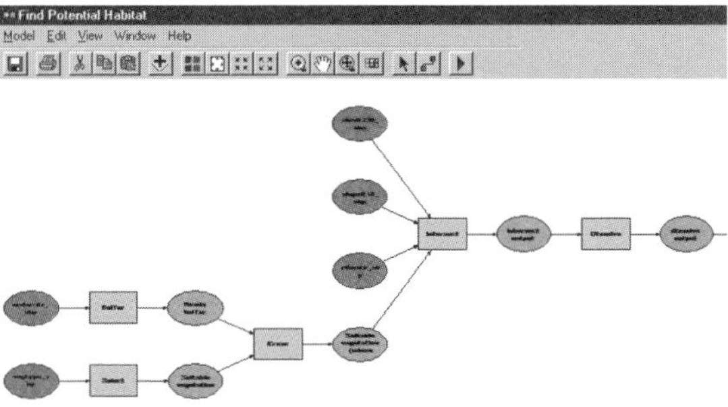

Clicking the Full Extent tool displays the entire model.

- The preset zoom levels on the View menu (25%, 50%, 100%, 200%, 400%) zoom to fixed percentages of the actual size.

To avoid losing track of where you are in a large model, you can open an Overview window. The Overview window displays the entire model at all times. Your current location in the model window is marked by a rectangle in the Overview window. When you navigate in the ModelBuilder window, this rectangle moves correspondingly.

To use the Overview window as a zoom tool, click and drag a rectangle in the Overview window. The model window zooms in to the area covered by the rectangle. You can click and drag to create a new zoom rectangle at any time. You can also resize a rectangle by clicking and dragging its black selection handles.

To use the Overview window as a pan tool, click inside the existing rectangle in the Overview window and drag it to a new location.

You can resize the Overview window and move it outside the ModelBuilder window. ▸

Using the overview window

1. Click the Windows menu and click Overview Window.
2. Drag, shrink, or expand the box in the Overview window to zoom/move the model to the area covered by the box.

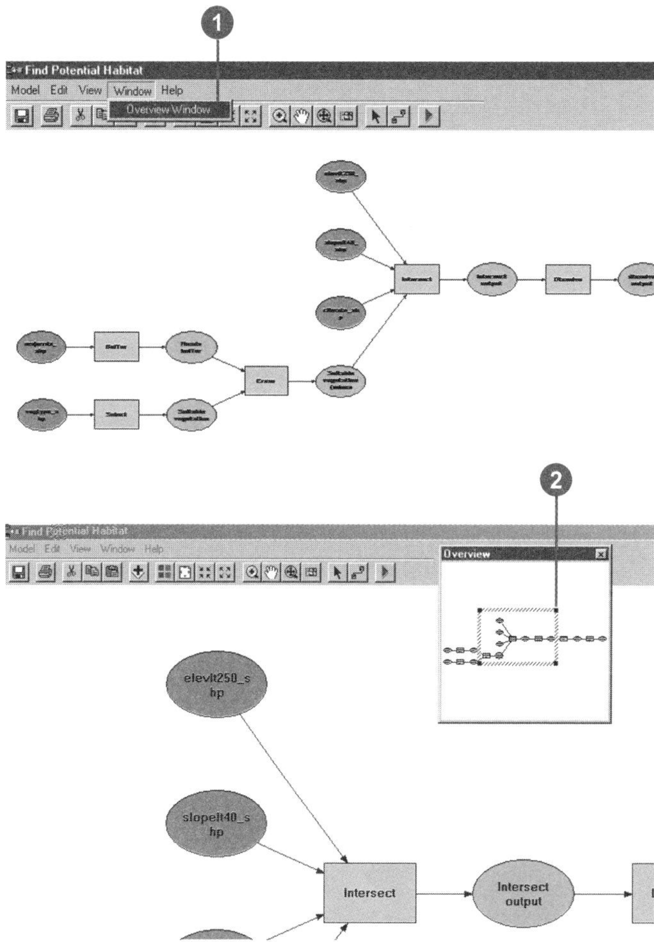

The display window is zoomed to the area shown by the rectangle in the Overview window. Dragging the overview window to a new location pans the model.

The Navigate tool allows you to navigate through the flow of a model, from one element to the next, either forward or backward. It can be useful if you want to step through a model, element by element.

To move forward, click the beginning of a connector line. To move backward, click the arrowhead of a connector line.

Navigating forward through the model

1. To navigate through the model, click the Navigate tool.
2. To move forward to the next element, click the beginning of the connector line.
3. Continue to follow step 2 to navigate forward through the entire model.

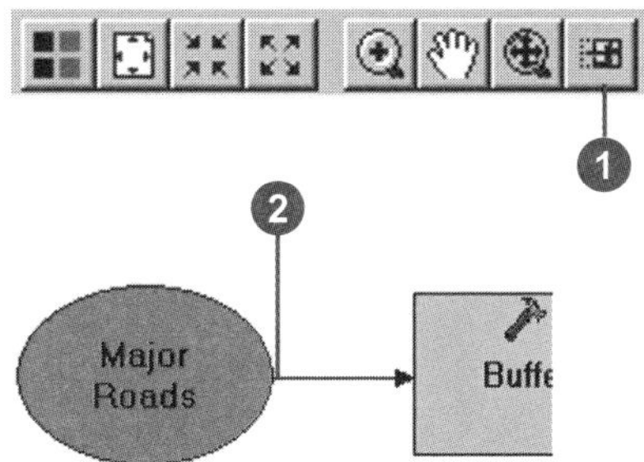

Navigating backward through the model

1. To navigate through the model, click the Navigate tool.
2. To move backward to the preceding element, click the arrowhead of the connector line.
3. Continue to follow step 2 to navigate backward through the entire model.

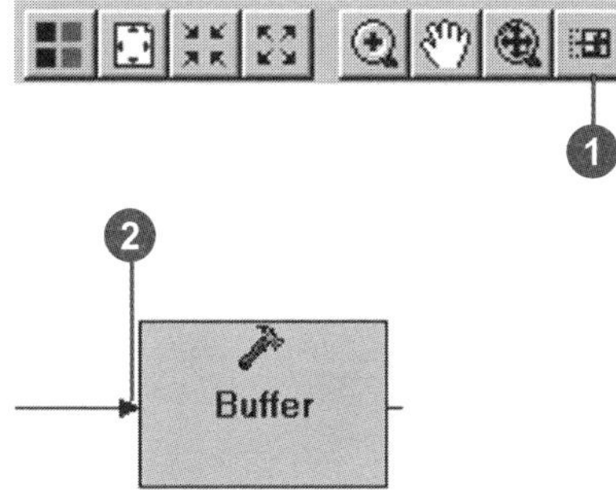

Validating a model

When you open an existing model inside a ModelBuilder window, processes that were run earlier will retain that already run state. However, if a parameter value has been altered, such as the dataset that is referenced, such changes will not be immediately apparent when you open the model.

By validating the entire model, you can verify that all parameter values are valid before you run the model. Note that validating an already run model will return its processes to a ready-to-run state. Parameter values may become invalid for several reasons. For example, the input dataset referenced no longer exists in its location on disk, or it was renamed. Alternatively, a property of a dataset might have changed, for example, a field has been added or deleted.

After you validate a model, if any of its processes change to a not-ready-to-run state, one or more of the parameter values is invalid and must be repaired.

See Also

See 'Repairing a model' in this chapter for information on repairing parameter values.

1. Click the Model menu and click Validate Entire Model.

   All parameter values set for each element are validated. If a parameter value set is invalid, the element and all those that depend on it will become not ready to run. You have to reset the value for invalid parameters to return affected processes to a ready-to-run state.

2. After running validation, messages display information about invalid parameter values. Right-click an element and click View Messages to identify problems.

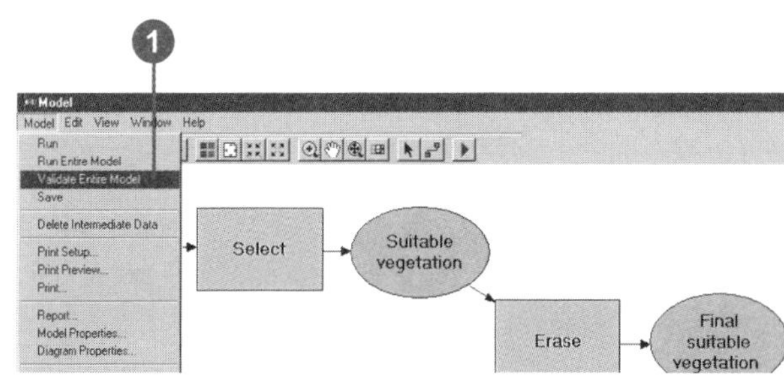

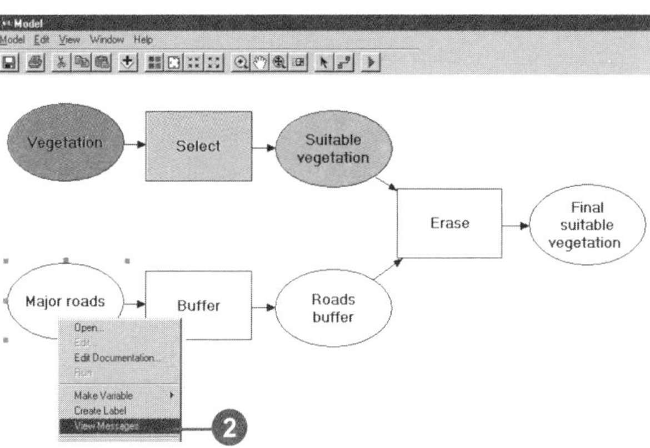

Repairing a model

There are two main reasons to repair a model:

- After validating a model, processes become not ready to run. There is an invalid parameter value set.
- A red 'x' (icon) appears over your model icon in the ArcCatalog tree or the ArcToolbox window. There may be a tool referenced in the model that no longer exists in its toolbox on disk, or it was renamed. You may have a COM tool inside the model, and the DLL is unregistered. Alternatively, a parameter name for a model or a script added to the model might have changed.

To fix invalid parameter values, go back through the model to identify the first process not in a ready-to-run state. Open the tool of this process and repair ▸

Tip

Automatic repair of parameter values

If you add a model you have created to a second model and update a parameter value in the first model, the parameter value will automatically be repaired in the second model.

Repairing the parameter value for input data elements

1. Right-click the input data element for which you want to repair the parameter value and click Open.

   Alternatively, right-click the tool element and click Open.

2. Click the Browse button for the input data parameter.
3. Locate the correct input data, and click Add.
4. Click OK on the input data element properties dialog box or the tool element dialog box.

   The input data element, as well as any other elements with valid parameter values that depend on it, will become ready to run.

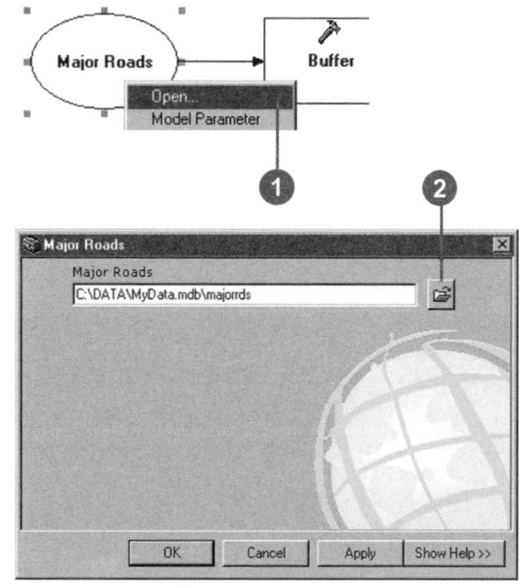

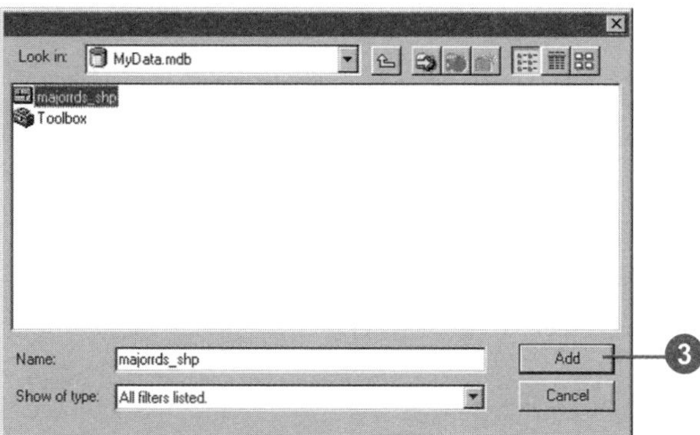

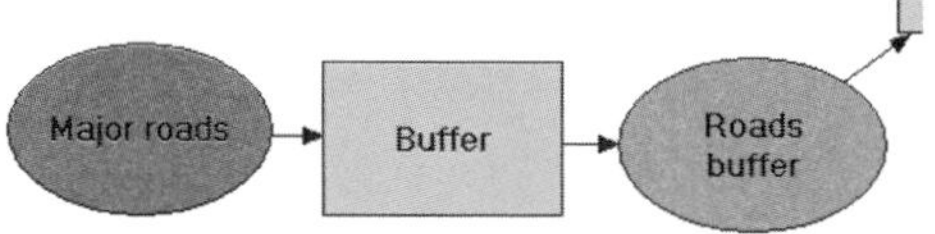

the invalid value set for the parameter. Once the parameter value is valid, the process will return to ready-to-run.

You can repair the tool element by right-clicking it and clicking Open. This opens a Browse dialog box so you can browse to the tool's disk location that the tool element references.

If the problem is an unregistered DLL for a COM tool you have inside your model, relocating the tool will not solve this problem. You must register the DLL within your operating system to resolve the problem.

If parameter names in a model or a script that is added to the model have changed, click Model, then Save to update the model with the changes.

Tip

Setting relative paths

You can avoid repairing paths in a model by setting relative paths before sending or moving a model, along with other relevant sources. See 'Storing relative pathnames' in Chapter 5 and 'Sharing your geoprocessing work' in Chapter 3 for more information.

Repairing a tool

1. Right-click the invalid tool and click Open.

   A browse dialog box will open.

2. Click the Look in dropdown arrow and navigate to the location of the tool.

3. Click Add.

   The tool is repaired. The next time you right-click the tool and click Open, the tool's dialog box will open normally.

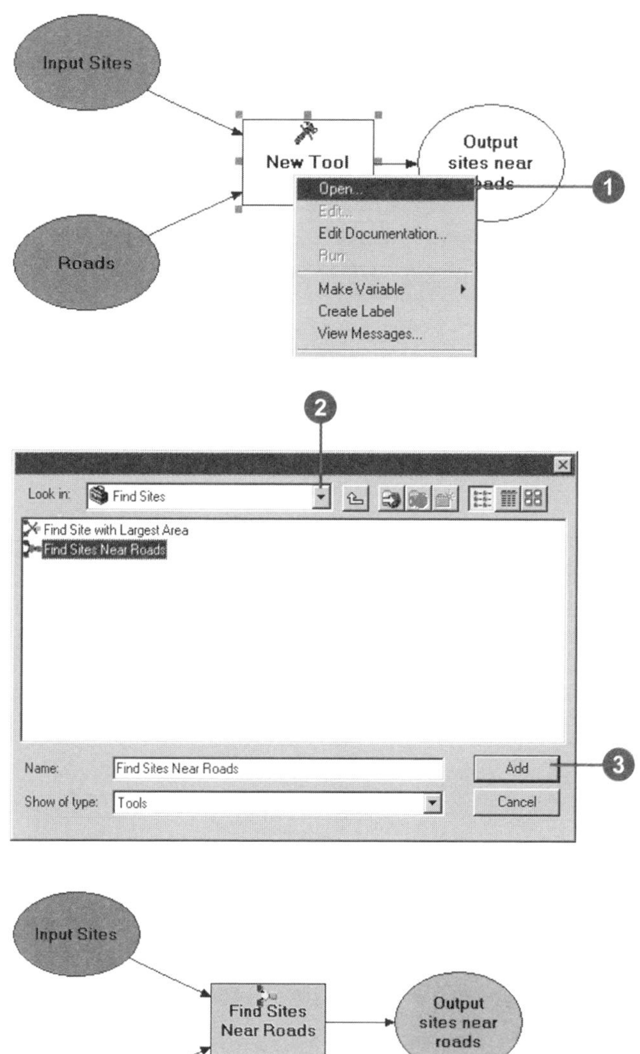

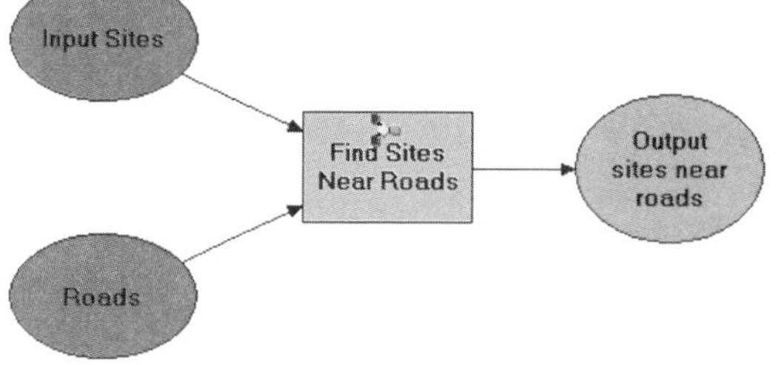

Print setup

The Print Setup dialog box lets you set printer options for a model.

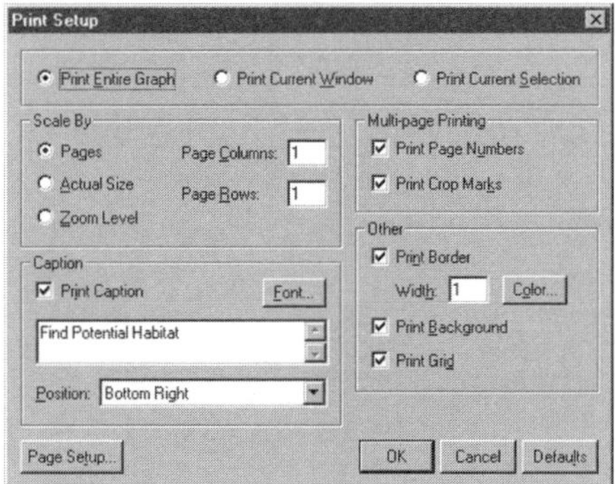

You can print the entire diagram (graph) regardless of whether you are zoomed in on the diagram; the current window, which is the portion of the diagram that is visible in the display window; or the current selection, which is the currently selected elements or connection lines within the model diagram.

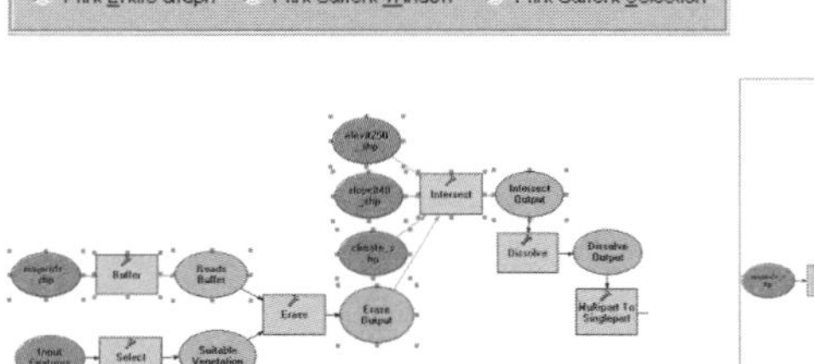

Selected elements and connector lines in the model diagram

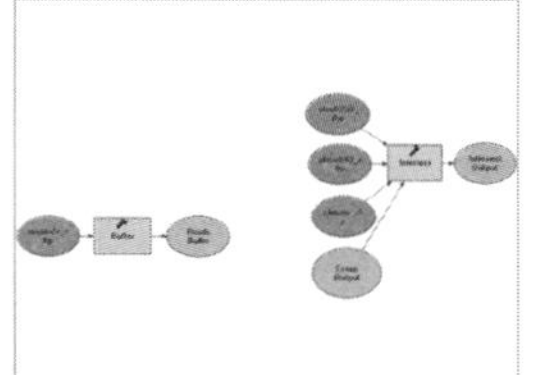

Print preview, with Print Current Selection set on the Print Setup dialog box

When the Print Border option is checked, a single-width border, also known as a "neatline", is drawn along the margin edges. By default, a border will be printed with a border width of 1. The width and color of the line can be altered.

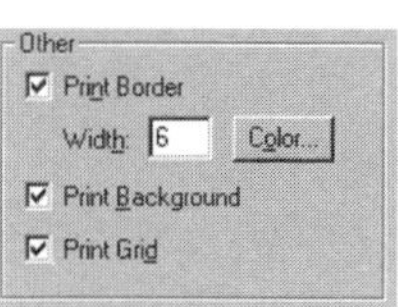

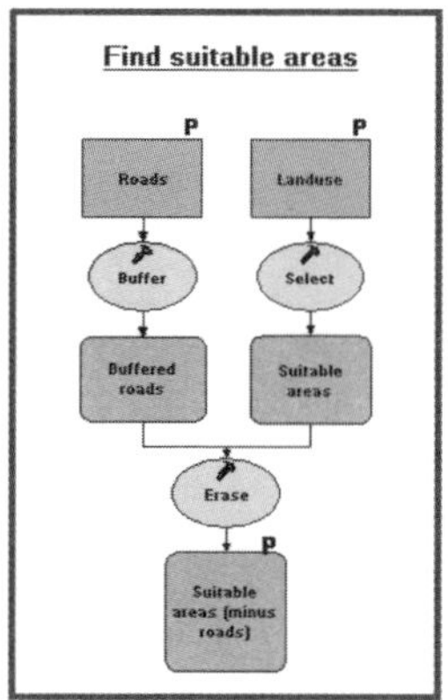

If you need help lining up elements, you can show a grid over the model diagram. Access the Diagram Properties dialog box via the Model menu. On the General tab, check Show Grid. You can choose whether or not to print this grid by checking or unchecking the Print Grid check box.

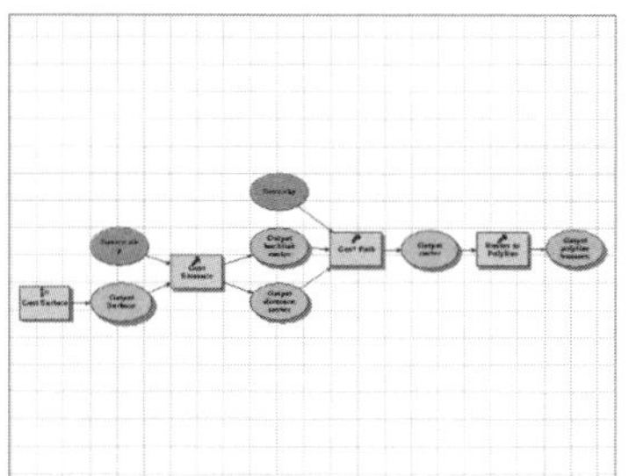

Print preview, with the Print Grid option checked

Scale by

You can scale by pages, by actual size, or by zoom level. If you scale by pages, the model diagram prints on the number of pages specified in the page column and page row boxes. If both boxes are set to 1, the entire model diagram will print on one page. If both boxes are set to 2, the model diagram will be scaled to print on four pages. If you scale by actual size, the model diagram prints on as many pages as necessary. If you scale by zoom level, the model diagram will be printed at the scale of the display.

Caption

By default, a caption is printed in the lower right corner of the page using the model's display name. You can change the text that will print by typing the desired text into the text input box, and you can change the position of the text by clicking the Position dropdown arrow and selecting a position for the text—bottom right, bottom center, bottom left, top right, top center, or top left. Click the Font button to change the font of the text that will be displayed. If you do not want to see a caption displayed, uncheck the Print Caption check box.

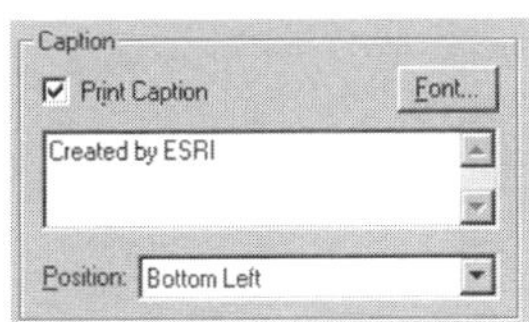

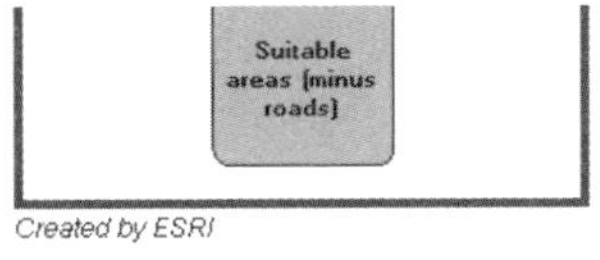

Multipage printing

For each printed page you can print the page number and add crop marks.

Page numbers

When Print Page Numbers is checked, page numbers are printed. No page number appears if the model is printed on a single page. Page numbers are centered in the margins on the sides of pages that are adjacent to other pages. Pages are numbered in the form [Row, Column] - Location, where Row is the row page number; Column is the column page number; and Location is top, bottom, left, or right. Page numbers are printed by default. The diagram that follows shows the location of page numbers.

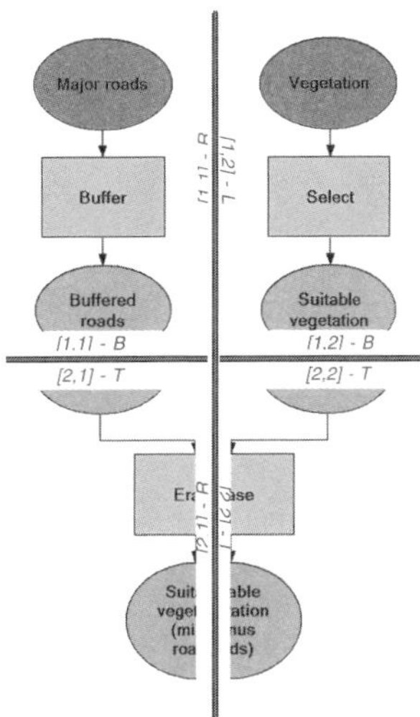

The model has been printed on four pages. The page displaying the upper left corner of the model is numbered [1,1]; the upper right corner is [1,2]; the lower left corner is [2,1]; and the lower right corner is [2,2]. The top, bottom, left, and right edges of adjacent pages are labeled T, B, L, and R.

Crop marks

Crop marks help you align models that are printed on several pages. They are single-width lines about one-half inch long printed along the margins of pages that are adjacent to other pages. They are printed by default. No crop marks appear if the model is printed on a single page. The following diagram shows a model printed with crop marks. To align the printed model, you would cut along the crop marks.

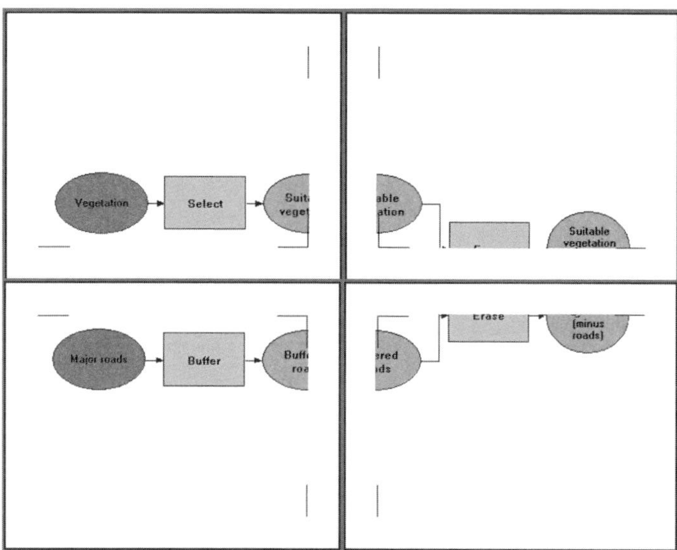

Crop marks help you align models that are printed on several pages.

Page setup

Click the Page Setup button to change the default page settings. You can change the orientation of the page (from Portrait to Landscape) and set the paper size and the paper source. Click the Printer button to change the name of the printer.

Setting up and printing a model

The Print Setup dialog box lets you specify the area to print; the number of pages on which the model will print; the width and color of the border; and other options, such as page numbering, crop marks, and a caption.

By default, the model prints on a single page. You can change this by setting new values in the Page Columns and Page Rows boxes. A setting of 2 by 2, for instance, will print the model on four pages.

By default, the page margin is 0.5 inches. This margin is uniform for all sides of the page.

To print the model, click the File menu and click Print. Here you can control options, such as the number of copies to print and the page orientation (Portrait or Landscape).

Tip

Setting up print options

You can set up print options before you print. On the Print Setup dialog box, click the Page Setup button to change the paper size and source, the orientation of the page, the page margins, and the printer you want to use.

1. Click the Model menu and click Print Setup.
2. Change the default options to more appropriate settings for your model.
3. Click OK.
4. Click the Model menu and click Print Preview to see what the model will look like when it is printed. ►

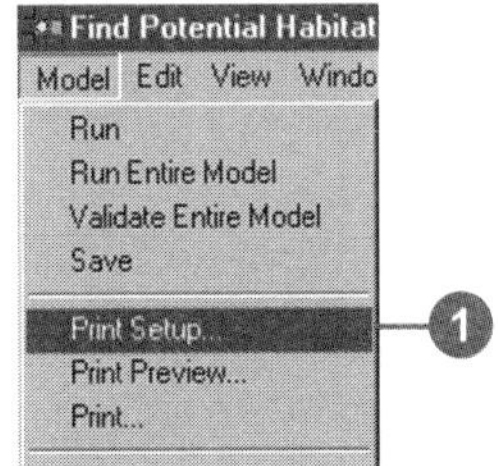

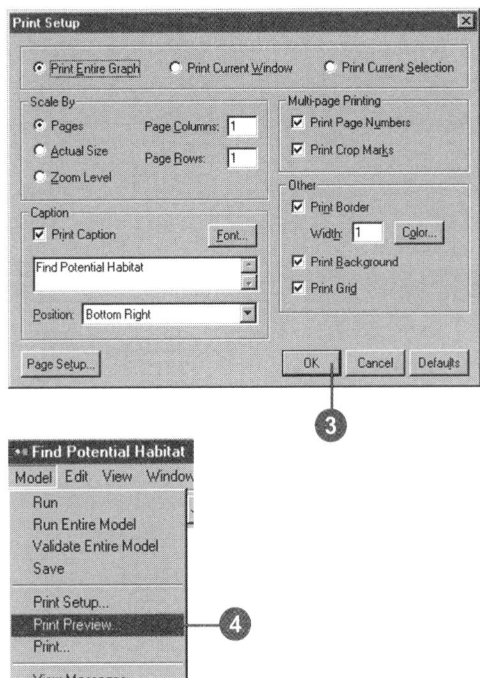

Tip

Printing a model

You can access the Print dialog box from the Model menu or from the Print Preview dialog box.

5. Click Print.
6. Set the options you want in the Print dialog box.
7. Click OK.

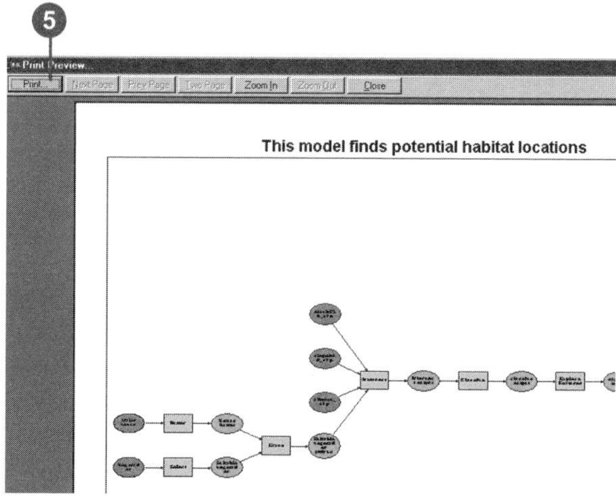

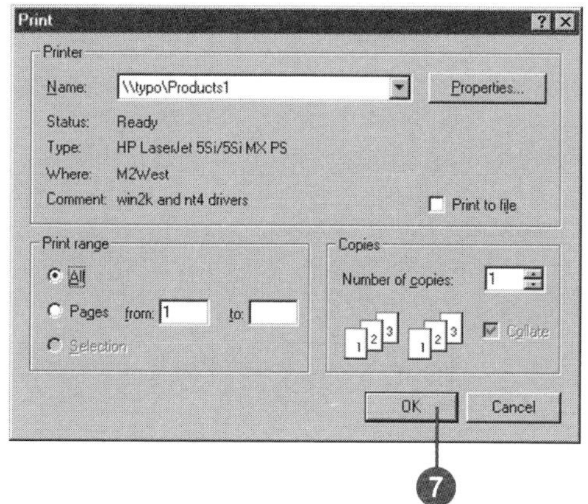

Generating a report

You can view or save a report for your model.

The report documents everything contained in the model, including the time and date it was generated, the variables used—including their data types and values—and the processes created.

The documentation for processes includes the tool name and the location (source) of the tool used in the process, the parameter values and their properties, and any messages. The messages explain the state of each process. For example, they inform you if a parameter value is invalid, and they supply execution messages to let you know if the process ran successfully.

Tip

Viewing messages

You don't have to generate a report to view messages. You can view messages in the message section of the Command Line window, or you can right-click a tool element in the model and click View Messages to view the messages for a particular tool.

Viewing a report

1. Click the Model menu and click Report.
2. Check View report in a window, then click OK.

   A Model Report window opens so you can view the report.
3. Click OK to close the window.

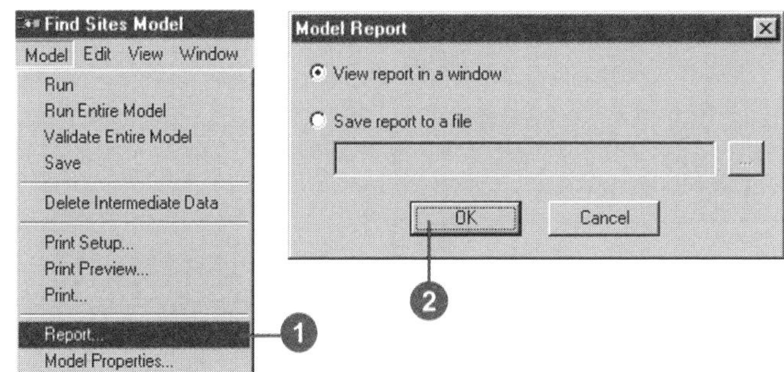

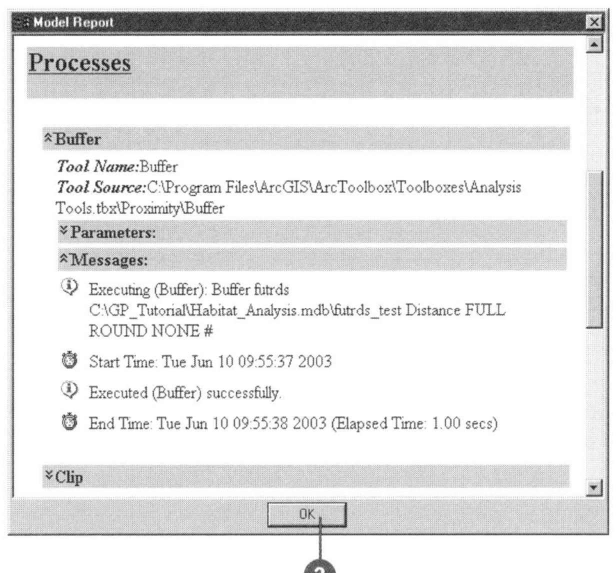

Saving a report

1. Click the Model menu and click Report.
2. Check Save report to a file, then click the Browse button.
3. Click the Save in dropdown arrow and navigate to the location at which to save the XML report.
4. Click the File name text box and type a name for the XML file.
5. Click Save.
6. Click OK on the Model Report dialog box.

   You can view the report at any time by opening it from its location on disk.

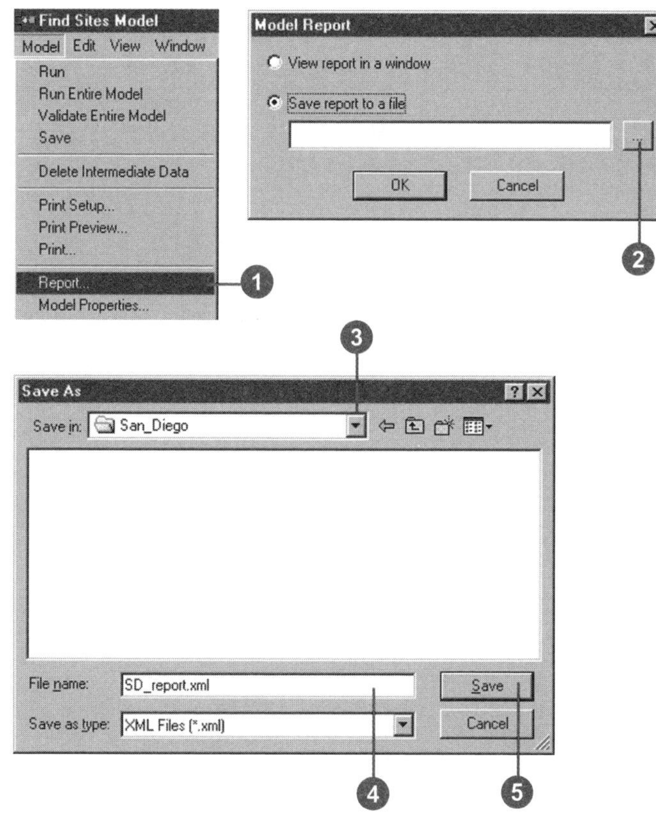

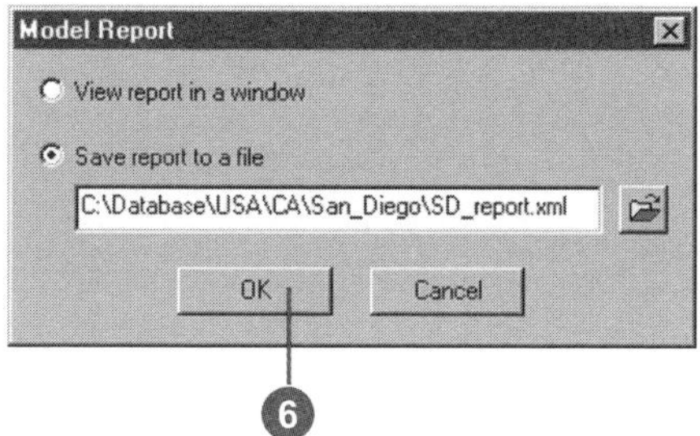

Documenting a process

Processes in your model can be documented similarly to how you document tools or toolboxes.

The documentation you add might describe what the process does. You can add paragraphs, bullet items, hyperlinks, illustrations, subsections, or indented text. The items you add can be rearranged in the Contents list using the up and down buttons, or content can be deleted using the Delete button. The order of the items in the Contents list is reflected in the Help page.

Tip

Indenting items

Any item can be indented. Click the Indent button to add an Indent item to the Contents list, click the Indent item in the Contents list, then click the button corresponding to the item you want indented.

See Also

For information on documenting toolboxes and tools, see 'Adding documentation to toolboxes' in Chapter 4 and 'Adding documentation to tools' in Chapter 5.

Adding documentation to a process

1. Right-click the tool of the process for which you want to add documentation and click Edit Documentation.

   The documentation window opens so you can begin to document the process.
2. To add a paragraph of text, click the tool name in the Contents list and click the Paragraph button.
3. Type the text for the paragraph in the right side of the dialog box.
4. Continue adding paragraphs, bullet items, links, illustrations, subsections, and indented text as appropriate.
5. Click OK.

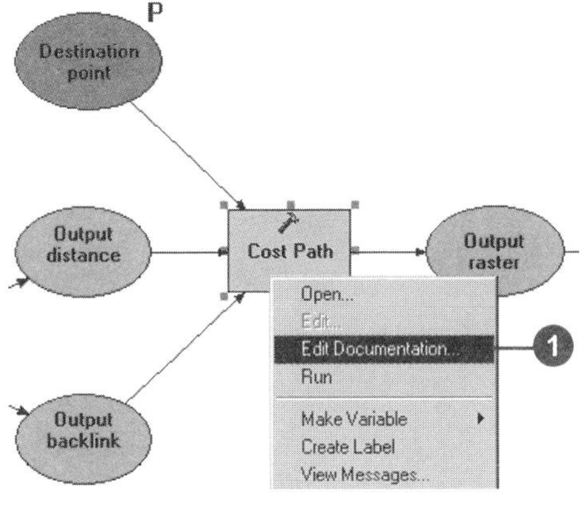

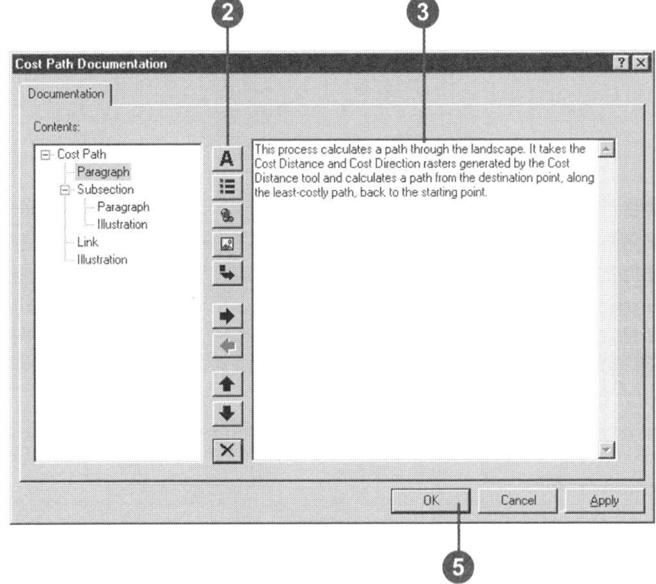

Viewing documentation for a process

1. In the ArcToolbox window, right-click the tool containing a documented process and click Help.
2. Click the Model section to expand its contents and navigate to the documented process.

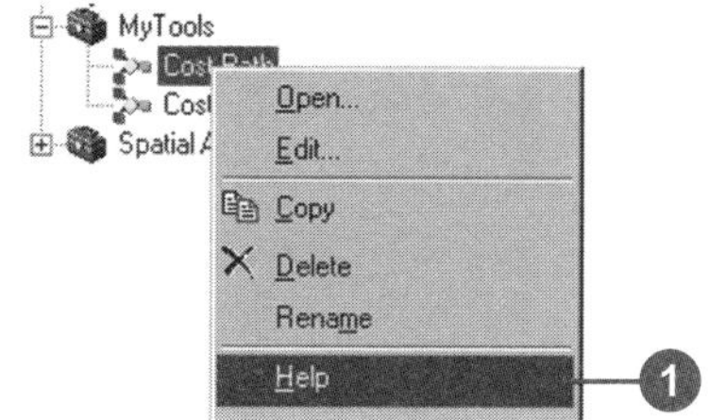

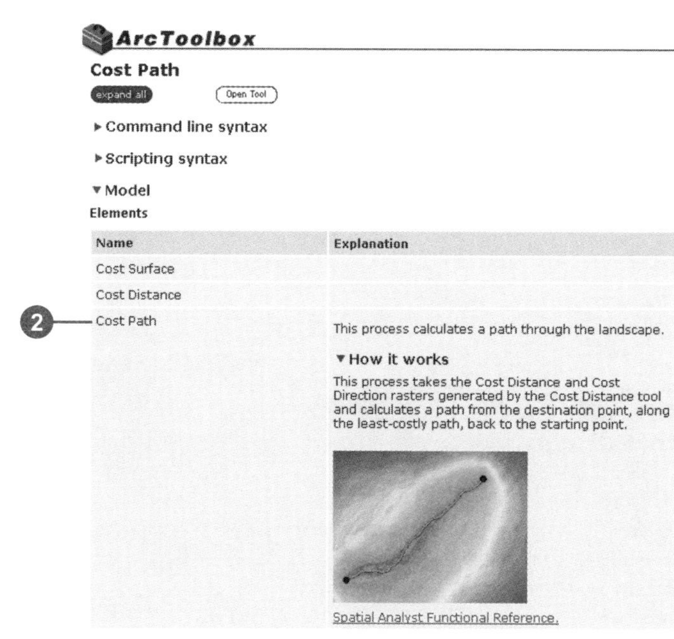

Importing a model from ArcView GIS 3

You can import existing models, which were created using ArcView GIS 3 ModelBuilder, into the ModelBuilder window if you have installed the ArcGIS Spatial Analyst extension. Each .xmd file can be converted to an ArcGIS model.

If you are working in ArcMap and you have an ArcView GIS 3 project, import the ArcView GIS 3 project, then import the ArcView GIS 3 model. However, you do not have to import the project. In any ArcGIS application with a display, layers added to the table of contents will be matched to the data referenced in the ArcView GIS 3 model. Layer names need to be the same as the theme name in the ArcView GIS 3 model. If a matching layer is not found for project data, the model will still be created. After importing the model in any ArcGIS application, including ArcCatalog, you can edit the project data variables to reference the project data. ▸

Importing an ArcView project

1. Start ArcMap and open a blank document.
2. Click the File menu and click Import from ArcView project.
3. Navigate to and select the ArcView project you would like to import to ArcMap. If you are importing a layout, choose the layout you would like to import from the Layouts dropdown list. The views that are in the layout will be automatically selected. Views that are not in the selected layout may be selected interactively.
4. Click OK to import the project. The selected import layout is now the layout view in ArcMap. Some graphic and text adjusting may be required to get the desired look for the map layout. The imported views are now separate data frames in the table of contents.

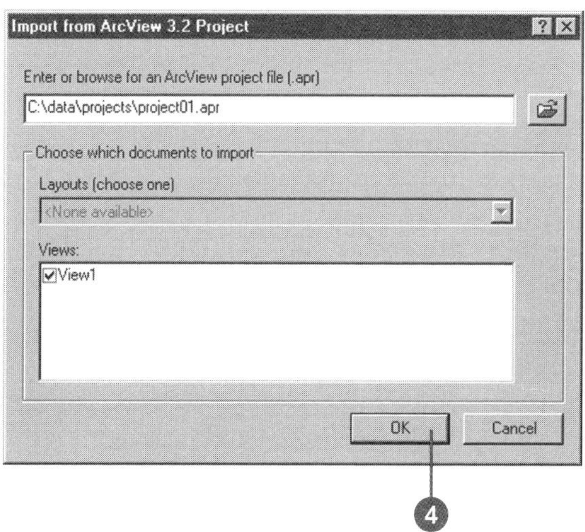

Although most functions convert directly to ArcGIS tools, some functions will change their name and some will convert to multiple tools.

Vector Conversion will become Feature to Raster. Point Interpolation will become IDW or Spline. DEM Conversion will change to DEM to Raster. Reclass will become Reclassify. Buffer will become a combination of Euclidean Distance and Reclassify. Arithmetic Overlay will become a combination of Plus, Minus, Times, Divide, and Reclassify.

See Also

See 'Creating models and adding scripts' in Chapter 5 for information on creating a new model inside a toolbox.

Importing an ArcView GIS 3 model

1. Create a new model inside a toolbox.
2. Click the Model menu, point to Import, and click ArcView 3 ModelBuilder.
3. Navigate to the location of the .xmd file.
4. Click the file and click Open.

   The ArcView GIS 3 model is imported.

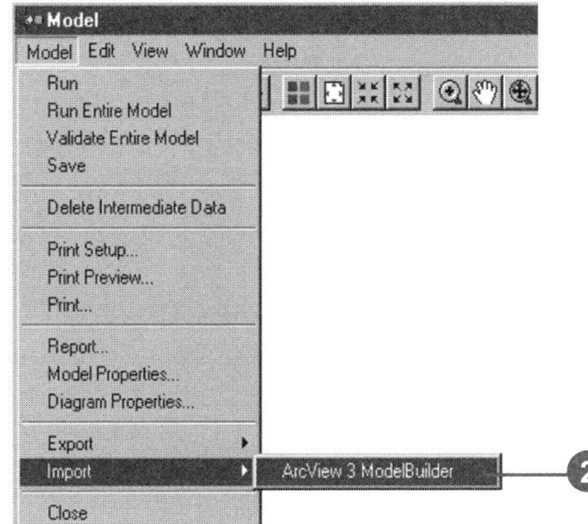

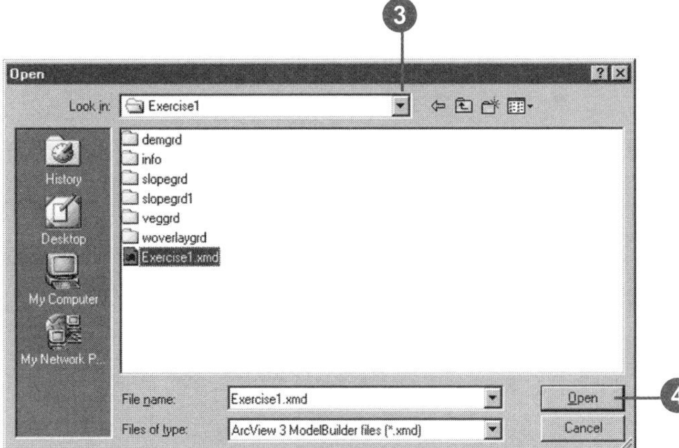

Exporting a model

You can export a model to a script or to a graphic.

An easy way to start writing scripts is to export a model you have created to a script, then start to modify the exported script.

You can export to three different script formats: Python, JScript, and VBScript.

The script can be shared via e-mail or over a network, and it can be added to a toolbox and run as any other tool.

In the example, the Buffer tool has been added to the ModelBuilder window and its parameter values have been set. You may have several inputs that you need to buffer. Rather than using multiple buffer tools inside the model, you can export the simple model that contains one process to a script, then modify the script to perform batch processing to include a loop that will buffer all the inputs inside a workspace. ►

See Also

See 'Introducing geoprocessing methods' in Chapter 3 for more information on modifying a script to perform batch processing.

Exporting a model to a script

1. Create a model in the display window.
2. Click the Model menu, point to Export, then To Script, and click the preferred script type.
3. Click the Save in dropdown arrow and navigate to the location to which you want to save your script.
4. Type a File name for the script.
5. Click Save.

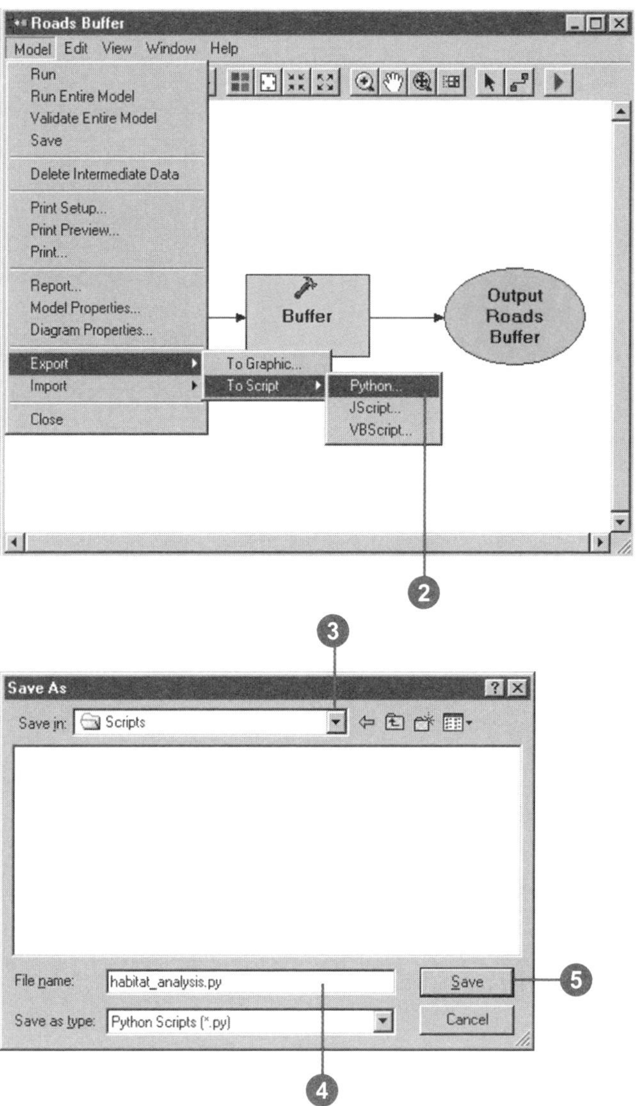

A model can be exported to a graphic. It can be useful to place a graphic of the model in the documentation for the model, reference the graphic inside a model's stylesheet so the graphic of the model is displayed on the model's dialog box, or place the graphic in a report.

See Also

See 'Adding documentation to tools' in Chapter 5 for more information on attaching a graphic of a model to the model's documentation. See 'Setting a modified stylesheet for a tool' in Chapter 5 for more information on setting a graphic as a background image for a tool.

Exporting a model to a graphic

1. Create a model in the display window.
2. Click the Model menu, point to Export, and click To Graphic.
3. Set the desired options in the Save As Image dialog box.
4. Click OK.

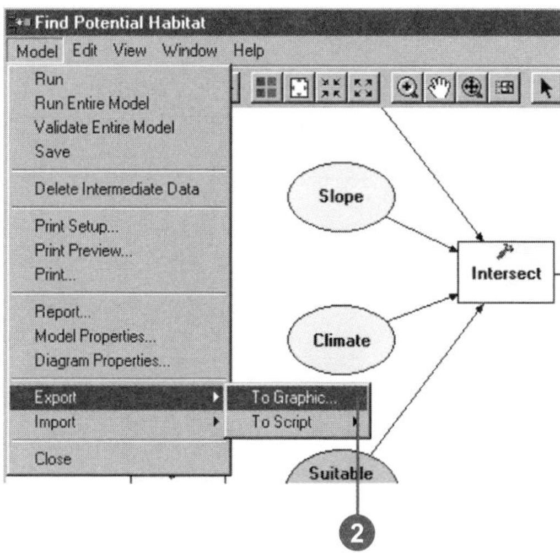

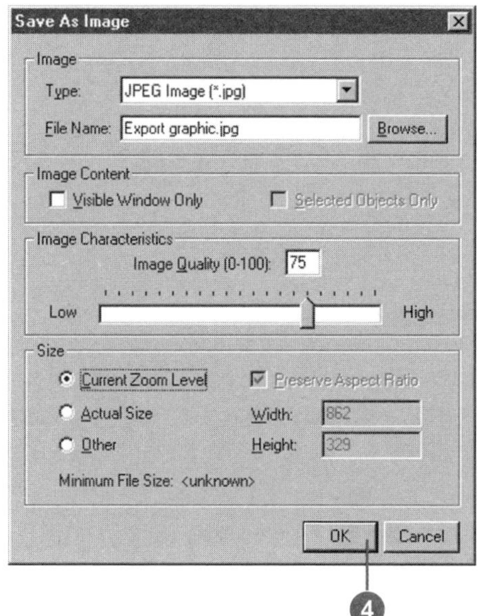

Appendix

IN THIS APPENDIX

- **How are default stylesheets applied?**
- **Customizing the default stylesheets**
- **Options for customizing stylesheets**

The dialog boxes of tools (system tools or custom tools) are similar in style because they have a default Extensible Stylesheet Language (XSL) stylesheet applied to them—they have the same background image, font, colors, and layout. Stylesheets are composed of display instructions, or XSL operations, which operate on XML statements. The result of these XSL operations is a set of HTML files that, when displayed, becomes the user interface that is interacted with whenever a tool is executed via its dialog box.

You can alter the default stylesheets to change the look of your tools, depending on personal preference. Two examples are provided to show you the types of changes you can make. Note that the code in the XSL files used in the examples can be copied from the online Help system if desired.

The first example focuses on how the default colors, background, and layout can be changed and explains how to add hyperlinks. The second example illustrates the creation of an expandable panel for optional parameters. In both examples new stylesheets are created that import the ESRI default stylesheets for those display characteristics that are not overwritten in the custom stylesheets. The alternative is to copy the default stylesheets locally to the location of the toolbox, modify your local copy, then override the default stylesheet applied to the tool's dialog box with your local copy by setting the path. The example in the section 'Setting a modified stylesheet for a tool' in Chapter 5 describes this process.

How are default stylesheets applied?

When the dialog box of a tool is opened, the XML statements representing the user interface elements for the tool are created. The XML statements do not indicate how these elements should be displayed; they only define which elements are involved. The XML file is called MdElements.xml and is created or re-created in the Documents and Settings\<user name>\Application Data\ESRI\ArcToolbox\Dlg folder each time the dialog box of a tool is opened. Unless the tool has been linked to custom stylesheets, the default stylesheets are applied to the MdElements.xml file. The following default stylesheets are located in your ArcGIS\ArcToolbox\Stylesheets folder on the drive where you installed ArcGIS:

- MdDlgMain.xsl
- MdDlgContent.xsl
- MdDlgHelp.xsl

These XSL stylesheets transform the XML statements (MdElements.xml) into the set of HTML files that represent the tool's dialog box. Three HTML files are created or re-created in the Documents and Settings\<user name>\Application Data\ESRI\ArcToolbox\Dlg folder:

- MdDlgMain.htm
- MdDlgContent.htm
- MdDlgHelp.htm

These HTML files are then displayed as the tool's dialog box.

Customizing the default stylesheets

You can override the default MdDlgContent.xsl and MdDlgHelp.xsl stylesheets that are applied to any tool's dialog box, provided the tool does not reside inside a read-only toolbox.

Example 1

In this example, the font, background image, and background color are altered for the dialog box of the Standard Distance tool and a hyperlink is added. This tool is located inside the Spatial Statistics system toolbox.

Without customization, the dialog box for the Standard Distance tool resembles the following graphic.

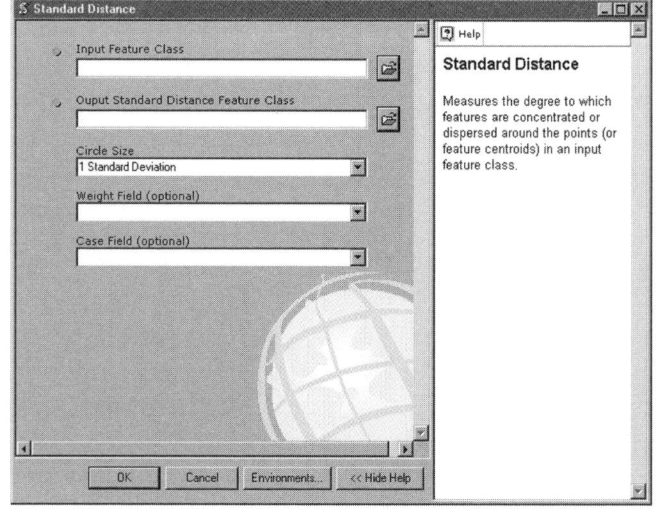

Using a custom stylesheet, the look of the dialog box can be altered dramatically, as the following graphic shows.

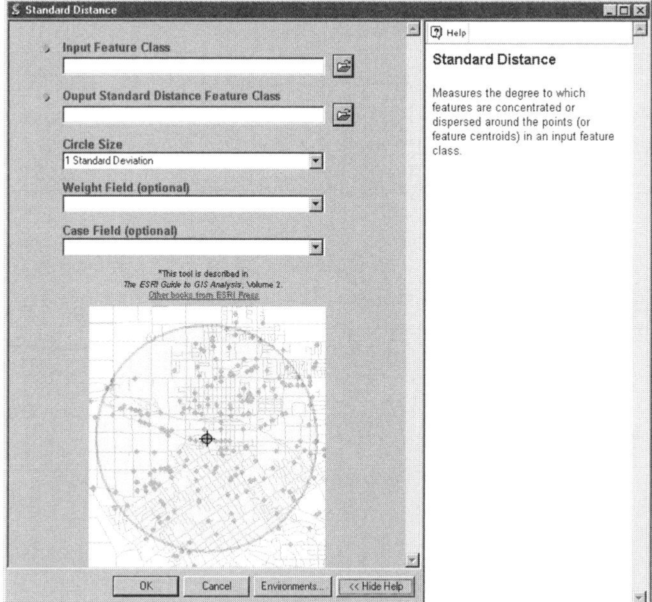

A custom stylesheet is used to alter the background image, font, and color of the dialog box and to add a hyperlink.

The default stylesheet that is applied to a tool's dialog box can be changed on the General tab of the tool's properties dialog box. Right-click the tool and click Properties to open the properties dialog box.

In this example, the name of a custom XSL file called SSContent.xsl is entered in the Stylesheet text box on the General tab of the tool's properties dialog box. By simply typing the name of the custom stylesheet, the path to the location of the default stylesheets (ArcGIS\ArcToolbox\Stylesheets) is used. To follow this technique, place your custom stylesheet or stylesheets in this location.

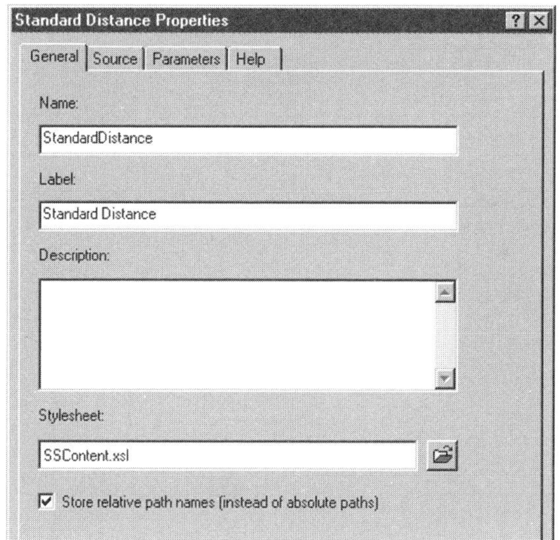

The contents of SSContent.xsl are displayed in the example that follows. In this example, this file is used in place of MdDlgContent.xsl. It imports the settings of MdDlgContent.xsl if they are not changed in SSContent.xsl. There is no need for a path to the default MdDlgContent.xsl file in this example, as both

SSContent.xsl and MdDlgContent.xsl are placed in the ArcGIS\ArcToolbox\Stylesheets location. The background image (mygraphic.jpg) can be placed in the DLG folder (\Documents and Settings\username000\Application Data\ESRI\ArcToolbox\ Dlg or \WINNT\Profiles\username000\Application Data\ESRI\ArcToolbox\Dlg [on Windows NT]). As this is the default location where the HTML files are generated from the XSL files, the graphic will be found in this location if just the name of the graphic is entered, for example, url(mygraphic.jpeg). Alternatively, a hard-coded path can be entered to the location of the graphic, such as url(C:/mydata/mygraphic.jpeg). A third alternative is displayed in the code below. The graphic can be placed in a location that is relative to the location of the Common folder (\ArcGIS\ArcToolbox\Common), such as in the Stylesheets folder.

```
<?xml version="1.0"?>
<xsl:stylesheet xmlns:xsl="http://www.w3.org/1999/XSL/Transform"
                version="1.0">

<xsl:import href = "MdDlgContent.xsl" />
<xsl:output method="html"/>

<!-- Overwrite Select Variable Definitions -->
<xsl:variable name="BackgroundColor">honeydew</xsl:variable>
<xsl:variable name="BackgroundImage">url(<xsl:value-of select="MdElementDialogInfo/CommonPath"/>/../
Stylesheets/mygraphic.jpg)</xsl:variable>
<xsl:variable name="BackgroundPosition">bottom center</xsl:variable>
<xsl:variable name="CaptionFont">arial,verdana</xsl:variable>
<xsl:variable name="CaptionSize">10pt</xsl:variable>
<xsl:variable name="CaptionColor">DarkGreen</xsl:variable>
<xsl:variable name="CaptionWeight">Bold</xsl:variable>

<!-- Add Web link -->
<xsl:template match="MdElementDialogInfo">
   <xsl:apply-imports/>
   <table border="0" cellspacing="0" cellpadding="0" width="98%"
onmousedown="parent.ShowHelpTopic('Intro');">
      <tr align="center">
       <td><span style="font-family:arial; font-size:7pt;">*This tool is described in</span></td></tr>
```

```
      <tr align="center">
       <td><span style="font-family:arial; font-size:7pt;"><i>The ESRI Guide to GIS Analysis</i>, Volume 2.</span></td></tr>
      <tr align="center">
       <td><span style="font-family:arial; font-size:7pt; color:blue;"><a href="http://gis.esri.com/esripress/display/index.cfm" target="esripress">Other books from ESRI Press</a></span></td></tr>
    </table>
</xsl:template>
</xsl:stylesheet>
```

Variables correspond to the background color, font type, size, color, image, and image position. Display instructions are also appended to the main MdElementDialogInfo template. Once all other XML elements have been processed, this extra XSL code adds the text, "This tool is described in *The ESRI Guide to GIS Analysis*, Volume 2", and a Web link to ESRI Press.

Example 2

You can change the stylesheet applied to a dialog box so that parameters are grouped into expandable sections within the tool's dialog box.

In the example below, the tool has multiple optional parameters. You can apply a custom stylesheet to group all the optional parameters into an expandable section.

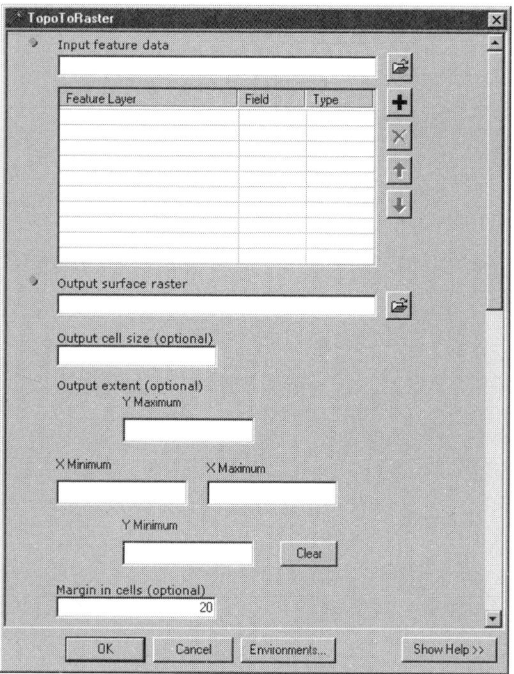

Note that system tools are read only, but you can still modify them by first copying them to your own toolbox. See Chapter 4, 'Working with toolboxes', for information on creating your own toolboxes, and see Chapter 5, 'Working with toolsets and tools', for information on copying a tool from one toolbox to another.

The modified code within a custom stylesheet (expand.xsl) that can be applied is displayed in the section that follows. This stylesheet only overrides certain settings, and the default stylesheet (MdDlgContent.xsl) is imported to take care of the rest. The default stylesheet resides in your ArcGIS\ArcToolbox\Stylesheets folder on the drive where you installed ArcGIS, so provided your custom stylesheet is placed in the same location, you can just type the name of the default stylesheet to import, with a .xsl extension within your custom stylesheet.

```
<?xml version="1.0"?>
<xsl:stylesheet xmlns:xsl="http://www.w3.org/1999/XSL/Transform"
                version="1.0">

<xsl:import href = "MdDlgContent.xsl" />
<xsl:output method="html"/>

<!-- Overwrite Select Templates -->
<xsl:template match="PropertyGroup">
   <TR valign="top"><TD>
       <DIV ID="GEN" STYLE="cursor: hand;">
           <!-- Process required parameters -->
           <xsl:for-each select="Property">
                <xsl:choose>
                  <xsl:when test='not(contains(PropertyLabel,"(optional)"))'>
                       <xsl:apply-templates select="." />
                   </xsl:when>
                </xsl:choose>
            </xsl:for-each>

           <!-- Add expanding section for optional parameters -->
          <TABLE onclick="parent.clicker({PropertyGroupName},{PropertyGroupName}Image);"
STYLE="cursor:hand;" border="0" cellspacing="0" cellpadding="0" width="94%">
<TR valign="top" bgcolor="menu">
                    <TD colspan="2">
                       <TABLE border="0" width="100%" cellpadding="0" cellspacing="0">
                            <TR bgcolor="menu">
                                <TH align="left">
```

```
                                   <IMG ID="{PropertyGroupName}Image" SRC="{../../CommonPath}/
triangle.gif" ALT="*" ALIGN="MIDDLE" BORDER="0" WIDTH="11" HEIGHT="11"/>
                                   <SPAN class="caption" STYLE="color:menutext;">Optional Parameters</
SPAN>
                                 </TH>
                             </TR>
                         </TABLE>
                     </TD>
                 </TR>
                 <TR valign="top">
                     <TD colspan="2">
                        <DIV ID="{PropertyGroupName}" STYLE="display:'none';"
onclick="window.event.cancelBubble = true;">
                            <TABLE border="0" cellspacing="1" cellpadding="4" width="90%">
                                <xsl:for-each select="Property">
                                    <xsl:choose>
                                      <xsl:when test='contains(PropertyLabel,"(optional)")'>
                                          <xsl:apply-templates select="." />
                                       </xsl:when>
                                    </xsl:choose>
                                </xsl:for-each>
                            </TABLE>
                        </DIV>
                     </TD>
                 </TR>
             </TABLE>
         </DIV>
     </TD></TR>
</xsl:template>
</xsl:stylesheet>
```

In this example, the stylesheet expand.xsl is specified in the Stylesheet text box on the General tab of the tool's properties dialog box. Since there is no path, this file should be saved in the location \ArcGIS\ArcToolbox\Stylesheets.

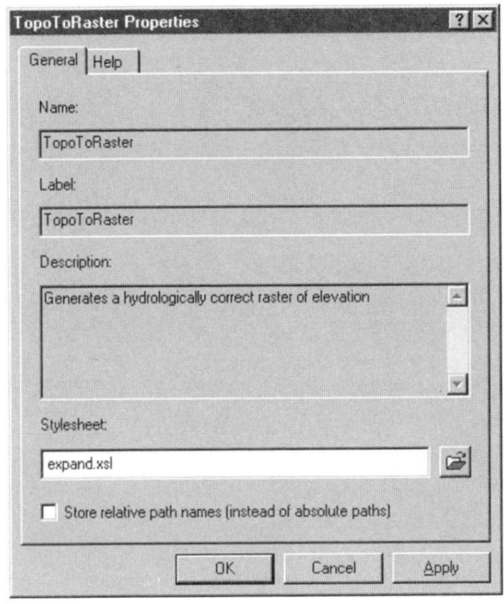

By opening the tool's dialog box you'll see the applied stylesheet. Notice how the optional parameters are collected into the Optional Parameters expandable section.

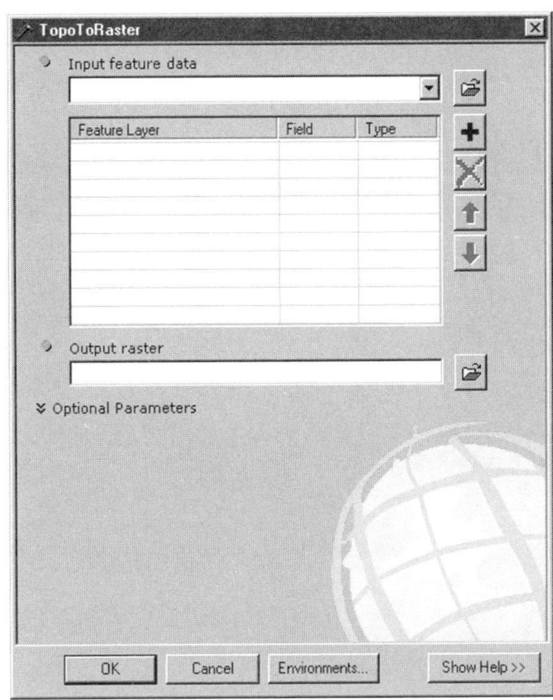

Clicking the arrow to the left of Optional Parameters expands the section to allow values for optional parameters to be supplied.

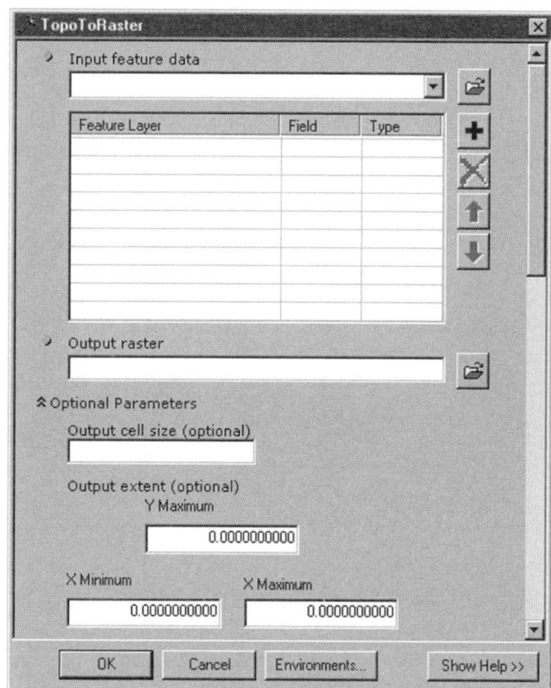

If you examine the code for expand.xsl, you'll notice that it overwrites the PropertyGroup template of MdDlgContent.xsl and uses <xsl:choose> statements to process optional and required parameters differently.

Options for customizing stylesheets

There are a number of options you can implement so others will see your stylesheet customizations when you distribute your tools, including the options that follow.

Option 1

- Copy the default stylesheets—MdDlgContent.xsl, MdDlgHelp.xsl, or both—from your \ArcGIS\ArcToolbox\Stylesheets folder to another location on your local drive. This is usually the same location as the toolbox or the geodatabase if the toolbox is stored inside a geodatabase.
- Modify the stylesheets as desired, then use them in place of the default stylesheets by specifying the location to the files in the Stylesheet text box on the General tab of the tool's properties dialog box. For more information on specifying custom stylesheets in the Stylesheet text box, see 'Setting a modified stylesheet for a tool' in Chapter 5.
- Set relative paths on the General tab of the tool's properties dialog box.
- Zip the folder containing the toolbox and the data, then send the ZIP file. Refer to 'Sharing your geoprocessing work' in Chapter 3 for more information on archiving your work.

Option 2

- Type the name of your custom stylesheet or stylesheets in the Stylesheets text box on the General tab of the tool's properties dialog box.
- Put your custom stylesheets in the same location on disk as the default stylesheets: \ArcGIS\ArcToolbox\Stylesheets.

- Instruct users of your tools to place your custom stylesheets in the Stylesheets folder on their machine. From then on, your stylesheets will always be applied when your tools are run.

Option 3

- Type the name and the path to your custom stylesheet or stylesheets in the Stylesheets text box on the General tab of the tool's properties dialog box.
- Set relative paths on the General tab of the tool's properties dialog box.
- In your custom stylesheet files, hard code the location to import the default stylesheet from—for example, <xsl:import href= "C:\Program Files\ArcGIS\ArcToolbox\Stylesheets\MdDlgContent.xsl" />. This path may have to be modified by receivers of your tool if they installed ArcGIS in a different location.

Glossary

alias

In geoprocessing, an alternate name for a toolbox. Toolbox aliases can be used to avoid confusion when working with tools with the same name that are stored in different toolboxes. For example, tools in the Analysis Tools toolbox can be differentiated from similar tools in the Spatial Analyst Tools toolbox by adding "_analysis" to their names at the command line, as in "clip_analysis."

AML

See ARC Macro Language (AML).

ARC Macro Language (AML)

A proprietary high-level algorithmic language for generating end user applications in ArcInfo Workstation. Features include the ability to create onscreen menus, use and assign variables, control statement execution, and get and use map or page unit coordinates. AML includes an extensive set of commands that can be used interactively or in AML programs (macros) as well as commands that report on the status of ArcInfo environment settings.

ArcCatalog tree

See Catalog tree.

ArcSDE geodatabase

A remote geodatabase in a relational database management system (RDBMS) served to client applications by ArcSDE. An ArcSDE geodatabase can be used as a workspace for geoprocessing tasks.

ArcToolbox window

A dockable window used to display, manage, and use the contents of toolboxes in ArcGIS. It provides a shortcut to frequently used tools contained within toolboxes that may be stored in folders or geodatabases on disk.

batch processing

Executing a series of noninteractive jobs at the same time. In geoprocessing, these system tools allow for batch processing: the Feature Class to Geodatabase tool, the Feature Class to Shapefile tool, the Table to Geodatabase tool, and the Table to dBASE tool.

CAD dataset

See CAD feature dataset.

CAD drawing dataset

The pictorial representation of an entire CAD file that can be viewed in any ArcGIS application with a display. The CAD drawing dataset is a vector data source of a mixed feature type in which the symbology is set to mimic that of the originating CAD application. The graphic properties of a CAD drawing dataset's objects can be identified, but the dataset is not usable for feature class-based queries or analysis.

CAD feature class

A read-only member of a CAD feature dataset, comprised of one of the following: polylines, points, polygons, multipatch, or annotation. The feature attribute table of a CAD feature class is a virtual table comprised of select CAD graphic properties and any existing field attribute values.

CAD feature dataset

The feature representation of a CAD file in a geodatabase-enforced schema. A CAD feature dataset is comprised of five read-only feature classes: points, polylines, polygons, multipatch, and annotation. Formats supported in ArcGIS include DWG (AutoCAD), DXF (AutoDesk Drawing Exchange Format), and DGN (the default MicroStation file format).

CAD file

The digital equivalent of a drawing, figure, or schematic created using a CAD system. CAD files are the data source for CAD drawing datasets, feature datasets, and feature classes. ArcGIS software-supported formats include DWG (AutoCAD), DXF (AutoDesk Drawing Exchange Format), and DGN (the default MicroStation file format). A CAD file is represented in ArcCatalog with a CAD feature dataset and a CAD drawing dataset.

Catalog tree

In ArcCatalog, a hierarchical view of folder connections that provides access to GIS data stored on local disks or shared on a network, and allows users to manage connections to databases and GIS servers.

cell size

The area on the ground covered by a single cell in an image, measured in map units.

cluster tolerance

In geodatabase feature classes, a definition for the minimum tolerated distance between vertices in the topology. Vertices that fall within the set cluster tolerance will be snapped together during the validate topology process.

COM

See Component Object Model (COM).

command

An instruction to a computer program, usually one word or concatenated words or letters, issued by the user from a control device, such as a keyboard, or read from a file by a command interpreter.

command line

An onscreen interface in which the user types in commands at a prompt. In geoprocessing, any tool added to the ArcToolbox window can be run from the command line.

Command Line window

In geoprocessing, a window that provides a command line for running tools, and a message window for viewing the status messages created when running those tools.

Component Object Model (COM)

A binary standard that enables software components to interoperate in a networked environment regardless of the language in which they were developed. Developed by

Microsoft, COM technology provides the underlying services of interface negotiation, life cycle management (determining when an object can be removed from a system), licensing, and event services (putting one object into service as the result of an event that has happened to another object).

compression

A reduction of file size for data handling and storage. Examples of such methods include quadtrees, run-length encoding, and wavelet.

connector

A visual representation of the relationship between elements in a model. Connectors join elements together to create processes. Typical processes connect an input data element, a tool element, and a derived data element.

coordinate system

A fixed reference framework superimposed onto the surface of an area to designate the position of a point within it; a reference system consisting of a set of points, lines, and/or surfaces; and a set of rules, used to define the positions of points in space in either two or three dimensions. The Cartesian coordinate system and the geographic coordinate system used on the earth's surface are common examples of coordinate systems.

coverage

A data model for storing geographic features using ArcInfo software. A coverage stores a set of thematically associated data considered to be a unit. It usually represents a single layer, such as soils, streams, roads, or land use. In a coverage, features are stored as both primary features (points, arcs, polygons) and secondary features (tics, links, annotation). Feature attributes are described and stored independently in feature attribute tables.

coverage feature class

In ArcInfo, a classification describing the format of geographic features and supporting data in a coverage. Feature classes include point, arc, node, route system, route, section, polygon, and region. One or more coverage features are used to model geographic features; for example, arcs and nodes can be used to model linear features, such as street centerlines. The tic, annotation, link, and boundary feature classes provide supporting data for coverage data management and viewing.

current workspace

A user-specified path to a container for file-based geographic data, set in the Environment Settings dialog box. Data from the current workspace can be accessed from any tool dialog box, including script or model dialog boxes, or the command line simply by typing its name.

See also scratch workspace.

custom tool

In geoprocessing, a tool created by a user and added to a toolset or a toolbox. Custom tools may only be added to custom toolsets or toolboxes.

custom toolset

In geoprocessing, a subset of a toolbox created by a user to hold custom tools or a group of frequently used tools.

data

Any collection of related facts arranged in a particular format; often, the basic elements of information that are produced, stored, or processed by a computer.

data source

Any geographic data. Data sources may include coverages, shapefiles, rasters, or feature classes.

data type

The attribute of a variable, field, or column in a table that determines the kind of data it can store. Common data types are character, integer, decimal, single, double, and string.

database management system (DBMS)

A set of computer programs that organizes the information in a database according to a conceptual schema and provides tools for data input, verification, storage, modification, and retrieval.

dataset

Any organized collection of data with a common theme.

DBMS

See database management system (DBMS).

derived data

Data created by running a geoprocessing operation on existing project data. Derived data from one process can serve as input data for another process.

derived value

Nongeographic data created by running certain geoprocessing tools. Derived values from one process in a model can serve as input values for other processes in the model.

See also intermediate data.

dialog box

In geoprocessing, a form consisting of a tool's parameters.

Documentation Editor

In geoprocessing, the interface used to write documentation for tools, toolsets, toolboxes, and processes within a model.

double precision

The level of coordinate exactness based on the possible number of significant digits that can be stored for each coordinate. Datasets can be stored in either single or double precision. Double-precision geometries store up to 15 significant digits per coordinate (typically 13 to 14 significant digits), retaining the accuracy of much less than one meter at a global extent.

See also single precision.

element

A component of a model. Elements can be variables, such as input and derived data, or tools.

environment settings

Settings that can apply to all tools within the application, all tools within a model or script, or a particular process within a model or script. Environment settings include current workspace, output spatial reference, output spatial grids, cell size, and tile size. They are generally set before running tools.

extent

The coordinate pairs defining the minimum bounding rectangle (xmin, ymin and xmax, ymax) of a data source. All coordinates for the data source fall within this boundary.

feature class

A collection of geographic features with the same geometry type (such as point, line, or polygon), the same attributes, and the same spatial reference. Feature classes can stand alone within a geodatabase or be contained within shapefiles, coverages, or other feature datasets. Feature classes allow homogeneous features to be grouped into a single unit for data storage purposes. For example, highways, primary roads, and secondary roads can be grouped into a line feature class named "roads". In a geodatabase, feature classes can also store annotation and dimensions.

feature dataset

A collection of feature classes stored together that share the same spatial reference; that is, they have the same coordinate system and their features fall within a common geographic area. Feature classes with different geometry types may be stored in a feature dataset.

folder

A location on a disk containing a set of files, other folders, or both.

geodatabase

An object-oriented data model introduced by ESRI that represents geographic features and attributes as objects and the relationships between objects, but is hosted inside a relational database management system. A geodatabase can store objects, such as feature classes, feature datasets, nonspatial tables, and relationship classes.

geodatabase raster band

See raster dataset band.

geodatabase raster dataset

See raster.

geographic information system (GIS)

An arrangement of computer hardware, software, and geographic data that people interact with to integrate, analyze, and visualize the data; identify relationships, patterns, and trends; and find solutions to problems. The system is designed to capture, store, update, manipulate, analyze, and display the geographic information. A GIS is typically used to represent maps as data layers that can be studied and used to perform analyses.

geoprocessing

A GIS operation used to manipulate data stored in a GIS workspace. A typical geoprocessing operation takes an input dataset, performs an operation on that dataset, and returns the result of the operation as an output dataset. Common geoprocessing operations are geographic feature overlay, feature selection and analysis, topology processing, and data conversion. Geoprocessing allows for definition, management, and analysis of information used to form decisions.

geoprocessing settings

Any settings that affect working with or running tools. Geoprocessing settings include the state of the ArcToolbox window, the state of the Environment Settings dialog box, and variables that have been created at the command line. In ArcMap, geoprocessing settings are saved with a map document. In ArcCatalog, geoprocessing settings are persisted with the application.

geostatistical layer files

A layer file created by the Geostatistical Analyst extension. The files can be exported to ESRI GRID for further geoprocessing.

GIS

See geographic information system (GIS).

history model

A model created, dated, and saved when the application is closed to document the tools and parameter values used for each session. The history model is contained within the History toolbox and can be viewed when the application is reopened.

See also model.

hyperlink

A reference (link) from some point in one electronic document or resource to some point in another document or resource or another place in the same document. Activating the link, usually by clicking it with the mouse, causes the browser to display the target of the link.

input data

Data that is entered into a computer, device, program, or process.

intermediate data

Any data, not set up as a model parameter, that is referenced by derived data variables in a model. Such data is part of a process and is not saved once the model has been successfully run from its dialog box. Input data is not classified as intermediate data because it existed before the model was executed.

See also derived data.

layer

A set of references to data sources such as a coverage, geodatabase feature class, raster, and so on that defines how the data should be displayed on a map. Layers can also define additional properties, such as which features from the data source are included. Layers can also be used as inputs to geoprocessing tools. Layers can be stored in map documents (.mxd) or saved individually as layer files (.lyr).

m-value

1. Vertex attributes that are stored with x,y point coordinates in ESRI's Geometry Engine. Every type of geometry (point, polyline, polygon, and so on) can have attributes for every vertex.
2. In linear referencing, measure values that may be added to linear features to perform dynamic segmentation. In linear referencing, m-values are used on vertices to imply a measurement along a linear feature. The m-value allows a location along a line to be found.

map projection

See projection.

mask

A means of performing raster analysis on a selected set of cells in a raster dataset. Cells that fall within the extent of the mask are processed. All other cells are characterized as NoData.

measure value

See m-value.

metadata

Information about the content, quality, condition, and other characteristics of data. Metadata for geographical data may document its subject matter; how, when, where, and by whom the data was collected; accuracy of the data; availability and distribution information; its projection, scale, resolution, and accuracy; and its reliability with regard to some standard. Metadata consists of properties and documentation. Properties are derived from the data source—for example, the coordinate system and projection of the data—while documentation is entered by a person (for example, keywords used to describe the data).

model

A set of rules and procedures for representing a phenomenon or predicting an outcome. In geoprocessing, a model consists of one process or a sequence of processes connected together. It is created in a toolbox and built in a ModelBuilder window. A model can be exported to a script file.

model parameter

A type of parameter exposed in a model that displays in a model's dialog box and allows for input.

ModelBuilder window

The interface used to build and edit models in ArcGIS.

My Toolboxes folder

In ArcCatalog, a folder that contains nonsystem toolboxes created in the ArcToolbox window or directly in the My Toolboxes folder in the ArcCatalog Tree. The My Toolboxes folder points to a location on disk that can be changed in the Geoprocessing tab of the Options dialog box.

output data

Data that is the result of a program or process.

See also input data.

parameter

In geoprocessing, a characteristic of a tool. Values set for parameters define a tool's behavior during run time.

personal geodatabase

A geodatabase that stores data in a single-user relational database management system. A personal geodatabase can be read simultaneously by several users, but only one user at a time can write data into it.

precision

Refers to the number of significant digits used to store coordinate values. Precision is important for accurate feature representation, analysis, and mapping. ArcInfo supports single and double precision.

process

A tool and its parameter values. One process, or multiple processes connected together, creates a model.

project data

Geographic input data that exists before a geoprocessing tool is run.

projection

A method by which the curved surface of the earth is portrayed on a flat surface. This generally requires a systematic mathematical transformation of the earth's graticule of lines of longitude and latitude onto a plane. It can be visualized as a transparent globe with a light bulb at its center casting lines of latitude and longitude onto a sheet of paper. Generally, the paper is either flat and placed tangent to the globe (a planar or azimuthal projection) or formed into a cone or cylinder and placed over the globe (cylindrical and conical projections). Every map projection distorts distance, area, shape, direction, or some combination thereof.

pyramid

In raster datasets, a reduced resolution layer that copies the original data in decreasing levels of resolution to enhance performance. The coarsest level of resolution is used to quickly draw the entire dataset. As the display zooms in, layers with finer resolutions are drawn; drawing speed is maintained because fewer pixels are needed to represent the successively smaller areas.

raster

A spatial data model that defines space as an array of equally sized cells arranged in rows and columns. Each cell contains an attribute value and location coordinates. Unlike a vector structure, which stores coordinates explicitly, raster coordinates are contained in the ordering of the matrix. Groups of cells that share the same value represent geographic features.

raster band

See raster dataset band.

raster catalog

A collection of raster datasets defined in a table of any format, in which the records define the individual raster datasets that are included in the catalog. A raster catalog is used to display adjacent or overlapping raster datasets without having to mosaic them together into one large file.

raster dataset

See raster.

raster dataset band

One layer in a raster dataset that represents data values for a specific range in the electromagnetic spectrum, such as ultraviolet, blue, green, red, infrared, or radar, or other values derived by manipulating the original image bands. A raster dataset can contain more than one band. For example, satellite imagery commonly has multiple bands representing different wavelengths of energy from along the electromagnetic spectrum.

relative path

In computing, the location of a computer file given in relation to the current working directory. For example, in ArcMap, a path to the data source of a layer contained in a map document (.mxd file) may be set relative to the location of the map document; in geoprocessing, a path to a source of information referenced by a tool in a toolbox may be set relative to the location of the toolbox.

scratch workspace

A path to a container for file-based geographic data that can be set in the Environment Settings dialog box or at the command line, into which all automatically generated outputs will be placed.

script

A set of instructions in plain text, usually stored in a file and interpreted, or compiled, at run time. In geoprocessing, scripts can be used to automate tasks, such as data conversion, or generate geodatabases and can be run from their scripting application or added to a toolbox. Geoprocessing scripts can be written in any COM-compliant scripting language, such as Python, JScript, or VBScript.

SDC dataset

A collection of Smart Data Compression (SDC) feature classes sharing attribute information with different geometries. SDC format is used by ESRI to provide StreetMap™ data. An SDC dataset is stored in a set of related files and contains multiple features classes.

SDC feature class

A highly compressed, read-only data structure that can store spatial geometry (points, lines, and polygons) and attribute data. The SDC structure supports geocoding, routing, and most spatial operations. SDC is the core data structure used in StreetMap, ArcIMS® RouteServer, RouteMAP™, IMS, ArcGIS Business Analyst, and BusinessMAP®.

shapefile

A vector data storage format for storing the location, shape, and attributes of geographic features. A shapefile is stored in a set of related files and contains one feature class.

single precision

Refers to a level of coordinate exactness based on the number of significant digits that can be stored for each coordinate. Single-precision numbers store up to seven significant digits for each coordinate, retaining a precision of ±5 meters in an extent of 1,000,000 meters. Datasets can be stored as either single- or double-precision coordinates.

See also double precision.

snap raster

An option in the Environment Setting dialog box that ensures that the cell alignment of the extent will match accurately with an existing raster. This is done by snapping the lower left corner of the specified extent to the lower left corner of the nearest cell in the snap raster, and snapping the upper right corner of the specified extent to the upper right corner of the nearest cell in the snap raster.

spatial domain

For a spatial dataset, the defined precision and allowable range for x and y coordinates and for m-values and z-values if present. The spatial domain must be specified by the user when creating a geodatabase feature dataset or standalone feature class.

spatial grid

A two-dimensional grid system that spans a feature class. It is used to quickly locate features in a feature class that might match the criteria of a spatial search.

spatial reference

The coordinate system used to store a spatial dataset. For feature classes and feature datasets within a geodatabase, the spatial reference also includes the spatial domain.

stylesheet

A file or form that provides style and layout information, such as margins, fonts, and alignment, for tagged content within an XML or HTML document. Stylesheets are frequently used to simplify XML and HTML document design, since one stylesheet may be applied to several documents. Transformational stylesheets may also contain code to transform the structure of an XML document and write its content into another document.

syntax

The structural rules for using statements in a command or programming language.

system tool

In geoprocessing, a tool installed with ArcGIS. System tools are stored in system toolsets and can be copied to custom toolsets and toolboxes.

system toolbox

In geoprocessing, a default toolbox that is installed with ArcGIS. System toolboxes contain system tools, organized into toolsets for ease of access.

system toolset

In geoprocessing, a subset of a toolbox that holds system tools.

table

A set of data elements arranged in rows and columns. Each row represents an individual entity, record, or feature, and each column represents a single field or attribute value. A table has a specified number of columns but can have any number of rows.

table view

A representation of tabular data for viewing and editing purposes. The table view created when a table is added to ArcMap is a copy of the actual table data stored in memory.

TIN dataset

A dataset containing a triangulated irregular network (TIN). The TIN dataset includes topological relationships between points and neighboring triangles.

tool

An entity in ArcGIS that performs such specific geoprocessing tasks as clip, split, erase, or buffer. A tool can belong to any number of toolsets or toolboxes.

toolbox

An object that contains toolsets and geoprocessing tools. It takes the form of a .tbx file on disk, or a table in a geodatabase.

toolset

In geoprocessing, a group of tools that perform similar tasks.

See also custom tool, system tool.

topology

1. In geodatabases, a set of governing rules applied to feature classes that explicitly define the spatial relationships that must exist between feature data.
2. In an ArcInfo coverage, the spatial relationships between connecting or adjacent features in a geographic data layer (for example, arcs, nodes, polygons, and points). Topological relationships are used for spatial modeling operations that do not require coordinate information.

usage

The way in which statements in a command or programming language are actually used. In geoprocessing, usage for a tool or environment setting can be viewed at the command line.

value

A measurable quantity which a function may take that is either assigned or determined by calculation.

variable

A symbol or placeholder that represents a changeable value or a value that has not yet been assigned. A variable has a quantity that can be measured, and it can be used to represent different types of data in expressions.

VPF

Vector Product Format. A U.S. Department of Defense military standard that defines a format, structure, and organization for large geographic databases. VPF data is read-only in ArcCatalog.

VPF dataset

See VPF.

VPF feature class

See feature class.

workspace

A container for geographic data. A workspace can be a folder that contains shapefiles, an ArcInfo workspace that contains coverages, a geodatabase, or a feature dataset.

z-value

The value for a given surface location that represents an attribute other than position. In an elevation or terrain model, the z-value represents elevation; in other kinds of surface models, it represents the density or quantity of a particular attribute.

Index

D

E

F

G

H

I

P

R

S

T

U

V

W

X

Z